AMS/IP
Studies in
Advanced
Mathematics

Volume 40

Geometric Analysis on the Heisenberg Group and Its Generalizations

Ovidiu Calin
Der-Chen Chang
Peter Greiner

American Mathematical Society • International Press

Shing-Tung Yau, General Editor

2000 *Mathematics Subject Classification.* Primary 53C17, 53C22, 35H20;
Secondary 46E25, 20C20.

Library of Congress Cataloging-in-Publication Data

Calin, Ovidiu.

Geometric analysis on the Heisenberg group and its generalizations / Ovidiu Calin, Der-Chen Chang, Peter Greiner.

p. cm. — (AMS/IP studies in advanced mathematics, ISSN 1089-3288 ; v. 40)

Includes bibliographical references and index.

ISBN-13: 978-0-8218-4319-2 (alk. paper)

ISBN-10: 0-8218-4319-2 (alk. paper)

1. Nilpotent Lie groups. I. Chang, Der-Chen E. II. Greiner, P. C. (Peter Charles), 1938– III. Series.

2007060760

Softcover ISBN: 978-0-8218-4688-9

∞ The paper used in this book is acid-free and falls within the guidelines
established to ensure permanence and durability.
Visit the AMS home page at http://www.ams.org/
Visit the International Press home page at URL: http://www.intlpress.com/

10 9 8 7 6 5 4 3 2 1 13 12 11 10 09 08

This book is dedicated to Professor Yum-Tong Siu
for his teaching, friendship and encouragement.

Contents

Preface

This book deals with a detailed analysis of the Heisenberg group and its extensions to superior steps, and their applications to Physics. The approach is using the complex and real Hamiltonian and Lagrangian formalism developed in the subRiemannian manifolds context. The exposition is enriched with chapter exercises which facilitate the reader's understanding.

An overview for the Reader: This work is a text for a course or seminars designed for graduate students interested in the most recent developments in the subRiemannian manifolds and sub-elliptic operators theory. It is useful for both pure and applied mathematicians and theoretical physicists working in the quantum mechanics area.

Scientific Outline: This book deals with the study of subRiemannian manifolds, which are manifolds with the Heisenberg principle built in. This brings the hope that Heisenberg manifolds (step 2 subRiemannain manifolds) will play a role in Quantum Mechanics in the future, similar to the role played by the Riemannian manifolds in Classical Mechanics. Some people also speculate that superior steps subRiemannian manifolds may play a similar role in Quantum Field Theory.

The subRiemannian geodesics behavior, which is very different than in the Riemannian case, plays an important role in finding heat kernels and propagators for the Schrödinger equation, as well as in finding fundamental solutions for subelliptic operators using a very ingenious geometric method involving complex modified action.

One of the novelties of this book is to introduce the complex Hamiltonian mechanics techniques and apply them in describing the fundamental solutions and heat propagators. The fact that the propagator depends on each subRiemannian distance is an expected fact in Quantum Mechanics.

Our method is quite different and new. The fact that each path has a contribution to the propagator led Feynman to introduce the path integrals. In this work we can manage without them, just working in the proper geometric framework.

For macro objects, like airplanes, planets or bullets, only the classical path is important and their motion is described by a unique geodesic in the context of Riemannian geometry. The situation for the sub-atomic particles is completely different and this book indicates the correct framework to study their motion, which is the subRiemannian geometry. The main issue which is the base of this interpretation is the fact that local and global in subRiemannian context are the same. Hence there are points arbitrary close which can be joined by infinitely many geodesics with distinct lengths and which are used in the asymptotic expansion of the heat kernel on the Heisenberg group.

The first chapter contains a detailed description of the geometry of the Heisenberg group which is the prototype of the step 2 subRiemannian manifolds. It shows that the Carnot-Carathéodory distance plays a very important role in finding the fundamental solution of the Heisenberg operator which retrieves the well-known Folland's formula.

The second chapter deals with a step 4 subRiemannian manifold and its geometry. The tools used to describe the geodesics involve complicated and explicit computations involving elliptic functions.

Chapter 3 deals with the geometry of a step $2(k+1)$ subRiemannian manifolds. As in chapter 2, the closed form solutions require the use of hypergeometric functions. We provide also a qualitative approach and we manage to prove global connectivity and geodesic completeness.

Chapter 4 is concerned with the non-isotropic multi-dimensional Heisenberg group. In this case, the group law will be

$$(x,t)\circ(x',t') = \Big(x + x', t + t' + 2\sum_{j=1}^{n} a_j\big[x_{2j}x'_{2j-1} - x_{2j-1}x'_{2j}\big]\Big).$$

In the isotropic case $a_1 = a_2 = \cdots = a_n$, the results of Chapter 1 carry over with no change, except that each family of geodesics from the point $(0,t)$ to the origin is parametrized by the $(2n-1)$-sphere. However, on a general non-isotropic Heisenberg group, the analysis is more laborious. More precisely, suppose that

$$0 < a_1 \le a_2 \le \cdots \le a_p < a_{p+1} = \cdots = a_n.$$

Then every point $(0,t)$ is connected to the origin by an infinite number of geodesics. Every point (x,t) with $x'' = (x_{2p+1}, x_{2p+2}, \ldots, x_{2n}) \neq 0$ is connected to the origin by a finite number of geodesics. On the other hand, there are points (x,t), with $x \neq 0$, but $x'' = 0$, which are connected to the origin by an infinite number of geodesics. If (x,t), $x \neq 0$, is connected to the origin by an infinite number of geodesics, then the infinity of the number of geodesics connecting (x,t) to $(0,0)$ is "smaller" than the infinity of the number of geodesics connecting $(0,t)$ to $(0,0)$.

The technique of complex Hamiltonian mechanics is introduced in **chapter 5**. It consists in defining a quantum Hamiltonian and introducing a complex modified action. The latter is used in obtaining the lengths of real geodesics and in finding a geometric formula for the fundamental solution of sub-elliptic operators.

Chapter 6 is dedicated to applications in Quantum Mechanics and Electromagnetism. The main application is the fundamental solution and the Schrödinger kernel (propagator) for the Heisenberg operator by means of complex Hamiltonian formalism.

Acknowledgments: This work owes much to the generous help and encouragement of many people. First, we would like to thank Professors Richard Beals, Bernard Gaveau, Stere Ianus, Adam Koranyi, Duong H. Phong, Yum-Tong Siu, and Elias M. Stein for their advice and help. Heartfelt thanks to the Mathematics Departments at Eastern Michigan University, Georgetown University and the University of Toronto for providing excellent research environments for us. We also owe Sandy Becker and Benjamin Jellema many thanks for technical support and help with technology.

Last but not least, we would like to express our gratitude to International Press, especially Professor Shing-Tung Yau, series editor in making this endeavor possible.

The Authors

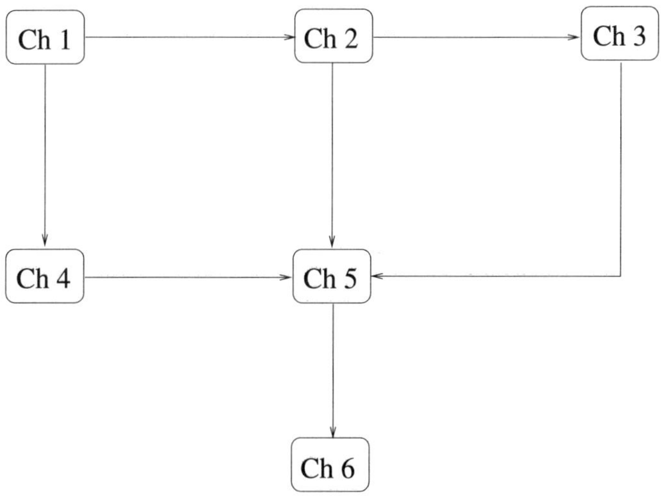

CHAPTER 1

Geometric Mechanics on the Heisenberg Group

The Heisenberg group and its sub-Laplacian are at the cross-roads of many domains of analysis and geometry: nilpotent Lie groups theory, hypoelliptic second order partial differential equations, strongly pseudoconvex domains in complex analysis, probability theory of degenerate diffusion process, subRiemannian geometry, control theory and semiclassical analysis of quantum mechanics, see *e.g.*, [**13**], [**16**], [**30**], [**31**], and [**44**].

Here we give a detailed discussion of the behavior of the subRiemannian geodesics on the Heisenberg group, which is the paradigm of the theory. Research on the subRiemannian geometry induced by the sub-Laplacian and its analytic consequences has been studied extensively in the past ten years (see *e.g.*, [**3**], [**13**], [**4**], [**19**], [**18**] and [**26**]). To the best of our knowledge, the first time these geodesics have been computed by Koranyi, see [**39**], [**38**], and [**40**]. In this chapter we shall provide a systematic method to handle problems along this direction.

The Heisenberg group is the prototype for the subRiemannian manifolds of step 2, which makes this case very special in the class of subRiemannian manifolds.

There are a few ways to introduce the Heisenberg group. In the first section we shall show that all of these are equivalent.

1.1. Definitions for the Heisenberg group

Let $\mathbf{G}$ be a noncommutative group. If $h, k \in \mathbf{G}$ are two elements, define the commutator of h and k by $[h,k] = hkh^{-1}k^{-1} = hk(kh)^{-1}$. If $[h,k] = e$, we say that h and k commute (e denotes the unit element of $\mathbf{G}$). The set of elements which commute with all other elements is called the center of the group $Z(\mathbf{G}) = \{g \in \mathbf{G}; [g,k] = e, \forall k \in G\}$. If $\mathbf{K} \lhd \mathbf{G}$ is a subgroup of $\mathbf{G}$ then let $[\mathbf{K}, \mathbf{G}]$ be the group generated by all commutators $[k,g]$ with $k \in \mathbf{K}$ and $g \in \mathbf{G}$. When $\mathbf{K} = \mathbf{G}$, then $[\mathbf{G}, \mathbf{G}]$ is called the commutator subgroup of $\mathbf{G}$.

DEFINITION 1.1. Let $\mathbf{G}$ be a group. Define the sequence of groups $\big(\Gamma_n(\mathbf{G})\big)_{n\geq 1}$ by $\Gamma_0(\mathbf{G}) = \mathbf{G}$, $\Gamma_{n+1}(\mathbf{G}) = [\Gamma_n(\mathbf{G}), \mathbf{G}]$. $\mathbf{G}$ is called nilpotent if there is an $n \in \mathbb{N}$ such that $\Gamma_n(\mathbf{G}) = e$. The smallest integer n with the above property is called the class of nilpotence of $\mathbf{G}$.

The subset of $\mathcal{M}_3(\mathbb{R})$ given by

$$\mathbf{G} = \left\{ \begin{pmatrix} 1 & a & c \\ 0 & 1 & b \\ 0 & 0 & 1 \end{pmatrix}; a, b, c \in \mathbb{R} \right\} \tag{1.1}$$

defines a noncommutative group with the usual matrix multiplication. Consider the matrices

$$A = \begin{pmatrix} 1 & a_1 & a_3 \\ 0 & 1 & a_2 \\ 0 & 0 & 1 \end{pmatrix}, \qquad B = \begin{pmatrix} 1 & b_1 & b_3 \\ 0 & 1 & b_2 \\ 0 & 0 & 1 \end{pmatrix}. \tag{1.2}$$

Then

$$AB = \begin{pmatrix} 1 & a_1 + b_1 & a_3 + b_3 + a_1 b_2 \\ 0 & 1 & a_2 + b_2 \\ 0 & 0 & 1 \end{pmatrix}, \tag{1.3}$$

$$A^{-1} = \begin{pmatrix} 1 & -a_1 & a_1 a_2 - a_3 \\ 0 & 1 & -a_2 \\ 0 & 0 & 1 \end{pmatrix}, \qquad B^{-1} = \begin{pmatrix} 1 & -b_1 & b_1 b_2 - b_3 \\ 0 & 1 & -b_2 \\ 0 & 0 & 1 \end{pmatrix} \tag{1.4}$$

The commutator

$$[A, B] = ABA^{-1}B^{-1} = \begin{pmatrix} 1 & 0 & a_1 b_2 - b_1 a_2 \\ 0 & 1 & 0 \\ 0 & 0 & 1 \end{pmatrix},$$

and hence the commutator subgroup is

$$\Gamma_1(\mathbf{G}) = [\mathbf{G}, \mathbf{G}] = \langle [A, B]; A, B \in \mathbf{G} \rangle = \left\{ \begin{pmatrix} 1 & 0 & k \\ 0 & 1 & 0 \\ 0 & 0 & 1 \end{pmatrix}; k \in \mathbb{R} \right\}.$$

Let

$$C = \begin{pmatrix} 1 & 0 & k \\ 0 & 1 & 0 \\ 0 & 0 & 1 \end{pmatrix}.$$

Then

$$AC = \begin{pmatrix} 1 & a & c+k \\ 0 & 1 & b \\ 0 & 0 & 1 \end{pmatrix} = CA,$$

and therefore $[A, C] = AC(AC)^{-1} = I_3$. Hence $\Gamma_2(\mathbf{G}) = [\Gamma_1(\mathbf{G}), \mathbf{G}] = I_2 = e$, and the group $\mathbf{G}$ is nilpotent of class 2. $\mathbf{G}$ is called the Heisenberg group with 3 parameters.

The nilpotence class measures the noncommutativity of the group. In the following we shall associate with this group a noncommutative geometry of step 2. This geometry will have the Heisenberg principle built in.

The bijection $\phi : \mathbb{R}^3 \to \mathcal{M}_3(\mathbb{R})$,

$$\phi(x_1, x_2, t) = \begin{pmatrix} 1 & x_1 & t \\ 0 & 1 & x_2 \\ 0 & 0 & 1 \end{pmatrix}$$

induces a noncommutative group law structure on $\mathbb{R}^3$

$$(x_1, x_2, t) \circ (x_1', x_2', t') = (x_1 + x_1', x_2 + x_2', t + t' + x_1 x_2'). \tag{1.5}$$

The zero element is $e = (0, 0, 0)$ and the inverse of (x_1, x_2, t) is $(-x_1, -x_2, x_1 x_2 - t)$.

$\mathbb{R}^3$ together with the group law (1.5) will be called the **nonsymmetric 3-dimensional Heisenberg group**. This group can be regarded also as a Lie group. The left translation $L_a : \mathbf{G} \to \mathbf{G}$, $L_a g = ag$, $\forall g \in \mathbf{G}$ is an analytic diffeomorphism with inverse $L_a^{-1} = L_{a^{-1}}$. A vector field X on $\mathbf{G}$ is called left invariant if

$$(L_a)_*(X_g) = X_{ag}, \qquad \forall a, g \in \mathbf{G}.$$

The set of all left invariant vector fields form the Lie algebra of $\mathbf{G}$, denoted by $L(\mathbf{G})$. The Lie algebra of $\mathbf{G}$ has the same dimension as $\mathbf{G}$ and it is isomorphic to the tangent space $T_e\mathbf{G}$. We shall use this result in the following proposition in order to compute a basis for the Lie algebra of the Heisenberg group.

PROPOSITION 1.2. The vector fields

$$X = \partial_{x_1}, \quad Y = \partial_{x_2} + x_1\partial_t, \quad T = \partial_t \tag{1.6}$$

are left invariant with respect to the Lie group law (1.5) on $\mathbb{R}^3$.

PROOF. Consider the notation $x_3 = t$. In this case the left translation is

$$L_{(a_1,a_2,a_3)}(x_1, x_2, x_3) = (a_1 + x_1, a_2 + x_2, a_3 + x_3 + a_1x_2).$$

Let X be a left invariant vector field. Then $\forall a = (a_1, a_2, a_3) \in \mathbf{G}$, $X_a = (L_a)_* X_e$. In local coordinates $X_a = \sum_i X_a^i \partial_{x_i}$. The components are given by

$$X_a^i = X_a(x_i) = (L_a)_* X_e(x_i) = X_e(x_i \circ L_a), \tag{1.7}$$

where x_i is the i-th coordinate and X_e is the value of the vector field X at the origin. Let $b = (b_1, b_2, b_3) \in \mathbf{G}$. Then

$$\begin{aligned}
(x_1 \circ L_a)(b) &= x_1(L_a b) = x_1(ab) = a_1 + b_1 = x_1(a) + x_1(b),\\
(x_2 \circ L_a)(b) &= x_2(L_a b) = x_2(ab) = a_2 + b_2 = x_2(a) + x_2(b),\\
(x_3 \circ L_a)(b) &= x_3(L_a b) = x_3(ab) = a_3 + b_3 + a_1b_2 = x_3(a) + x_3(b) + x_1(a)x_2(b).
\end{aligned}$$

Dropping b, yields

$$\begin{aligned}
x_1 \circ L_a &= x_1(a) + x_1,\\
x_2 \circ L_a &= x_2(a) + x_2,\\
x_3 \circ L_a &= x_3(a) + x_3 + x_1(a)x_2.
\end{aligned}$$

Substituting in equation (1.7) and using $X_e = \xi^1\partial_{x_1} + \xi^2\partial_{x_2} + \xi^3\partial_{x_3}$, yields

$$\begin{aligned}
X_a^1 &= X_e\big(x_1(a) + x_1\big) = \xi^1,\\
X_a^2 &= X_e\big(x_2(a) + x_2\big) = \xi^2,\\
X_a^3 &= X_e\big(x_3(a) + x_3 + x_1(a)x_2\big) = \xi^3 + x_1(a)\xi^2.
\end{aligned}$$

Hence, the left invariant vector field X depends on the parameters ξ^i

$$\begin{aligned}
X &= \xi^1\partial_{x_1} + \xi^2\partial_{x_2} + (\xi^3 + x_1\xi^2)\partial_{x_3}\\
&= \xi^1\partial_{x_1} + \xi^2(\partial_{x_2} + x_1\partial_{x_3}) + \xi^3\partial_{x_3}\\
&= \xi^1 X + \xi^2 Y + \xi^3 T,
\end{aligned}$$

and the Lie algebra is generated by the linear independent vector fields X, Y and T. □

In the following we shall show that the nonsymmetric model can always be reduced to a symmetric model, using a coordinate transformation.

PROPOSITION 1.3. Under the change of coordinates

$$y_1 = x_1, \quad y_2 = x_2, \quad \tau = 4t - 2x_1x_2,$$

the vector fields

$$X = \partial_{x_1}, \quad Y = \partial_{x_2} + x_1\partial_t, \quad T = \partial_t,$$

are transformed into

$$X = \partial_{y_1} - 2y_2\partial_\tau, \quad Y = \partial_{y_2} + 2y_1\partial_\tau, \quad T = 4\partial_\tau. \tag{1.8}$$

PROOF. The proof follows from the following relationships

$$\begin{aligned} \partial_t &= 4\partial_\tau, \\ \partial_{x_2} &= \partial_{y_2} - 2y_1\partial_\tau, \\ \partial_{x_1} &= \partial_{y_1} - 2y_2\partial_\tau. \end{aligned}$$

□

Making $y_1 = x_1$, $y_2 = x_2$, $\tau = -t$ in (1.8) we consider the vector fields $X_1 = \partial_{x_1} + 2x_2\partial_t$, $X_2 = \partial_{x_2} - 2x_1\partial_t$, $X_3 = \partial_t$ on $\mathbb{R}^3 = \mathbb{R}^2_x \times \mathbb{R}_t$. We are interested in a Lie group law on $\mathbb{R}^3$ such that X_1, X_2 and X_3 are left invariant. This shall be done using the Campbell-Hausdorff formula. The constants of structure are denoted by c_{ij}^k and are defined by

$$[X_i, X_j] = \sum_{k=1}^{3} c_{ij}^k X_k.$$

From $[X_1, X_2] = -4\partial_t$ and $[X_1, \partial_t] = [X_2, \partial_t] = 0$, the constants of structure are

$$c_{12}^1 = c_{12}^2 = 0, \qquad c_{12}^3 = -4,$$

$$c_{13}^j = c_{23}^j = 0, \qquad j = 1, 2, 3.$$

If $x = (x_1, x_2, x_3)$ and $y = (y_1, y_2, y_3)$, a locally Lie group structure is given by the Campbell-Hausdorff formula:

$$(x \circ y)_i = x_i + y_i + \frac{1}{2}\sum_{j,k} c_{jk}^i\, x_j\, y_k + \frac{1}{12}\sum_{k,s,p,j} x_p\, y_j\, x_s\, c_{pj}^k\, c_{ks}^i + \dots \tag{1.9}$$

In our case, we obtain a globally defined group structure

$$\begin{aligned} (x \circ y)_1 &= x_1 + y_1, \\ (x \circ y)_2 &= x_2 + y_2, \\ (x \circ y)_3 &= x_3 + y_3 - \frac{1}{2} \cdot 4(x_1y_2 - x_2y_1), \end{aligned}$$

because the term $x_p\, y_j\, x_s\, c_{pj}^k\, c_{ks}^3 = 0$. We arrived at the following result.

PROPOSITION 1.4. The vector fields $X_1 = \partial_{x_1} + 2x_2\partial_t$, $X_2 = \partial_{x_2} - 2x_1\partial_t$ are left invariant with respect to the following Lie group law on $\mathbb{R}^3$

$$(x_1, x_2, t) \circ (x_1', x_2', t') = \big(x_1 + x_1', x_2 + x_2', t + t' - 2(x_1x_2' - x_2x_1')\big). \tag{1.10}$$

The Lie group $\mathbf{H}_1 = (\mathbb{R}^3, \circ)$ is called the symmetric three dimensional **Heisenberg group**. The unit element is $e = (0,0,0)$ and the inverse is $(x_1, x_2, t)^{-1} = (-x_1, -x_2, -t)$.

We shall study the subRiemannian geometry associated with this model. The geometry has the Heisenberg uncertainty principle

$$[X_1, X_2] = -4\partial_t \tag{1.11}$$

built in. This brings the hope, that Heisenberg manifolds (step 2 subRiemannian manifolds) will play a role for Quantum Mechanics in the future, similar to the role played by the Riemannian manifolds for Classical Mechanics. (see *e.g.* [**2**], [**1**], [**14**]). In Quantum Mechanics, the states of a quantum particle (position, momentum) are described by differential operators. It is known that two states, which cannot be measured simultaneously correspond to operators, which do not commute (see [**45**]).

For instance, if $\mathbf{x}$ and $\mathbf{p}$ are the position and the momentum of a particle, then one cannot measure them simultaneously and hence, we write $[\mathbf{x}, \mathbf{p}] \neq 0$. The state of the particle is measured using a radiation beam sent towards the particle. The radiation is reflected partially back. Using the variation of frequency between the sent and reflected beams, the Doppler-Fizeau formula will provide the speed of the particle. This method will provide accurate results if the radiation will not significantly change the speed of the particle *i.e.*, its kinetic energy $K = mv^2/2$. This means the radiation has a low energy and hence, a low frequency, because $E_{radiation} = h\nu$. Therefore the wave length of the radiation $\lambda = 1/\nu$ will be large. In this case the position of the particle cannot be measured accurately (see [**32**]).

In order to measure the position accurately, the radiation wave length has to be as small as possible. In this case the frequency ν is large as will be the energy $E_{radiation}$. This will change the kinetic energy of the particle and hence its velocity. Hence, one cannot measure accurately both the position and the speed of the particle.

The Heisenberg uncertainty principle, fundamental in the study of quantum particles, can be found also in other examples at the large scale structure. Let's assume that you are watching high-street traffic from an airplane. You will notice the position of the cars but you cannot say too much about their speed. They look like they are not moving at all. A policeman on the road will see the picture completely differently. For him, the speed of the cars will make more sense than their position. The latter is changing too fast to be noticed accurately.

The Heisenberg group is also closely linked with theory of functions in several complex variables. Let us start with the unit disc $\{w \in \mathbb{C} : |w| \le 1\}$. Consider the Cayley transform:

$$w = \frac{i - z}{i + z}. \tag{1.12}$$

It is known that the transformation (1.12) maps the real line $\mathrm{Im}(z) = 0$ together with the point "∞" into the unit circle $|w| = 1$, and the upper half plane $\{z = x + iy \in \mathbb{C} : \mathrm{Im}(z) > 0\}$ into the disc $|w| < 1$. From the point of view of Fourier analysis, there are two groups of holomorphic self-mappings on it which are of importance to us. The first one is the translation group, $z = x + iy \to (x + h) + iy$,

$h \in \mathbb{R}$, which is isomorphic to $\mathbb{R}$. This group acts transitively on each level set $\{z = x + iy \in \mathbb{R}^2_+ : y = c, c \geq 0\}$, which is responsible for the fact that many important operators, *e.g.*, Hilbert transform and Cauchy operator, are convolution operators. The second one is the dilation group, $z \to \delta z$, $\delta > 0$, which determines the homogeneity of the kernels of these operators. The analogues of these groups are equally significant.

A complex analogue of the upper-half plane $\mathbb{R}^2_+$ is the Siegel upper-half space

$$\mathcal{U} = \Big\{(z_1, z_2) \in \mathbb{C}^2 : \mathrm{Im}(z_2) > |z_1|^2\Big\}.$$

Its boundary is the "paraboloid"

$$\partial\mathcal{U} = \big\{(z_1, z_2) \in \mathbb{C}^2 : \mathrm{Im}(z_2) = |z_1|^2\big\}.$$

The domain $\mathcal{U}$ is biholomorphically equivalent with the unit ball

$$B_2 = \big\{(w_1, w_2) \in \mathbb{C}^2 : |w_1|^2 + |w_2|^2 < 1\big\}$$

via the generalized Cayley transform:

$$w_1 = \frac{2iz_1}{i + z_2}, \qquad w_2 = \frac{i - z_2}{i + z_2}.$$

In this correspondence the boundary $\partial\mathcal{U}$, together with the "point at ∞" maps onto the unit sphere in $\mathbb{C}^2$.

It is known that the Heisenberg group $\mathbf{H}_1$ gives the translations of the domain $\mathcal{U}$. The action of an element $(z, t) \in \mathbf{H}_1$ on the domain $\mathcal{U}$ is given by

$$(\xi_1, \xi_2) \to \big(\xi_1 + z, \xi_2 + t + i(2\xi_1 \cdot \bar{z}_1 + |z_1|^2)\big). \tag{1.13}$$

Since $|\xi_1 + z_1|^2 - |z_1|^2 = \mathrm{Im}\{i(2\xi_1 \cdot \bar{z}_1 + |z_1|^2)\}$, this mapping preserves each "level surface"

$$\mathrm{Im}(\xi_2) - |\xi_1|^2 = \text{constant}.$$

It follows that the mapping (1.13) is simply transitive on the boundary $\partial\mathcal{U}$: the boundary $\partial\mathcal{U}$ therefore may be identified as the orbit of the "origin" $(0, 0)$ under the action of the group $\mathbf{H}_1$:

$$\mathbf{H}_1 \in (z, t) \mapsto (z, t + i|z|^2) \in \partial\mathcal{U}.$$

The second group consists of dilations. The action on the domain $\mathcal{U}$ is given by

$$(\xi_1, \xi_2) \to (\delta\xi_1, \delta^2\xi_2), \qquad \delta > 0.$$

It is easy to see that these are automorphisms of the group structure of $\mathbf{H}_1$. Summarizing our discussion, the group $\mathbf{H}_1$ is isomorphic to the boundary $\partial\mathcal{U}$. Similar to analysis in $\mathbb{R}^2$, we may use the Heisenberg group and the Siegel upper-half space to "approximate" a bounded strongly pseudoconvex domain in $\mathbb{C}^{n+1}$. This phenomena indeed happened in many situations, see Folland-Stein [**24**] , Phong-Stein [**49**], [**48**], and Greiner-Stein [**28**].

1.2. The horizontal distribution

Unlike on Riemannian manifolds, where one may measure the velocity and distances in all directions, on Heisenberg manifolds there are directions where we cannot say anything using direct methods.

On the Heisenberg group, an important role is played by the distribution $\mathcal{H}$ generated by the linearly independent vector fields X_1 and X_2:

$$x \to \mathcal{H}_x = \text{span}_x\{X_1, X_2\},$$

called the *horizontal distribution.* As $[X_1, X_2] = -4\partial_t \notin \mathcal{H}$, the horizontal distribution $\mathcal{H}$ is not involutive, and hence, by Frobenius theorem, it is not integrable, *i.e.*, there is no surface locally tangent to $\mathcal{H}$.

A vector field V on $\mathbb{R}^3$ is called *horizontal* if and only if $V_x \in \mathcal{H}_x$, $\forall x$. A curve $c : [0,1] \to \mathbb{R}^3$ is called horizontal if the velocity vector $\dot{c}(s)$ is a horizontal vector field along $c(s)$. Horizontality is a constraint on the velocities and it is called a *non-holonomic constraint.*

In this chapter we shall construct many horizontal objects, *i.e.*, geometrical objects which can be constructed directly from the horizontal distribution and the subRiemannian metric defined on it. The main goal is to recover the external structure of the space, such as the missing direction ∂_t by means of horizontal objects.

PROPOSITION 1.5. A curve $c = (x_1, x_2, t)$ is horizontal if and only if

$$\dot{t} = 2(\dot{x}_1 x_2 - x_1 \dot{x}_2). \tag{1.14}$$

PROOF. The velocity vector can be written as

$$\begin{aligned} \dot{c} &= \dot{x}_1\partial_{x_1} + \dot{x}_2\partial_{x_2} + \dot{t}\partial_t \\ &= \dot{x}_1\big(\partial_{x_1} + 2x_2\partial_t\big) - 2\dot{x}_1 x_2\partial_t \\ &\quad + \dot{x}_2\big(\partial_{x_2} - 2x_1\partial_t\big) + 2x_1\dot{x}_2\partial_t + \dot{t}\partial_t \\ &= \dot{x}_1 X_1 + \dot{x}_2 X_2 + \big(\dot{t} + 2x_1\dot{x}_2 - 2x_2\dot{x}_1\big)\partial_t. \end{aligned}$$

Hence, $\dot{c} \in \mathcal{H}$ iff the coefficient of ∂_t vanishes *i.e.*, (1.14) holds. □

COROLLARY 1.6. A curve $c = (x_1, x_2, t)$ is horizontal if and only if

$$\dot{c} = \dot{x}_1 X_1 + \dot{x}_2 X_2. \tag{1.15}$$

In the following we shall give a geometrical interpretation for the t component of a horizontal curve. This will be used in the proof of the connectivity theorem later. Using the polar coordinates $x_1 = r\cos\phi$, $x_2 = r\sin\phi$, equation (1.14) becomes

$$\begin{aligned} \dot{t} &= 2(\dot{x}_1 x_2 - x_1\dot{x}_2) \\ &= -2r^2\dot{\phi}(\sin^2\phi + \cos^2\phi) \\ &= -2r^2\dot{\phi}. \end{aligned}$$

In differential notation

$$dt = -2r^2\, d\phi. \tag{1.16}$$

Let $r = r(\phi)$ be the equation in polar coordinates of the projection of the horizontal curve on the x-plane. The area of an infinitesimal triangle with vertices at the origin,

$(r(\phi), \phi)$ and $(r(\phi + d\phi), \phi + d\phi)$ is $\frac{1}{2}r(\phi)r(\phi + d\phi)d\phi \approx \frac{1}{2}r^2(\phi)d\phi$. Integrating, we obtain the area swept by the vectorial radius between the initial angle ϕ_0 and ϕ, see Figure 1.1.

$$\mathcal{A} = \frac{1}{2}\int_{\phi_0}^{\phi} r^2(\phi)\, d\phi.$$

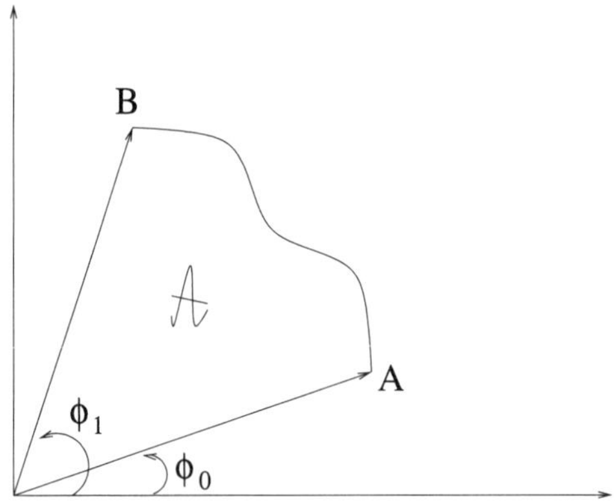

Figure 1.1 : The area swept by the vectorial radius between two points in the plane.

Taking the derivative, we obtain

$$\frac{d\mathcal{A}}{d\phi} = \frac{1}{2}r^2(\phi),$$

or

$$dA = \frac{1}{2}r^2\, d\phi. \tag{1.17}$$

Dividing the equations (1.16) and (1.17) yields

$$\frac{dt}{d\mathcal{A}} = -4, \tag{1.18}$$

which says that the t-component is roughly the area swept by the vectorial radius on the x-plane, up to a multiplication factor. The negative sign in (1.18) shows that when t is increasing, the rotation in the x-plane is done clock-wise. The equation (1.18) is valid only if t is not constant.

The *areal velocity* is defined as

$$\alpha = \frac{d\mathcal{A}}{ds}.$$

Areal velocity also appears in Kepler's laws: all of the planets have plane trajectories, which are ellipses with the sun in one of the focuses. The areal velocity α is constant along the motion.

THEOREM 1.7. *A curve c in $\mathbb{R}^3$ is horizontal if and only if the rate of change of the t-component is equal to 4α, i.e., $\dot{t} = 4\alpha$.*

PROOF. $c = (x_1, x_2, t)$ is a horizontal curve iff

$$\dot{t} = 2(\dot{x}_1 x_2 - x_1\dot{x}_2),$$

which in polar coordinates is

$$\dot{x}_1 x_2 - x_1\dot{x}_2 = r^2\dot{\phi},$$

and hence

$$\alpha = \frac{1}{2}(\dot{x}_1 x_2 - x_1 \dot{x}_2) = \frac{1}{4}\dot{t}.$$

□

The characterization of horizontal curves with t constant is given in the following result.

PROPOSITION 1.8. A smooth curve $c(s)$ is horizontal with $t(s) = t$ constant if and only if $c(s) = (as, bs, t)$, with $a, b \in \mathbb{R}$, $a^2 + b^2 \neq 0$.

PROOF. If $c(s)$ is horizontal with t constant, the equation $\dot{t} = -2r^2\dot{\phi}$ yields ϕ =constant. Hence the projection on the x-space is a line which passes through the origin.
If $c(s) = (as, bs, t)$, with t constant, then

$$2(\dot{x}_1 x_2 - x_1 \dot{x}_2) = 2(abs - abs) = 0 = \dot{t}$$

and hence the horizontality condition (1.14) holds. □

The following proposition shows that the left translation of a horizontal curve is a horizontal curve.

PROPOSITION 1.9. If $c(s)$ is a horizontal curve, then $\bar{c}(s) = L_a c(s)$ is a horizontal curve, for any $a \in \mathbf{H_1}$.

PROOF. If $c = (c_1, c_2, c_3)$ and $\bar{c} = (\bar{c}_1, \bar{c}_2, \bar{c}_3)$, then

$$\begin{aligned} \bar{c}_1 &= a_1 + c_1 \Longrightarrow \dot{\bar{c}}_1 = \dot{c}_1 \\ \bar{c}_2 &= a_2 + c_2 \Longrightarrow \dot{\bar{c}}_2 = \dot{c}_2 \\ \bar{c}_3 &= a_3 + c_3 - 2(a_1 c_2 - a_2 c_1) \Longrightarrow \dot{\bar{c}}_3 = \dot{c}_3 - 2(a_1 \dot{c}_2 - a_2 \dot{c}_1). \end{aligned} \tag{1.19}$$

Using that c is horizontal, equation (1.14) yields

$$\begin{aligned} \dot{\bar{c}}_3 &= \dot{c}_3 - 2(a_1 \dot{c}_2 - a_2 \dot{c}_1) \\ &= 2(\dot{c}_1 c_2 - c_1 \dot{c}_2) - 2(a_1 \dot{c}_2 - a_2 \dot{c}_1) \\ &= 2(\dot{c}_1(a_2 + c_2) - \dot{c}_2(a_1 + c_1)) \\ &= 2(\dot{\bar{c}}_1 \bar{c}_2 - \dot{\bar{c}}_2 \bar{c}_1). \end{aligned}$$

From equation (1.14) it follows that $\bar{c}(s)$ is horizontal. □

COROLLARY 1.10. The velocity of the horizontal curve $\bar{c}(s) = L_a c(s)$ is

$$\dot{\bar{c}}(s) = (L_a)_* \dot{c}(s) = \dot{c}_1(s) X_{1|\bar{c}(s)} + \dot{c}_2(s) X_{2|\bar{c}(s)}. \tag{1.20}$$

PROOF. As $\bar{c}(s)$ is horizontal, from equations (1.15) and (1.19) we have

$$\begin{aligned} \dot{\bar{c}}(s) &= \dot{\bar{c}}_1(s) X_{1|\bar{c}(s)} + \dot{\bar{c}}_2(s) X_{2|\bar{c}(s)} \\ &= \dot{c}_1(s) X_{1|\bar{c}(s)} + \dot{c}_2(s) X_{2|\bar{c}(s)} \\ &= \dot{c}_1(s) (L_a)_* X_{1|c(s)} + \dot{c}_2(s) (L_a)_* X_{2|c(s)} \\ &= (L_a)_* (\dot{c}_1(s) X_{1|c(s)} + \dot{c}_2(s) X_{2|c(s)}) \\ &= (L_a)_* \dot{c}(s). \end{aligned}$$

□

1.3. Horizontal connectivity theorem

On the Heisenberg group $\mathbf{H_1}$, the vector fields X_1, X_2 and $[X_1, X_2]$ generate the tangent space of $\mathbb{R}^3$ at every point. Such a subRiemannian manifold is called step 2. In the case of a general subRiemannian manifold, the number of brackets needed to generate all directions +1 is called the step of the manifold. The higher the step, the more noncommutative and the harder it will be to study the geometry. The step 1 corresponds to Riemannain geometry, which is the commutative case. The step condition was used independently by Chow [**22**] and Hörmander [**33**] to study connectivity and hypoellipticity, respectively. Using Hörmander's theorem, the Heisenberg Laplacian $\Delta_H = \frac{1}{2}(X_1^2 + X_2^2)$ is hypoelliptic, *i.e.*, $\Delta_H u = f \in C^\infty \Longrightarrow u \in C^\infty$.

In the following we shall prove Chow's connectivity theorem (see [**22**]) in the particular case of the Heisenberg group $\mathbf{H_1}$.

PROPOSITION 1.11. *Any two points in $\mathbf{H_1}$ can be joined by a piece-wise horizontal curve, i.e., a curve tangent to the horizontal distribution.*

PROOF. Let P and Q be two points in $\mathbb{R}^3$. Let t_P and t_Q be the t-coordinates of P and Q. We distinguish between the following two cases:
case (i): $t_P \neq t_Q$
Consider the number $\alpha = t_P - t_Q \neq 0$. Let P_1 and Q_1 be the projections on the x-plane of the points P and Q. Consider in the x-plane a curve $\bar{\phi} : [0,1] \to \mathbb{R}^2$ which joins P_1 and Q_1, such that the area situated between the graph of $\bar{\phi}$ and the segments OP_1 and OQ_1 is equal to $\alpha/4$. The area will be considered positive in the case of a counter clock-wise rotation of the curve $\bar{\phi}$ between P_1 and Q_1, see Figure 1.2. If $\bar{\phi}(s) = (x_1(s), x_2(s))$, then consider the function

$$t(s) = t_P + 2\int_0^s \Big(x_2(u)\dot{x}_1(u) - x_1(u)\dot{x}_2(u)\Big)\, du. \tag{1.21}$$

We claim that $\phi : [0,1] \to \mathbb{R}^3$ defined as $\phi(s) = \big(\bar{\phi}(s), t(s)\big)$ is a horizontal curve joining P and Q. Differentiating in (1.21) we obtain the horizontality condition $\dot{t}(s) = 2\big(x_2(s)\dot{x}_1(s) - x_1(s)\dot{x}_2(s)\big)$ and hence, $\phi(s)$ is a horizontal curve. We shall check that ϕ joins the points P and Q.

$$\phi(0) = (\bar{\phi}(0), t(0)) = (x(P_1), t_P) = P.$$

Using $t(0) = t_P$, integrating between 0 and 1 in equation (1.18) yields

$$\begin{aligned} t(1) &= t(0) - 4\big(\mathcal{A}(1) - \mathcal{A}(0)\big) \\ &= t(0) - \alpha = t_P - (t_P - t_Q) \\ &= t_Q, \end{aligned}$$

and hence, $\phi(1) = (\bar{\phi}(1), t(1)) = (x(Q_1), t_Q) = Q$.

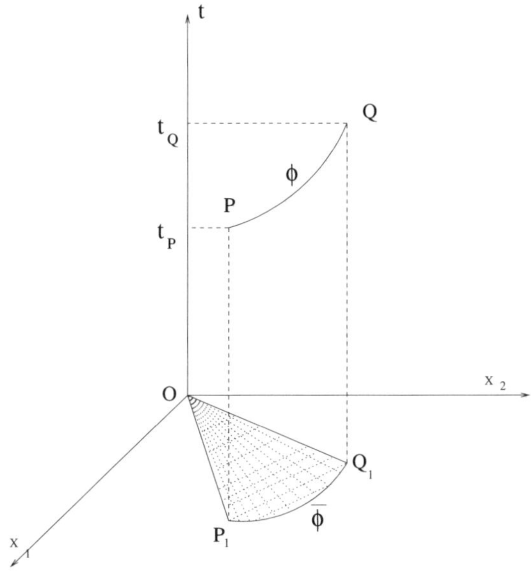

Figure 1.2: The projection of a horizontal curve.

case (ii): $t_P = t_Q = t$
Let $R = (0, 0, t)$. From Proposition 1.8 the curves $c_i : [0, 1] \to \mathbb{R}^3$

$$c_1(s) = \big(sx_1(P), sx_2(P), t\big),$$

$$c_2(s) = \big(sx_1(Q), sx_2(Q), t\big)$$

are horizontal and join the points R with P and R with Q, respectively. Then the piece-wise defined curve

$$\phi(s) = \begin{cases} c_1(1 - 2s), & 0 \le s \le \frac{1}{2} \\ c_2(2s - 1), & \frac{1}{2} \le s \le 1 \end{cases}$$

is horizontal with $\phi(0) = c_1(1) = P$ and $\phi(1) = c_2(1) = Q$. □

The piece-wise condition in the above proposition can be dropped. The proof can be modified such that any two given points can be connected by a horizontal smooth curve. In order to do that, we should take advantage of the group law.

Given the points $P(x_1, y_1, t_1)$ and $Q(x_2, y_2, t_2)$, a translation to the left by $(-x_1, -y_1, -t_1)$ will transform them into $O(0, 0, 0)$ and $S(x', y', t')$, with

$$x' = x_2 - x_1, \quad y' = y_2 - y_1, \quad t' = t_2 - t_1 - 2(y_1 x_2 - x_1 y_2).$$

If $t' \neq 0$, applying case (i) of the previous proposition, we get a smooth horizontal curve $c(s)$ joining O and S.
If $t' = 0$, then $c(s) = (sx', sy', 0)$ is horizontal and joins O and S, see Proposition 1.8. Translating to the left by (x_1, y_1, t_1), the points O and S are sent into P and Q, respectively. Applying Proposition 1.9, the curve $c(s) = (c_1(s), c_2(s), c_3(s))$ is sent into the horizontal smooth curve between P and Q

$$(1.22)\quad (x_1, y_1, t_1) \circ c(s) = \Big(x_1 + c_1(s), y_1 + c_2(s), t_1 + c_3(s) - 2(x_1 c_2(s) - y_1 c_1(s)\Big).$$

COROLLARY 1.12. By a left translation, the t-axis $c(s) = (0, 0, s)$ is transformed into

$$L_{(x_1,y_1,t_1)}c(s) = (x_1, y_1, t_1 + s).$$

PROOF. It is an obvious consequence of equation (1.22). □

In the following we shall find an explicit smooth horizontal curve between the origin and a given point $P(\mathbf{x_1}, \mathbf{x_2}, \mathbf{t})$.

LEMMA 1.13. *For any $a, b, \alpha, \beta \in \mathbb{R}$, the curve $c(s) = \Big(x_1(s), x_2(s), t(s)\Big)$,*

$$\begin{aligned} x_1(s) &= as + bs^2 \\ x_2(s) &= \alpha s^2 + \beta s^3 \\ t(s) &= -\frac{2}{3}\alpha a s^3 - a\beta s^4 - \frac{2}{5}b\beta s^5 \end{aligned}$$

is horizontal and passes through the origin.

PROOF. A computation shows that

$$\begin{aligned} \dot{x}_1 x_2 - x_1 \dot{x}_2 &= (a + 2bs)(\alpha s^2 + \beta s^3) - (as + bs^2)(2\alpha a + 3\beta s^2) \\ &= -\alpha a s^2 - 2a\beta s^3 - b\beta s^4 = \frac{1}{2}\dot{t} \end{aligned}$$

and hence the curve is horizontal. For $s = 0$ the curve passes through the origin. □

We shall show that given $\mathbf{x_1}, \mathbf{x_2}, \mathbf{t}$, we can chose the constants a, b, α, β such that the curve given by Lemma 1.13 passes through the point $P(\mathbf{x_1}, \mathbf{x_2}, \mathbf{t})$ for $s = 1$. This means that a, b, α and β are solutions of the following system

$$\begin{aligned} \mathbf{x_1} &= a + b \\ \mathbf{x_2} &= \alpha + \beta \\ \mathbf{t} &= -\frac{2}{3}\alpha a - a\beta - \frac{2}{5}b\beta. \end{aligned}$$

By computation

$$\begin{aligned} 15\mathbf{t} &= -10\alpha a - 15a\beta - 6b\beta = -10a(\alpha + \beta) - 5a\beta - 6b\beta \\ &= -10a\mathbf{x_2} - (5a + 6b)\beta = -10a\mathbf{x_2} - (5\mathbf{x_1} + b)\beta \\ &= -10(\mathbf{x_1} - b)\mathbf{x_2} - 5\mathbf{x_1}\beta - b\beta = -10\mathbf{x_1}\mathbf{x_2} + 10b\mathbf{x_2} - (5\mathbf{x_1} + b)\beta. \end{aligned}$$

Hence

$$15\mathbf{t} + 10\mathbf{x_1}\mathbf{x_2} - 10b\mathbf{x_2} + (5\mathbf{x_1} + b)\beta = 0.$$

Choose b such that $5\mathbf{x_1} + b \neq 0$. Then

$$\beta = \frac{-15\mathbf{t} - 10\mathbf{x_1}\mathbf{x_2} + 10b\mathbf{x_2}}{5\mathbf{x_1} + b},$$

and

$$\alpha = \mathbf{x_2} - \beta, \qquad a = \mathbf{x_1} - b.$$

Hence we have constructed a family of horizontal curves between the origin and P, which depends on the parameter b.

1.4. Hamiltonian formalism on the Heisenberg group

The Heisenberg group is a good environment to apply the Hamiltonian formalism. Consider the Hamiltonian $H : T^*\mathbb{R}^3_{(x,t)} \to \mathbb{R}$ given by

$$H(\xi, \theta, x, t) = \frac{1}{2}(\xi_1 + 2x_2\theta)^2 + \frac{1}{2}(\xi_2 - 2x_1\theta)^2, \tag{1.23}$$

which is the principal symbol of the Heisenberg Laplacian

$$\Delta_H = \frac{1}{2}(X_1^2 + X_2^2), \tag{1.24}$$

where $X_1 = \partial_{x_1} + 2x_2\partial_t$, $X_2 = \partial_{x_2} - 2x_1\partial_t$. In Quantum Mechanics, the procedure of obtaining the operator (1.24) from the Hamiltonian (1.23) is called *quantization*.

It is natural to consider the Hamiltonian system

$$\begin{aligned} \dot{x} &= \partial H/\partial \xi \\ \dot{t} &= \partial H/\partial \theta \\ \dot{\xi} &= -\partial H/\partial x \\ \dot{\theta} &= -\partial H/\partial t. \end{aligned} \tag{1.25}$$

The solutions $c(s) = \big(x(s), t(s), \xi(s), \theta(s)\big)$ of the system (1.25) are called *bicharacteristics*.

DEFINITION 1.14. Given two points $P(x_0, t_0)$, $Q(x_1, t_1) \in \mathbb{R}^3$, a geodesic between P and Q is the projection on the (x, t)-space of a bicharacteristic $c : [0, \tau] \to \mathbb{R}^3$, which satisfies the boundary conditions:

$$(x(0), t(0)) = (x_0, t_0), \qquad (x(\tau), t(\tau)) = (x_1, t_1). \tag{1.26}$$

The most basic questions we shall answer in this chapter are:

- Given any two points, can we join them by a geodesic?
- How many geodesics are between any two given points?

We shall start by proving the following result.

PROPOSITION 1.15. Any geodesic is a horizontal curve.

PROOF. Let $c(s) = \big(x_1(s), x_2(s), t(s)\big)$ be a geodesic. From the Hamiltonian system (1.25)

$$\dot{x}_1 = \xi_1 + 2x_2\theta, \qquad \dot{x}_2 = \xi_2 - 2x_1\theta,$$

and then

$$\begin{aligned} \dot{t} &= \frac{\partial H}{\partial \theta} \\ &= 2x_2(\xi_1 + 2x_2\theta) - 2x_1(\xi_2 - 2x_1\theta) \\ &= 2x_2\dot{x}_1 - 2x_1\dot{x}_2, \end{aligned}$$

which is the horizontality condition (1.14). Hence, any geodesic is a horizontal curve. □

Solving the Hamiltonian system. We shall solve the Hamiltonian system explicitly. We start with the observation that H does not depend on t. Then

$$\dot{\theta} = -\frac{\partial H}{\partial t} = 0$$

and hence, $\theta =$ constant. The equations

$$\dot{x}_1 = \frac{\partial H}{\partial \xi_1}, \qquad \dot{x}_2 = \frac{\partial H}{\partial \xi_2}$$

become

$$\dot{x}_1 = \xi_1 + 2x_2\theta, \qquad \dot{x}_2 = \xi_2 - 2x_1\theta. \tag{1.27}$$

Differentiating, yields

$$\ddot{x}_1 = \dot{\xi}_1 + 2\dot{x}_2\theta, \qquad \ddot{x}_2 = \dot{\xi}_2 - 2\dot{x}_1\theta. \tag{1.28}$$

Using $\dot{\xi} = -\partial H/\partial x$ and the system (1.27) we obtain

$$\begin{aligned} \dot{\xi}_1 &= 2\theta(\xi_2 - 2x_1\theta) = 2\theta\dot{x}_2 \\ \dot{\xi}_2 &= -2\theta(\xi_1 + 2x_2\theta) = -2\theta\dot{x}_1. \end{aligned} \tag{1.29}$$

(1.30)

From systems (1.28) and (1.29)

$$\ddot{x}_1 = 4\theta\dot{x}_2, \qquad \ddot{x}_2 = -4\theta\dot{x}_1 \tag{1.31}$$

with constant θ. The system (1.31) can be written as

$$\ddot{x}(s) = 4\theta J\dot{x}(s), \tag{1.32}$$

where

$$J = \begin{pmatrix} 0 & 1 \\ -1 & 0 \end{pmatrix}$$

and $x = \begin{pmatrix} x_1 & x_1 \end{pmatrix}$. The equation (1.32) describes the projection of the geodesic on the x-space. We shall show that this is a circle.

With the substitution $y(s) = \dot{x}(s)$ equation (1.32) becomes

$$\dot{y}(s) = 4\theta Jy(s), \tag{1.33}$$

with the solution

$$y(s) = e^{4\theta Js}y(0).$$

Therefore $\dot{x}(s) = e^{4\theta Js}y(0)$. Integrating and using that J and $e^{4\theta Js}$ commute, yields

$$\begin{aligned} x(s) &= x(0) + \int_0^s e^{4\theta Ju}y(0)\,du \\ &= x(0) + \frac{1}{4\theta}J^{-1}\,e^{4\theta Js}y(0)\Big|_{u=0}^{u=s} \\ &= x(0) - \frac{1}{4\theta}Je^{4\theta Js}y(0) + \frac{1}{4\theta}\mathcal{J}^{-1}y(0) \\ &= e^{4\theta Js}K + C, \end{aligned} \tag{1.34}$$

where $K = -Jy(0)/(4\theta)$ and $C = x(0) + K$. We need the following result.

LEMMA 1.16. *If* $J = \begin{pmatrix} 0 & 1 \\ -1 & 0 \end{pmatrix}$ *then* $e^{4\theta Js} = R_{4\theta s}$*, where* R_α *denotes the rotation by angle* α *in the* x*-plane.*

PROOF. A computation yields

$$\begin{aligned} e^{4\theta Js} &= \sum_{n=0}^{\infty} \frac{(4\theta s)^n J^n}{n!} = I \sum_{k=0}^{\infty} \frac{(4\theta s)^{4k}}{(4k)!} + J \sum_{k=0}^{\infty} \frac{(4\theta s)^{4k+1}}{(4k+1)!} \\ &\quad -I \sum_{k=0}^{\infty} \frac{(4\theta s)^{4k+2}}{(4k+2)!} - J \sum_{k=0}^{\infty} \frac{(4\theta s)^{4k+3}}{(4k+3)!} \\ &= \sum_{k=0}^{\infty} \begin{pmatrix} \frac{(4\theta s)^{4k}}{(4k)!} - \frac{(4\theta s)^{4k+2}}{(4k+2)!} & \frac{(4\theta s)^{4k+1}}{(4k+1)!} - \frac{(4\theta s)^{4k+3}}{(4k+3)!} \\ -\frac{(4\theta s)^{4k+1}}{(4k+1)!} + \frac{(4\theta s)^{4k+3}}{(4k+3)!} & \frac{(4\theta s)^{4k}}{(4k)!} - \frac{(4\theta s)^{4k+2}}{(4k+2)!} \end{pmatrix} \\ &= \begin{pmatrix} \cos(4\theta s) & \sin(4\theta s) \\ -\sin(4\theta s) & \cos(4\theta s) \end{pmatrix} \\ &= R_{4\theta s}. \end{aligned}$$

□

Using Lemma 1.4, the equation (1.34) becomes

$$x(s) = R_{4\theta s} K + C. \tag{1.35}$$

As $|x(s) - C| = |R_{4\theta s}K| = |K| =$ constant, $x(s)$ will describe a circle centered at C with the radius

$$|K| = \left| \frac{-Jy(0)}{4\theta} \right| = \frac{|y(0)|}{4|\theta|} = \frac{|\dot{x}(0))|}{4|\theta|}. \tag{1.36}$$

PROPOSITION 1.17. Consider a geodesic which joins the points $P(x_0, t_0)$ and $Q(x_1, t_1)$, with $t_0 \neq t_1$.
(i) The projection of the geodesic on the x-space is a circle or a piece of a circle with end points x_0 and x_1.
(ii) If the projection is one complete circle, with $x_0 = x_1$, let σ be its area. Then

$$\sigma = \frac{|t_1 - t_0|}{4}.$$

PROOF. (i) comes from the solution of the Hamiltonian system discussed above.
(ii) By Proposition 1.15, any geodesic is horizontal. From equation (1.18), the area of the projection on the x-plane and the t-component of a horizontal curve are related by

$$4\, d\mathcal{A} = -dt.$$

Integrating, yields

$$\begin{aligned} 4\sigma &= \int_0^\tau 4\, d\mathcal{A} = -\int_0^\tau dt \\ &= t_0 - t_1, \end{aligned}$$

where $x(\tau) = x_1$, $t(\tau) = t_1$. □

COROLLARY 1.18. The radius of the projection circle is $R = \sqrt{\frac{|t_1 - t_0|}{4\pi}}$.

If we let $\theta \to 0$ in (1.36), the radius of the circle $|K| \to \infty$. This corresponds to a circle with infinite radius, which is a line. This case is covered by the next result.

PROPOSITION 1.19. If $\theta = 0$, the projection of the geodesic onto the x-space is a line.

PROOF. If $\theta = 0$ the equation (1.32) yields $\ddot{x} = 0$, which corresponds to a line. $\square$

Proposition 1.15 claims that geodesics are horizontal curves. We shall show that the converse is false.

PROPOSITION 1.20. There are horizontal curves which are not geodesics.

PROOF. Consider $c(s) = \big(s^2/2, s, s^3/3\big)$. The curve is horizontal, because the horizontality condition holds

$$\dot{t}(s) = s^2 = 2\big(s^2 - s^2/2\big) = 2\big(\dot{x}_1 x_2 - x_1 \dot{x}_2\big).$$

On the other hand, the system (1.31) becomes $4\theta = 1$ and $0 = -4\theta s$, which leads to a contradiction. $\square$

The t-component

Using the Hamiltonian equation $\dot{t} = \partial H/\partial \theta$, and (1.36) yields

$$\begin{aligned}
\dot{t}(s) &= 2\big(x_2(s)\dot{x}_1(s) - x_1(s)\dot{x}_2(s)\big) \\
&= 2\langle x(s), J(\dot{x}(s))\rangle \\
&= 2\langle e^{4\theta Js}K + C,\ J(4\theta J e^{4\theta Js}K)\rangle \\
&= 2\langle e^{4\theta Js}K,\ -4\theta e^{4\theta Js}K\rangle + 2\langle C,\ -4\theta e^{4\theta Js}K\rangle \\
&= -8\theta|K|^2 - 8\theta\langle C, e^{4\theta Js}K\rangle.
\end{aligned}$$

Integrating we obtain

$$t(s) = \int \Big(-8\theta|K|^2 - 8\theta\langle C, e^{4\theta Js}K\rangle\Big)\, ds.$$

As

$$\begin{aligned}
\frac{d}{ds}\langle JC, e^{4\theta Js}K\rangle &= \langle JC, \frac{d}{ds}e^{4\theta Js}K\rangle = \langle JC, 4\theta J e^{4\theta Js}K\rangle \\
&= 4\theta\langle J^T JC, e^{4\theta Js}K\rangle = 4\theta\langle C, e^{4\theta Js}K\rangle
\end{aligned}$$

then

$$\int \langle C, e^{4\theta Js}K\rangle\, ds = \frac{1}{4\theta}\langle JC, e^{4\theta Js}K\rangle + const$$

Hence

$$t(s) = -8\theta|K|^2 s - 2\langle JC, e^{4\theta Js}K\rangle + C_1$$

where $C_1 = t(0) + 2\langle JC, K\rangle$.

The conservation of energy. Let $E = \frac{1}{2}(\dot{x}_1^2 + \dot{x}_2^2)$ be the kinetic energy. One may show that the Hamiltonian is equal to E along the geodesics, and hence E is a first integral for the Hamiltonian system. In the following proposition we shall give a direct proof.

PROPOSITION 1.21. The kinetic energy is preserved along the geodesics.

PROOF. Using equation (1.32)

$$\begin{aligned} \frac{dE}{ds} &= \frac{d}{ds}\frac{\dot{x}_1^2 + \dot{x}_2^2}{2} = \dot{x}_1\ddot{x}_1 + \dot{x}_2\ddot{x}_2 \\ &= \langle \dot{x}, \ddot{x} \rangle = 4\theta \langle \dot{x}, \mathcal{J}\dot{x} \rangle = 0. \end{aligned}$$

□

In the following we shall show that the projection of the solution is a circle, using the conservation of energy. Let $2E = R^2$ with $R > 0$ constant. For conservative systems with $\dot{x}_1(s)^2 + \dot{x}_2(s)^2 = R^2$, there is a function $\alpha = \alpha(s)$ such that

$$\dot{x}_1(s) = R\cos\alpha(s), \qquad \dot{x}_2(s) = R\sin\alpha(s). \tag{1.37}$$

Then

$$\ddot{x}_1(s) = -R\sin\alpha(s)\,\dot{\alpha}(s), \qquad \ddot{x}_2(s) = R\cos\alpha(s)\,\dot{\alpha}(s),$$

and the system (1.31)

$$\begin{aligned} \ddot{x}_1 &= 4\theta\dot{x}_2 \\ \ddot{x}_2 &= -4\theta\dot{x}_1 \end{aligned}$$

becomes

$$\begin{aligned} \sin\alpha(s)\big(4\theta + \dot{\alpha}\big) &= 0 \\ \cos\alpha(s)\big(4\theta + \dot{\alpha}\big) &= 0. \end{aligned}$$

Adding the squares, yields $\dot{\alpha}(s) = 4\theta$. Integrating, we obtain

$$\alpha(s) = -4\theta s - \alpha_0.$$

Hence equations (1.37) can be integrated

$$\begin{aligned} x_1(s) &= R\int_0^s \cos(-4\theta s - \alpha_0)\,ds \\ &= \frac{R}{4\theta}\sin(4\theta s + \alpha_0) + x_1(0). \\ x_2(s) &= R\int_0^s \sin(-4\theta s - \alpha_0)\,ds \\ &= \frac{R}{4\theta}\cos(4\theta s + \alpha_0) + x_2(0), \end{aligned}$$

where

$$x_1(0) = -\frac{R\sin\alpha_0}{4\theta}, \qquad x_2(0) = -\frac{R\cos\alpha_0}{4\theta}.$$

We obtain the solution

$$x(s) = \frac{R}{4\theta}\Big(\sin(4\theta s + \alpha_0), \cos(4\theta s + \alpha_0)\Big) + x(0),$$

which is the parametric equation of a circle of radius $\frac{R}{4\theta}$ centered at $x(0)$.

1.5. The connection form

Let $x \to \mathcal{H}_x = \text{span}_x\{X_1, X_2\}$ be the horizontal distribution on $\mathbb{R}^3$. A connection 1-form is a non-vanishing form $\omega \in T^*\mathbb{R}^3$ such that $\ker_x\omega = \mathcal{H}_x$. The form ω is unique up to a multiplicative factor. In this chapter we shall choose the *standard* 1-form with the property $\omega(\partial_t) = 1$, which is

$$\omega = dt - 2(x_2 dx_1 - x_1 dx_2). \tag{1.38}$$

DEFINITION 1.22. The curvature 2-form of the distribution $\mathcal{H}$ is defined as $\Omega : \mathcal{H} \times \mathcal{H} \to \mathcal{F}(\mathcal{R}^3)$

$$\Omega(U, V) = d\omega(U, V). \tag{1.39}$$

In our case $\Omega = 4dx_1 \wedge dx_2$. If the horizontal distribution $\mathcal{H}$ belongs to the intrinsic subRiemannian geometry, the form Ω describes the extrinsic geometry of the Heisenberg group. In general, the 2-form Ω describes the non-integrability of the horizontal distribution.

DEFINITION 1.23. The pair $(\mathbb{R}^3, \omega)$ is called a contact manifold if $\omega \wedge \Omega$ never vanishes.

In our case $\omega \wedge \Omega = 4dt \wedge dx_1 \wedge dx_2$ and hence, the Heisenberg group becomes a contact manifold. The following theorem shows that, locally, all contact manifolds are the same as the Heisenberg group, see Cartan [**21**].

THEOREM 1.24. *(Darboux) Each point p of a contact manifold admits a local coordinate system t, x_1, x_2 in a neighborhood U such that $\omega = dt - 2(x_2dx_1 - x_1dx_2)$.*

1.5.0.1. *The osculator plane.* Let $c(s) = \big(x_1(s), x_2(s), t(s)\big)$ be a curve. The *osculator* plane at $c(s)$ is defined by $\text{span}\{\dot c(s), \ddot c(s)\}$.

PROPOSITION 1.25. Let $c(s)$ be a curve. Then the curve $c(s)$ is horizontal if and only if the osculator plane at $c(s)$ is the horizontal plane $\mathcal{H}_{c(s)}$, for any s.

PROOF. If $\text{span}\{\dot c(s), \ddot c(s)\} = \mathcal{H}_{c(s)}$, then $\dot c(s) \in \mathcal{H}_{c(s)}$. Hence, the curve is horizontal.
If the curve $c(s)$ is horizontal, $\dot c(s) \in \mathcal{H}_{c(s)}$. It suffices to show $\ddot c(s) \in \mathcal{H}_{c(s)}$. From the horizontality condition (1.14),

$$\dot t = 2x_2\dot x_1 - 2x_1\dot x_2. \tag{1.40}$$

Differentiating in equation (1.40) yields

$$\begin{aligned} \ddot t &= 2\dot x_2\dot x_1 + 2x_2\ddot x_1 - 2\dot x_1\dot x_2 - 2x_1\ddot x_2 \\ &= 2x_2\ddot x_1 - 2x_1\ddot x_2. \end{aligned} \tag{1.41}$$

The acceleration vector along $c(s)$ is

$$\begin{aligned} \ddot c &= \ddot x_1\partial_{x_1} + \ddot x_2\partial_{x_2} + \ddot t\partial_t \\ &= \ddot x_1\big(\partial_{x_1} + 2x_2\partial_t\big) - 2x_2\ddot x_1\partial_t \\ &\quad + \ddot x_2\big(\partial_{x_2} - 2x_1\partial_t\big) + 2x_1\ddot x_2\partial_t + \ddot t\partial_t \\ &= \ddot x_1 X_1 + \ddot x_2 X_2 + \big(\ddot t - 2x_2\ddot x_1 + 2\ddot x_2 x_1\big)\partial_t \\ &= \ddot x_1 X_1 + \ddot x_2 X_2, \end{aligned}$$

since we used equation (1.41). Hence, $\ddot c \in \mathcal{H}_c$ and the osculator plane is horizontal. □

COROLLARY 1.26. For a horizontal curve c

$$\ddot{c} = \ddot{x}_1 X_1 + \ddot{x}_2 X_2, \tag{1.42}$$

$$\ddot{t} = 2x_2\ddot{x}_1 - 2\ddot{x}_2 x_1. \tag{1.43}$$

DEFINITION 1.27. Let $J : \mathcal{H} \to \mathcal{H}$ be defined by $J(X_1) = -X_2$, $J(X_2) = X_1$. J is called the complex structure of the horizontal plane.

We shall use J in order to write the equations for the geodesics on the Heisenberg group. The following result shows that the geodesics satisfy a Newton type equation. The left side is the acceleration, while the right side is the force, which keeps the distribution bent. As before, θ is a constant.

PROPOSITION 1.28. A curve c is a geodesic on the Heisenberg group if and only if
(i) c is a horizontal curve and
(ii) c satisfies

$$\ddot{c} = 4\theta J\dot{c}. \tag{1.44}$$

PROOF. If $c(s)$ is a geodesic, by Proposition 1.15, $c(s)$ is horizontal. Using Corollary 1.26 and the system (1.31),

$$\begin{aligned}
\ddot{c} &= \ddot{x}_1 X_1 + \ddot{x}_2 X_2 \\
&= 4\theta\dot{x}_2 X_1 - 4\theta\dot{x}_1 X_2 \\
&= 4\theta\dot{x}_2 J(X_2) + 4\theta\dot{x}_1 J(X_1) \\
&= 4\theta J\big(\dot{x}_1 X_1 + \dot{x}_2 X_2\big) \\
&= 4\theta J(\dot{c}).
\end{aligned}$$

We shall now prove the converse: if (i) and (ii) hold, then c is a geodesic. We shall use the definition 1.14. The horizontality condition (i) can be written as $\dot{t} = \partial H/\partial\theta$, which is the Hamiltonian equation for t. Using a similar computation as in the first part, equation (1.44) written in components becomes the system (1.31). Let $x_1(s)$ and $x_2(s)$ be solutions of this system. Define the following curve in the cotangent space

$$\gamma(s) = \big(x_1(s), x_2(s), t(s), \xi_1(s), \xi_2(s), \theta\big),$$

where

$$\xi_1 = \dot{x}_1 - 2x_2(s)\theta, \quad \xi_2 = \dot{x}_2 + 2x_1\theta,$$

with θ constant. Then $\gamma(s)$ satisfies the bicharacteristics system (1.25) for the Hamiltonian (1.23). Then the projection on the (x,t)-space is a geodesic and hence $c(s)$ is a geodesic. □

The subRiemannian metric.

DEFINITION 1.29. A non-degenerate, positive definite bilinear form $g_x : \mathcal{H}_x \times \mathcal{H}_x \to \mathcal{F}(\mathbb{R})$ at any point $x \in \mathbb{R}^3$, is called a subRiemannian metric.

We shall consider the subRiemannian metric in which the vector fields X_1, X_2 are orthonormal. The subRiemannian metric will become a Kaehler metric on the horizontal distribution, as we shall explain later. The following definitions can be found, for instance, in Kobayashy and Nomizu, see [**37**].

DEFINITION 1.30. A Hermitian metric on a real vector space V with a complex structure J is a non-degenerate, positive definite inner product h such that

$$h(JX, JY) = h(X, Y), \qquad \text{for } X, Y \in V.$$

We associate with each Hermitian metric of a vector space V a skew-symmetric bilinear form on V.

DEFINITION 1.31. The fundamental 2-form Φ is defined by

$$\Phi(X, Y) = h(X, JY), \qquad \text{for all vector fields } X \text{ and } Y.$$

A Hermitian metric on a vector space V with a complex structure J is called a Kaehler metric if its fundamental 2-form is closed.

The relationship with the subRiemannian metric is given in the following proposition.

PROPOSITION 1.32. The subRiemannian metric g in which $\{X_1, X_2\}$ are orthonormal is a Kaehler metric on $\mathcal{H}_x$, for any $x \in \mathbb{R}^3$. The fundamental 2-form satisfies $4\Phi = \Omega$. Hence

$$\Omega(U, V) = 4g(U, JV), \qquad \text{for all horizontal vectors } U \text{ and } V.$$

PROOF. Consider $V = \mathcal{H}_x$. We shall show first that g is a Hermitian metric. Let $U = U^1X_1 + U^2X_2$ and $V = V^1X_1 + V^2X_2$ be two horizontal vector fields. Using $JX_1 = -X_2$ and $JX_2 = X_1$, yields

$$JU = -U^1X_2 + U^2X_1,$$

$$JV = -V^1X_2 + V^2X_1.$$

Using the orthonormality of X_1 and X_2 we obtain

$$\begin{aligned} g(JU, JV) &= g(U^2X_1 - U^1X_2, V^2X_1 - V^1X_2) \\ &= U^1V^1 + U^2V^2 \\ &= g(U^1X_1 + U^2X_2, V^1X_1 + V^2X_2) \\ &= g(U, V), \end{aligned}$$

and hence g is invariant by J. The 2-form Ω is closed because it is exact $\Omega = d\omega$.

$$\begin{aligned}
\Omega(U,V) &= \Omega(U^1X_1 + U^2X_2, V^1X_1 + V^2X_2) \\
&= (U^1V^2 - U^2V^1)\Omega(X_1, X_2) \\
&= (U^1V^2 - U^2V^1)\big(X_1(\omega(X_2)) - X_2(\omega(X_1)) - \omega([X_1, X_2])\big) \\
&= 4(U^1V^2 - U^2V^1)\omega(\partial_t) \\
&= 4(U^1V^2 - U^2V^1) \\
&= 4g(U^1X_1 + U^2X_2, V^2X_1 - V^1X_2) \\
&= 4g(U, JV).
\end{aligned}$$

□

Using the skew-symmetry of Ω we obtain:

COROLLARY 1.33.

$$g(U, JU) = 0, \qquad \text{for any horizontal vector } U.$$

The geometrical interpretation of Ω is given below.

PROPOSITION 1.34. Let $\pi : \mathbb{R}^3_{(x,t)} \to \mathbb{R}^2_x$ be the projection which sends the horizontal plane onto the x-plane

$$\pi_*(X_1) = \partial_{x_1}, \qquad \pi_*(X_2) = \partial_{x_2}.$$

Then, for any horizontal vectors U and V, the area of the parallelogram generated by $\pi_*(U)$ and $\pi_*(V)$ is equal to $|\Omega(U,V)|/4$.

PROOF. We have

$$\begin{aligned}
\Omega(U,V) &= 4\big(dx_1 \wedge dx_2\big)(U,V) = 4\begin{vmatrix} dx_1(U) & dx_1(V) \\ dx_2(U) & dx_2(V) \end{vmatrix} \\
&= 4\begin{vmatrix} U(x_1) & V(x_1) \\ U(x_2) & V(x_2) \end{vmatrix} = 4\begin{vmatrix} U^1 & V^1 \\ U^2 & V^2 \end{vmatrix}.
\end{aligned}$$

Using the interpretation of the determinant as an area, the proof is complete. □

PROPOSITION 1.35. Let $c(s)$ be a geodesic curve. Then

$$\theta\Omega(U, \dot{c}) = g(U, \ddot{c}) \qquad \text{for any horizontal vector } U.$$

PROOF. Using the Kaehler property of the metric g and the geodesics equation (1.44)

$$\theta\Omega(U, \dot{c}) = 4\theta g(U, J\dot{c}) = g(U, \ddot{c}).$$

□

If $\theta = 0$, then $g(U, \ddot{c}) = 0$ for any horizontal vector U and then $\ddot{c} = 0$. Hence $\ddot{c}_1(s) = \ddot{c}_2(s) = 0$. We obtain the following result:

COROLLARY 1.36. Let $c(s)$ be a geodesic for which the momentum θ vanishes. Then

$$c(s) = (as + a_0, bs + b_0, t_0),$$

with a, a_0, b, b_0 and t_0 constants.

PROOF. The fact that $c_1(s) = as + a_0$ and $c_2(s) = bs + b_0$ is obvious. We shall show that $t(s)$ is constant. We have

$$\begin{aligned} \dot{t}(s) &= 2(x_2\dot{x}_1 - x_1\dot{x}_2) \\ &= 2(c_2\dot{c}_1 - c_1\dot{c}_2) \\ &= 2\Big((bs + b_0)a - (as + a_0)b\Big) \\ &= 0, \end{aligned}$$

i.e., t is constant. □

The following proposition deals with the metric properties of the velocity and acceleration.

PROPOSITION 1.37. Let $c(s)$ be a geodesic. Then
(i) The velocity $\dot{c}$ and the acceleration $\ddot{c}$ vector fields are perpendicular in the subRiemannian metric.
(ii) The magnitude of $\dot{c}$ in the subRiemannian metric is constant along the geodesic.
(iii) The magnitude of $\ddot{c}$ in the subRiemannian metric is constant along the geodesic.

PROOF. (i) For $U = \dot{c}$, Proposition 1.35 yields

$$g(\dot{c}, \ddot{c}) = \theta\Omega(\dot{c}, \dot{c}) = 0.$$

(ii) Differentiating and using (i) yields

$$\frac{d}{ds}g(\dot{c}, \dot{c}) = g(\ddot{c}, \dot{c}) + g(\dot{c}, \ddot{c}) = 0,$$

i.e. $g(\dot{c}(s), \dot{c}(s))$ is constant.
(iii) The vector field $\ddot{c}$ is horizontal, and using equation (1.44) and (ii) we have

$$g(\ddot{c}, \ddot{c}) = g(4\theta J\dot{c}, 4\theta J\dot{c}) = 16\theta^2 g(J\dot{c}, J\dot{c}) = 16\theta^2 g(\dot{c}, \dot{c}) = \text{ constant.}$$

□

In the classical theory of three dimensional curves, the curvature becomes along a curve $c(s)$ is a function defined by $k(s) = |\dot{T}(s)|$, where $T(s) = \dot{c}(s)/|\dot{c}(s)|$ is the unit tangent vector. If s is the arc length parameter, $|\dot{c}(s)| = 1$ and the curvature becomes $k(s) = |\ddot{c}(s)|$.

PROPOSITION 1.38. The curvature of a geodesic curve is constant, $k(s) = 4|\theta|$.

PROOF. Consider the geodesic $c(s)$ parametrized by the arc length. Then

$$k(s)^2 = |\ddot{c}|^2 = g(\ddot{c}, \ddot{c}) = 16\theta^2 g(\dot{c}, \dot{c}) = 16\theta^2.$$

□

The curvature of a geodesic depends on the momentum θ. As we shall show later, θ is a Lagrange multiplier which describes the number of rotations of the geodesics around the t-axis. Hence, we shall be able to prove a Gauss-Bonett type theorem for the geodesic curves.

The following proposition provides the complex structure of the horizontal distribution in function of Ω and the basic horizontal vector fields.

PROPOSITION 1.39. For any horizontal vector field U, we have

$$J(U) = \frac{1}{4}\Big(\Omega(X_1, U)X_1 + \Omega(X_2, U)X_2\Big).$$

PROOF. As both sides are linear, it suffices to check the relation only for the basic vector fields X_1 and X_2. Using $\Omega(X_1, X_2) = -\Omega(X_2, X_1) = 4$, and $\Omega(X_i, X_i) = 0$, yields

$$JX_1 = -X_2 = \frac{1}{4}\Omega(X_2, X_1)X_2 = \frac{1}{4}\Big(\Omega(X_1, X_1)X_1 + \Omega(X_2, X_1)X_2\Big)$$

$$JX_2 = X_1 = \frac{1}{4}\Omega(X_1, X_2)X_1 = \frac{1}{4}\Big(\Omega(X_1, X_2)X_1 + \Omega(X_2, X_2)X_2\Big).$$

□

1.6. Lagrangian formalism on the Heisenberg group

In this section we shall find the subRiemannian geodesics and characterize their lengths. The horizontality condition is a constraint on velocities, *i.e.*, they are *non-holonomic*. The Lagrangian which describes the geodesics has a non-holonomic constraint. This constraint can be expressed using the 1-form ω.

PROPOSITION 1.40. If $\phi(s)$ is a horizontal curve, then

$$\int_\phi \omega = 0. \tag{1.45}$$

PROOF. As $\phi^*\omega$ is a 1-form on $\mathbb{R}$, then $\phi^*\omega$ and ds are proportional

$$\phi^*\omega(s) = h(s)\, ds,$$

where the proportionality function $h(s)$ is

$$h(s) = \phi^*_{(s)}\omega\Big(\frac{d}{ds}\Big).$$

Let $\phi(s)$ be defined on $[0, 1]$. Then

$$\begin{aligned}
\int_\phi \omega &= \int_0^1 \phi^*\omega = \int_0^1 h(s)\, ds = \int_0^1 \phi^*_{(s)}(\omega)\Big(\frac{d}{ds}\Big)\, ds \\
&= \int_0^1 \omega\Big(\phi_*\Big(\frac{d}{ds}\Big)\Big)\, ds = \int_0^1 \omega(\dot\phi(s)) = 0,
\end{aligned}$$

because $\dot\phi \in \mathcal{H}$. □

We shall associate a Lagrangian $L : T\mathbb{R}^3 \to \mathbb{R}$ with the Hamiltonian (1.23). This will be done using a formal Legendre transform in $(\dot x, \dot t)$, see [**14**]. In the case of a quadratic Hamiltonian, the Lagrangian is given by the maximal distance between the hyperplane $< \xi, \dot x > + \theta \dot t$ and the convex surface given by the Hamiltonian in $\mathbb{R}^6$:

$$\begin{aligned}
L(x, t, \dot x, \dot t) &= \max_{\xi,\theta}\Big(\xi_1\dot x_1 + \xi_2\dot x_2 + \theta\dot t - H(\xi, \theta, x, t)\Big) \\
&= \max_{\xi,\theta} F(x, t, \dot x, \dot t, \xi, \theta).
\end{aligned}$$

In our case the Hamiltonian is degenerate, so that the equivalence between the Lagrangian and the Hamiltonian formalism of Classical Mechanics might not hold. We shall still proceed formally, and set

$$\frac{\partial F}{\partial \xi} = 0, \quad \frac{\partial F}{\partial \theta} = 0 \tag{1.46}$$

as in Classical Mechanics, where the partial derivatives vanish when the maximum is reached.

The equations (1.46) can be written as

$$\dot{x}_i = \frac{\partial H}{\partial \xi_i}, \quad \dot{t} = \frac{\partial H}{\partial \theta} = 0, \tag{1.47}$$

$$\dot{x}_1 = \xi_1 + 2x_2\theta, \quad \dot{x}_2 = \xi_2 - 2x_1\theta, \quad \dot{t} = 2x_2\dot{x}_1 - 2x_1\dot{x}_2. \tag{1.48}$$

Using relations (1.48), we formally define the Lagrangian by

$$\begin{aligned} L(x,t,\dot{x},\dot{t}) &= \xi_1\dot{x}_1 + \xi_2\dot{x}_2 + \theta\dot{t} - \frac{1}{2}(\xi_1 + 2x_2\theta)^2 - \frac{1}{2}(\xi_2 - 2x_1\theta)^2 \\ &= (\dot{x}_1 - 2x_2\theta)\dot{x}_1 + (\dot{x}_2 + 2x_1\theta)\dot{x}_2 + \theta\dot{t} - \frac{1}{2}(\dot{x}_1^2 + \dot{x}_2^2) \\ &= (\dot{x}_1^2 + \dot{x}_2^2) + \theta(\dot{t} - 2x_2\dot{x}_1 + 2x_1\dot{x}_2) - \frac{1}{2}(\dot{x}_1^2 + \dot{x}_2^2) \\ &= \frac{1}{2}(\dot{x}_1^2 + \dot{x}_2^2) + \theta(\dot{t} - 2x_2\dot{x}_1 + 2x_1\dot{x}_2). \end{aligned}$$

Using the 1-connection form ω, we write the Lagrangian as

$$L(c,\dot{c}) = \frac{1}{2}(\dot{x}_1^2 + \dot{x}_2^2) + \theta\omega(\dot{c}), \tag{1.49}$$

where $c = (x_1, x_2, t)$. Let the action integral be

$$S(c,\tau) = \int_0^\tau L(c,\dot{c})\,ds = \int_0^\tau \frac{1}{2}(\dot{x}_1^2 + \dot{x}_2^2)\,ds + \theta\int_c \omega. \tag{1.50}$$

The action has two parts: the kinetic energy and the non-holonomic constraint. In this case the constant θ is a Lagrange multiplier. The curves c, which are critical points for the action $S(c,\tau)$, satisfy the Euler-Lagrange equation

$$\frac{d}{ds}\Big(\frac{\partial L}{\partial \dot{c}}\Big) = \frac{\partial L}{\partial c}. \tag{1.51}$$

The partial equivalence between the Lagrangian and the Hamiltonian formalism is given in the following result.

PROPOSITION 1.41. *A solution of the Euler-Lagrange equation (1.51) is a geodesic if and only if it is a horizontal curve.*

PROOF. As any geodesic is a horizontal curve, it suffices to show that a horizontal solution of (1.51) is a geodesic. If $c = (x_1, x_2, t)$, the equation (1.51) becomes on components

$$\ddot{x}_1 = 4\theta\dot{x}_2, \quad \ddot{x}_2 = -4\theta\dot{x}_1, \quad \dot{\theta} = 0 \Longrightarrow \theta = \text{constant}.$$

Proposition 1.28 completes the proof. □

REMARK 1.42. The Euler-Lagrange equation (1.51) does not imply horizontality of solutions, which corresponds to the Hamiltonian equation $\dot{t} = \partial H/\partial \theta$. Hence the Hamiltonian and the Euler-Lagrange systems are not equivalent. This behavior is specific to subRiemannian geometry and has no analog in the Riemannian case.

Lagrangian symmetries. The symmetry of the Lagrangian influences the symmetry of the solutions of the Euler-Lagrange system and hence the geodesics. The study of the Lagrangian symmetries makes easier the understanding of geodesic behavior.

PROPOSITION 1.43. The Lagrangian (1.49) is left invariant with respect to the Heisenberg Lie group structure, *i.e.*, $L(\overline{c}, \dot{\overline{c}}) = L(c, \dot{c})$, where $\overline{c} = L_a(c)$, for any $a \in \mathbf{H}_1$.

PROOF. If $c = (c_1, c_2, c_3)$, $\overline{c} = (\overline{c}_1, \overline{c}_2, \overline{c}_3)$ and $a = (a_1, a_2, a_3)$, then

$$\begin{aligned} \overline{c}_1 &= a_1 + c_1 \Longrightarrow \dot{\overline{c}}_1 = \dot{c}_1 \\ \overline{c}_2 &= a_2 + c_2 \Longrightarrow \dot{\overline{c}}_2 = \dot{c}_2 \\ \overline{c}_3 &= a_3 + c_3 - 2(a_1 c_2 - a_2 c_1) \Longrightarrow \dot{\overline{c}}_3 = \dot{c}_3 - 2(a_1 \dot{c}_2 - a_2 \dot{c}_1) \end{aligned}$$

Then the kinetic energy is left invariant

$$\frac{1}{2}(\dot{c}_1^2 + \dot{c}_2^2) = \frac{1}{2}(\dot{\overline{c}}_1^{\,2} + \dot{\overline{c}}_2^{\,2}).$$

The horizontal constraint is also invariant

$$\begin{aligned} \dot{\overline{c}}_3 - 2\overline{c}_2\dot{\overline{c}}_1 + 2\overline{c}_1\dot{\overline{c}}_2 &= \dot{c}_3 - 2(a_1\dot{c}_2 - a_2\dot{c}_1) - 2(a_2 + c_2)\dot{c}_1 + 2(a_1 + c_1)\dot{c}_2 \\ &= \dot{c}_3 - 2c_2\dot{c}_1 + 2c_1\dot{c}_2. \end{aligned}$$

Hence, the Lagrangian is preserved by left translations. □

COROLLARY 1.44. The solutions of the Euler-Lagrange equation (1.51) are left invariant by the translations on $\mathbf{H}_1$.

In the following we shall use Noether's theorem approach to find first integrals of motion for the Lagrangian (1.49)

$$L(x, t, \dot{x}, \dot{t}) = \frac{1}{2}(\dot{x}_1^2 + \dot{x}_2^2) + \theta(\dot{t} - 2x_2\dot{x}_1 + 2x_1\dot{x}_2).$$

The following theorem can be found in Arnold [**2**]. For its generalizations on manifolds one may consult [**14**].

THEOREM 1.45. *(Noether) To every one-parameter group of diffeomorphisms $(h_s)_s$ of the coordinate space M of a Lagrangian system which preserves the Lagrangian, corresponds a first integral of the Euler-Lagrange equations of motion $I : TM \to \mathbb{R}$ given by*

$$I(q, \dot{q}) = \Big\langle \frac{\partial L}{\partial \dot{q}}, \frac{dh_s(q)}{ds}\Big|_{s=0} \Big\rangle, \tag{1.52}$$

where $\langle , \rangle$ denotes the standard inner product on $\mathbb{R}^3$.

In the case of the Heisenberg group, $M = \mathbb{R}^3$, $q = (x, t)$ and

$$\frac{\partial L}{\partial \dot{q}} = \Big(\frac{\partial L}{\partial \dot{x}_1}, \frac{\partial L}{\partial \dot{x}_2}, \frac{\partial L}{\partial \dot{t}}\Big) = \Big(\dot{x}_1 - 2\theta x_2, \dot{x}_2 + 2\theta x_1, \theta\Big).$$

Consider the following three independent one-parameter groups of diffeomorphisms

$$h_s(x,t) = L_{a(s)}(x,t) = (a_1(s) + x_1, a_2(s) + x_2, a_3(s) + t - 2(a_1(s)x_2 - a_2(s)x_1)),$$

with $a(s) \in \{(0,0,s), (0,s,0), (s,0,0)\}$. The associated vector field is

$$\frac{dh_s(q)}{ds}\Big|_{s=0} = \begin{cases} (0,0,1), & \text{for } a(s) = (0,0,s) \\ (0,1,2x_1), & \text{for } a(s) = (0,s,0) \\ (1,0,-2x_2), & \text{for } a(s) = (s,0,0). \end{cases}$$

Hence, formula (1.52) provides the following three functional independent first integrals

$$\begin{aligned} I_1 &= \theta = \text{constant}, \\ I_2 &= \dot{x}_2 + 4\theta x_1 = \text{constant}, \\ I_3 &= \dot{x}_1 - 4\theta x_2 = \text{constant}. \end{aligned}$$

Differentiating, we obtain the Euler-Lagrange system

$$\begin{cases} \ddot{x}_1 = 4\theta \dot{x}_2 \\ \ddot{x}_2 = -4\theta \dot{x}_1 \\ \theta = \text{constant}, \end{cases}$$

which is equivalent to (1.51).

The rotational symmetry of the Lagrangian will provide a non obvious first integral. The Lagrangian is invariant by the one-parameter group of rotations $h_s(x,t) = (R_s x, t)$. Using $R_s(x) = e^{Js}x$, we have $\frac{dR_s}{ds} = Je^{Js}$ and hence, the vector field generated by the rotation is

$$\frac{dh_s(q)}{ds}\Big|_{s=0} = \big(Jx, 0\big) = (x_2, -x_1, 0).$$

The first integral associated to the rotation vector field is the kinetic momentum with respect to the t-axis

$$\begin{aligned} I &= (\dot{x}_1 - 2\theta x_2)x_2 + (\dot{x}_2 + 2\theta x_1)(-x_1) \\ &= (\dot{x}_1 x_2 - \dot{x}_2 x_1) - 2\theta |x|^2. \end{aligned}$$

Using the horizontality condition $\dot{t} = 2\dot{x}_1 x_2 - 2\dot{x}_2 x_1$, we get the first integral

$$2I = \dot{t} - 4\theta |x|^2 = \text{constant}. \tag{1.53}$$

PROPOSITION 1.46. If c is a geodesic, for any $a \in \mathbf{H}_1$ we have
(i) the left translation $\overline{c} = L_a c$ is also a geodesic,
(ii) the geodesics c and $\overline{c}$ have the same length.

PROOF. *(i)* Let c be a geodesic and let $\overline{c} = L_a c$. From Proposition 1.41 the curve c is horizontal and solves the equation (1.51). From Corollary 1.44 and Proposition 1.9, the curve $\overline{c}$ is a solution of the Euler-Lagrange equation (1.51) and it is horizontal. Using Proposition 1.41, we find that $\overline{c}$ is a geodesic.
(ii) From Corollary 1.10

$$\dot{\overline{c}} = \dot{c}_1 X_{1|\overline{c}} + \dot{c}_2 X_{2|\overline{c}}$$

and hence $|\dot{c}|^2 = |\dot{\overline{c}}|^2$. Proposition 1.37, (ii) yields

$$\ell(c) = \int_0^1 |\dot{c}|\, ds = |\dot{c}| = |\dot{\overline{c}}| = \int_0^1 |\dot{\overline{c}}|\, ds = \ell(\overline{c}).$$

□

Connectivity by geodesics. Consider a geodesic joining the points P and Q. By a left translation, the point P can be transformed into $(0,0,0)$. By Proposition 1.46, the geodesic is transformed into another geodesic of the same length, which starts at the origin. Hence, it makes sense to study the metric properties and the connectivity only for geodesics starting from the origin.

Because of the Lagrangian rotational symmetry, it is useful to use polar coordinates $x_1 = r\cos\phi$, $x_2 = r\sin\phi$. The Lagrangian becomes

$$L = \frac{1}{2}(\dot{r}^2 + r^2\dot{\phi}^2) + \theta(\dot{t} + 2r^2\dot{\phi}). \tag{1.54}$$

The symmetries for the Lagrangian will provide symmetries for the geodesics. The Lagrangian (1.54) is invariant under the symmetry

$$S : (r, \phi, t; \theta) \to (r, -\phi, -t; -\theta). \tag{1.55}$$

If $(r(s), \phi(s), t(s))$ is a geodesic corresponding to θ, then $(r(s), -\phi(s), -t(s))$ is a geodesic corresponding to $-\theta$. Consequently, whatever statement is made for the geodesics joining the origin with the point $(r(s), \phi(s), t(s))$, when $\theta > 0$, it can also be made for the geodesics between the origin and $(r(s), -\phi(s), -t(s))$, when $\theta < 0$. This allows us to do the analysis only for the case $\theta > 0$ and $\theta = 0$.

1.6.0.2. *Euler-Lagrange equations.* A computation shows

$$\frac{d}{ds}\frac{\partial L}{\partial \dot{r}} = \ddot{r} \qquad \frac{\partial L}{\partial r} = r\dot{\phi}(\dot{\phi} + 4\theta), \tag{1.56}$$

$$\frac{\partial L}{\partial \dot{\phi}} = r^2\dot{\phi} + 2\theta r^2, \qquad \frac{\partial L}{\partial \phi} = 0, \tag{1.57}$$

$$\frac{\partial L}{\partial \dot{t}} = \theta, \qquad \frac{\partial L}{\partial t} = 0, \tag{1.58}$$

and hence $r(s)$, $\phi(s)$ and θ satisfy the Euler-Lagrange system

$$\begin{cases} \ddot{r} = r\dot{\phi}(\dot{\phi} + 4\theta) \\ r^2(\dot{\phi} + 2\theta) = C(\text{constant}) \\ \theta = \theta_0(\text{constant}). \end{cases} \tag{1.59}$$

If a geodesic starts at the origin, then $r(0) = 0$ and hence $C = 0$. The second equation of (1.59) yields $\dot{\phi} = -2\theta$. With the assumption $\theta > 0$, the argument angle ϕ will rotate clock-wise. The Euler-Lagrange system becomes

$$\begin{cases} \ddot{r} = -4\theta^2 r \\ \dot{\phi} = -2\theta \\ \theta = \theta_0(\text{constant}). \end{cases} \tag{1.60}$$

When $\theta = 0$, the system (1.60) becomes

$$\begin{cases} \ddot{r} = 0 \\ \dot{\phi} = 0 \\ \theta = 0. \end{cases} \tag{1.61}$$

PROPOSITION 1.47. Given a point $P(x, 0)$, there is a unique geodesic between the origin and P. It is a straight line in the plane $\{t = 0\}$ of length $|x|$, and it is obtained for $\theta = 0$.

PROOF. In the case $\theta = 0$, the Hamiltonian is $H(\xi) = \frac{1}{2}\xi_1^2 + \frac{1}{2}\xi_2^2$. From the Hamiltonian equation $\dot{t} = \partial H/\partial\theta = 0$, and hence $t(s)$ is constant. Since $t(0) = 0$, it follows that $t(s) = 0$ and the solution belongs to the x-plane. Using the equation $\ddot{r} = 0$, it follows the solution is a straight line. □

From now on, unless otherwise stated, we shall assume $\theta > 0$. From the system (1.60) we can arrive at a first integral of energy.

PROPOSITION 1.48. (i) The function $I(r, \dot{t}, \theta) = \dot{r}^2 + 4\theta^2 r^2$ is a first integral for the system (1.60).
(ii) $\frac{1}{2}I(r, \dot{t}, \theta)$ is equal to the energy of the system, *i.e.*, $I = \dot{x}_1^2 + \dot{x}_2^2$.

PROOF. (*i*) Differentiating

$$\frac{d}{ds}I(r, \dot{t}, \theta) = 2\dot{r}(\ddot{r} + 4\theta^2 r) = 0.$$

(*ii*) In polar coordinates, $\dot{x}_1^2 + \dot{x}_2^2 = \dot{r}^2 + r^2\dot{\phi}^2$. Using $\dot{\phi} = -2\theta$ completes the proof. □

Let $c(s) = \big(x_1(s), x_2(s), t(s)\big)$ be a geodesic. Since $c(s)$ is a horizontal curve,

$$\dot{c} = \dot{x}_1 X_1 + \dot{x}_2 X_2$$

Let g denote the subRiemannian metric in which X_1, X_2 are orthonormal. Then

$$|\dot{c}(s)|_g^2 = g(\dot{c}(s), \dot{c}(s)) = \dot{x}_1^2(s) + \dot{x}_2^2(s).$$

If s is the arc length along the geodesic, the curve $c(s)$ becomes unit speed

$$\dot{x}_1^2(s) + \dot{x}_2^2(s) = 1.$$

Proposition 1.48 yields

$$\dot{r}^2(s) + 4\theta^2 r^2(s) = 1. \tag{1.62}$$

Separating the variables, we have

$$\frac{dr}{\sqrt{1 - 4\theta^2 r^2}} = \pm ds,$$

and integrating, we obtain

$$\frac{1}{2\theta}\arcsin(2\theta r(s)) = \pm s,$$

which yields

$$r(s) = \pm\frac{1}{2\theta}\sin(2\theta s), \tag{1.63}$$

where we consider

$$+ \quad \text{sign for } 2n\pi \le 2\theta s \le (2n+1)\pi, \text{ and}$$
$$- \quad \text{sign for } (2n+1)\pi \le 2\theta s \le (2n+2)\pi.$$

This yields a circle of diameter $1/(2\theta)$ which passes through the origin.

LEMMA 1.49. *If $\phi(0) = \phi_0$, then*

$$r(\phi)^2 = \Big(\frac{1}{2\theta}\Big)^2 \sin^2(\phi - \phi_0), \tag{1.64}$$

$$t(\phi) - t(\phi_0) = \frac{1}{4\theta^2}\frac{\sin 2(\phi-\phi_0) - 2(\phi-\phi_0)}{2}. \tag{1.65}$$

PROOF. From the second equation of the system (1.60) we obtain $\phi(s) - \phi(0) = -2\theta s$. Substituting in (1.63) yields (1.64). One of the Hamiltonian equations yields

$$\dot t = \frac{\partial H}{\partial \theta} = 2(\dot x_1 x_2 - x_1 \dot x_2) = -2r^2\dot\phi.$$

Integrating between ϕ_0 and ϕ and using equation (1.64) we have

$$\begin{aligned} t(\phi) - t(\phi_0) &= -2\int_{\phi_0}^{\phi} r^2(\phi)\,d\phi = -2\Big(\frac{1}{2\theta}\Big)^2 \int_{\phi_0}^{\phi} \sin^2(\phi-\phi_0)\,d\phi \\ &= -\frac{2}{4\theta^2}\int_0^{\phi-\phi_0} \sin^2 u\,du = \frac{1}{4\theta^2}\frac{\sin 2(\phi-\phi_0) - 2(\phi-\phi_0)}{2}. \end{aligned}$$

□

1.6.0.3. *Boundary conditions.* We are interested in connecting the origin with a point $P(x, t)$ by a geodesic $c : [0, s_f] \to \mathbb{R}^3$. Consider the following boundary conditions:

$$x(0) = 0, \quad t(0) = 0, \quad \phi(0) = \phi_0, \tag{1.66}$$

$$\|x(s_f)\| = R, \quad t(s_f) = t, \quad \phi(s_f) = \phi_f. \tag{1.67}$$

Because of the rotational invariance around the t-axis, we may choose $\phi_0 = 0$. From $\dot t = -2r^2\dot\phi$ and $\dot\phi = -2\theta$, we get $\dot t = 4\theta r^2 > 0$. Hence $t(s)$ is increasing and if $t(0) = 0$, then $t(s_f) > 0$.

LEMMA 1.50. *The following relations among the boundary conditions take place:*

$$\phi_f = -2\theta s_f, \tag{1.68}$$

$$(\sin\phi_f)^2 = 4\theta^2 R^2, \tag{1.69}$$

$$t = \frac{1}{4\theta^2}\frac{\sin(2\phi_f) - 2\phi_f}{2}, \tag{1.70}$$

$$\frac{t}{R^2} = -\mu(\phi_f) = \mu(2\theta s_f), \tag{1.71}$$

where

$$\mu(x) = \frac{x}{\sin^2 x} - \cot x. \tag{1.72}$$

PROOF. Integrating in the second equation of the system (1.60) and using the boundary conditions yields (1.68). Equations (1.64) and (1.65) together with the boundary conditions (1.66)-(1.67) yield equations (1.69) and (1.70). Eliminating $4\theta^2$ from equations (1.69) and (1.70) yields (1.71). □

The behavior of the function μ given by (1.72) is very important in understanding the subRiemannian geometry of the Heisenberg group. The function μ was first used by Beals, Gaveau and Greiner in [**3**]. The graph of μ for $x \geq 0$ is given in Figure 1.2.

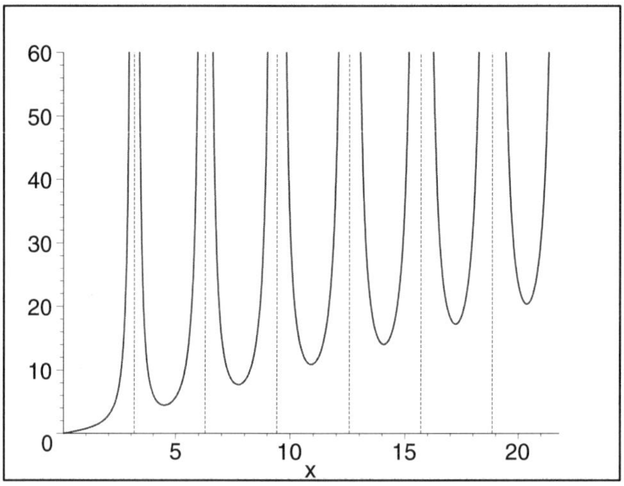

Figure 1.2: The graph of $\mu(x)$ for $x \geq 0$.

LEMMA 1.51. *μ is a monotone increasing diffeomorphism of the interval $(-\pi, \pi)$ onto $\mathbb{R}$. On each interval $(m\pi, (m+1)\pi)$, $m = 1, 2, \dots$, μ has a unique critical point x_m. On this interval μ decreases strictly from $+\infty$ to $\mu(x_m)$ and then increases strictly from $\mu(x_m)$ to $+\infty$. Moreover*

$$\mu(x_m) + \pi < \mu(x_{m+1}), \quad m = 1, 2, \dots \tag{1.73}$$

$$0 < \left(m + \frac{1}{2}\right)\pi - x_m < \frac{1}{m\pi}. \tag{1.74}$$

PROOF. As μ is an odd function, it suffices to show that it is a monotone increasing diffeomorphism of the interval $(0, \pi)$ onto $(0, +\infty)$. We note that $\sin x - x\cos x$ vanishes at $x = 0$ and it is increasing on $(0, \pi)$. Then

$$\frac{1}{2}\mu'(x) = \frac{\sin x - x\cos x}{\sin^3 x} = \begin{cases} = 1/3, & x = 0, \\ > 1/3, & x \in (0, \pi). \end{cases}$$

The first identity holds as an application of the l'Hospital rule:

$$\begin{aligned} \lim_{x \to 0} \frac{\sin x - x\cos x}{\sin^3 x} &= \lim_{x \to 0} \frac{\cos x - \cos x + x\sin x}{3\sin^2 x} \\ &= \frac{1}{3} \lim_{x \to 0} \frac{x}{\sin x} = \frac{1}{3}. \end{aligned}$$

The second inequality holds because

$$\frac{1}{2}\mu''(x) = \frac{x + 2x\cos^2 x - 3\cos x \sin x}{\sin^4 x} > 0.$$

The numerator vanishes at $x = 0$, and its derivative is

$$4\sin x(\sin x - x\cos x) > 0, \quad x \in (0, \pi).$$

Therefore μ is a diffeomorphism of the interval $(0, \pi)$ onto $(0, \infty)$. In the interval $(m\pi, (m+1)\pi)$ the function μ approaches $+\infty$ at the endpoints. In order to find the critical points, we set

$$\frac{1}{2}\mu'(x) = \frac{\sin x - x\cos x}{\sin^3 x} = \frac{1 - x\cot x}{\sin^4 x} = 0.$$

Hence the critical point x_m is the solution of the equation $x = \tan x$ on the interval $(m\pi, (m+1)\pi)$. Note that

$$\begin{aligned}
\mu(x+\pi) &= \frac{x+\pi}{\sin^2(x+\pi)} - \cot(x+\pi) \\
&= \frac{x}{\sin^2(x+\pi)} - \cot(x+\pi) + \frac{\pi}{\sin^2 x} \\
&= \mu(x) + \frac{\pi}{\sin^2 x},
\end{aligned}$$

so the successive minimum values increase by more than π. From Figure 1.3 we have

$$m\pi < x_m < m\pi + \frac{\pi}{2} = (m + \frac{1}{2})\pi. \tag{1.75}$$

Using $x_m = \tan x_m$, yields

$$\cot x_m = \frac{1}{x_m} < \frac{1}{m\pi}. \tag{1.76}$$

Let $f(x) = \cot x$. As $f'(x) = -\frac{1}{\sin^2 x} < -1$, there is a ξ between x and y such that

$$f(x) - f(y) = f'(\xi)(x - y) < -(x - y).$$

Hence $x - y < f(y) - f(x)$. Choosing $x = m\pi + \frac{\pi}{2}$, $y = x_m$ and using

$$f(m\pi + \frac{\pi}{2}) = \frac{\cos(m\pi + \frac{\pi}{2})}{\sin(m\pi + \frac{\pi}{2})} = 0,$$

and (1.76) yields

$$0 < (m + \frac{1}{2})\pi - x_m < \cot x_m < \frac{1}{m\pi}.$$

□

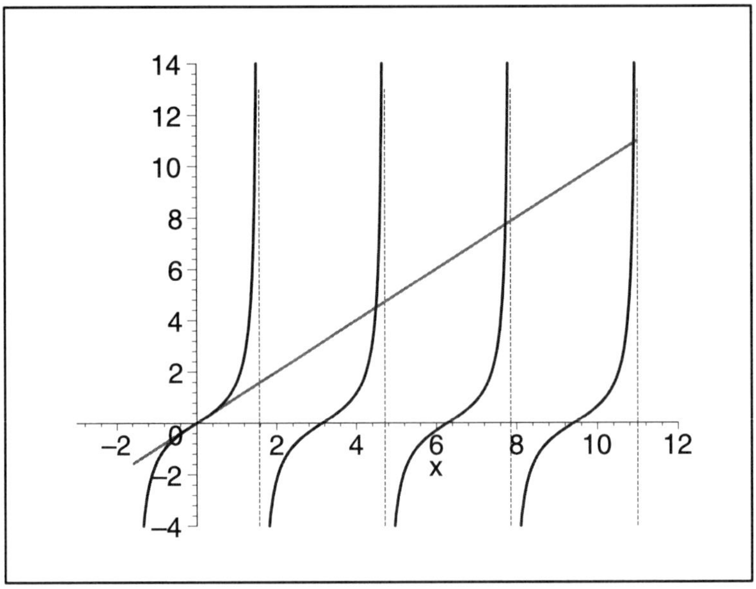

Figure 1.3*: Critical points of* μ *are solutions of* $\tan x = x$.

The number of geodesics which join the origin with an arbitrary given point is given in the following theorems, which can be found in [**3**].

THEOREM 1.52. *There are finitely many geodesics that join the origin to* (x, t) *if and only if* $x \neq 0$. *These geodesics are parametrized by the solutions* ζ *of*

$$\frac{|t|}{\|x\|^2} = \mu(\zeta). \tag{1.77}$$

There is exactly one such geodesic if and only if

$$|t| < \mu(x_1)\|x\|^2, \tag{1.78}$$

where x_1 *is the first critical point of* μ. *The number of geodesics increase without bound as* $|t|/\|x\|^2 \to \infty$. *Let* $\zeta_1 < \zeta_2 < \cdots < \zeta_N$ *be the solutions of (1.77). The square of the length of the geodesic associated with the solution* ζ_m *is*

$$s_m^2 = \left(\frac{\zeta_m}{\sin \zeta_m}\right)^2 \|x\|^2. \tag{1.79}$$

PROOF. The enumeration of geodesics follows from Lemma 1.51. The line $y = |t|/\|x\|^2$ intersects the graph of μ finitely many times. There is only one intersection if and only if inequality (1.78) holds. See the graph of μ in Figure 1.2. As the geodesic was parametrized by arc length, its length is given by the value of the parameter s_f. Dividing the square of the equation (1.68) by equation (1.69) yields the square of the length

$$s_f^2 = \left(\frac{\phi_f}{\sin \phi_f}\right)^2 R^2.$$

ϕ_f satisfies equation (1.71). For each $\phi_f = -\zeta_m$ we obtain the length described by formula (1.79). □

The natural dilations of the Heisenberg Laplacian operator

$$\frac{1}{2}\Big(\partial_{x_1} + 2x_2\partial_t\Big)^2 + \frac{1}{2}\Big(\partial_{x_2} - 2x_1\partial_t\Big)^2$$

are

$$(x_1, x_2, t) \to (\lambda x_1, \lambda x_2, \lambda^2 t), \quad \lambda > 0.$$

We are looking for a formula for the length of geodesics, different than (1.79), such that s_f is homogeneous of degree 1 with respect to the above dilations. This is done in the next result.

THEOREM 1.53. *Consider the geodesics joining the origin with the point (x,t), $x \neq 0$. Let $\zeta_1, \dots, \zeta_N$ be the solutions of equation (1.77). Then the square of the lengths is given by*

$$s_m^2 = \nu(\zeta_m)\Big(|t| + \|x\|^2\Big), \tag{1.80}$$

where

$$\nu(x) = \frac{x^2}{x + \sin^2 x - \sin x \cos x}.$$

PROOF. Equation (1.77) yields

$$|t| + \|x\|^2 = \mu(\zeta_m)\|x\|^2 + \|x\|^2 = (1 + \mu(\zeta_m))\|x\|^2,$$

and hence

$$\|x\|^2 = \frac{1}{1+\mu(\zeta_m)}\Big(|t| + \|x\|^2\Big).$$

Using equation (1.79) and the definition of μ given in (1.72)

$$\begin{aligned} s_m^2 &= \Big(\frac{\zeta_m}{\sin\zeta_m}\Big)^2\|x\|^2 = \Big(\frac{\zeta_m^2}{\sin^2\zeta_m}\cdot\frac{1}{1+\mu(\zeta_m)}\Big)\Big(|t| + \|x\|^2\Big) \\ &= \frac{\zeta_m^2}{\sin^2\zeta_m + \zeta_m - \sin^2\zeta_m\cot\zeta_m}\Big(|t| + \|x\|^2\Big) = \nu(\zeta_m)\Big(|t| + \|x\|^2\Big). \end{aligned}$$

For the graph of ν see Figure 1.4. □

PROPOSITION 1.54. The projection of the geodesics joining the origin and (x,t), $x \neq 0$ are circles or arcs of circle with diameters of at least $\|x\|$

$$\frac{1}{2\theta_m} = \frac{\|x\|}{|\sin\zeta_m|} \geq \|x\|. \tag{1.81}$$

PROOF. From the first equation of Lemma 1.49, the diameter of the circle is $1/2\theta$. Using relation (1.69) yields

$$\frac{1}{2\theta} = \frac{R}{|\sin\phi_f|}.$$

Replacing R by $\|x\|$ and ϕ_f by ζ_m, we arrive at relation (1.81). □

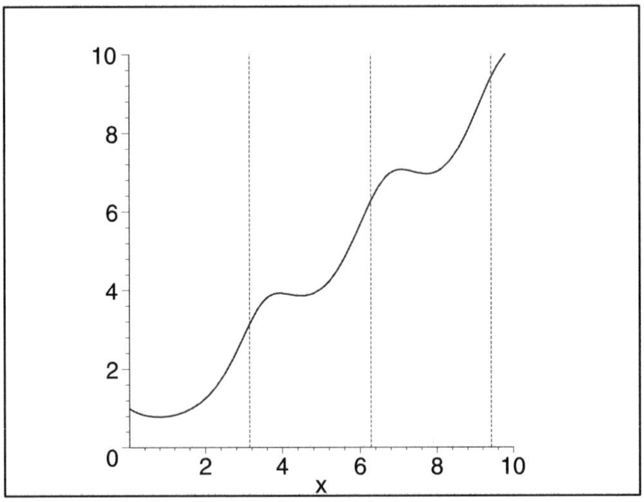

Figure 1.4: The graph of $\nu(x)$.

1.6.0.4. *Geodesics between the origin and points on the t-axis.* We have seen that the number of geodesics increases without bound as $|t|/\|x\|^2 \to \infty$. The following theorem deals with the limit case $\|x\| = 0$. This result was found by Gaveau (see [**26**] and [**3**]).

THEOREM 1.55. *The geodesics that join the origin to a point $(0,0,t)$ have lengths $s_1, s_2 \dots$, where*

$$s_m^2 = m\pi|t|. \tag{1.82}$$

For each length s_m, the geodesics of that length are parametrized by the circle $\mathbb{S}^1$.

PROOF. We shall treat the problem as the limit case $\|x\| \to 0$ of Theorem 1.53. Then the solutions $\zeta_m \to m\pi$. Relation (1.80) becomes

$$s_m^2 = \nu(m\pi)|t|,$$

with the coefficient given by

$$\nu(m\pi) = \frac{m^2\pi^2}{m\pi + \sin^2 m\pi - \sin m\pi \, \cos m\pi} = m\pi.$$

□

The above proof is using Theorem 1.53. In the following we shall provide a direct proof. The approach will use polar coordinates and the fact that the solution is given by

$$r(\phi) = r_{max}|\sin(\phi - \phi_0)|.$$

Consider the solution γ parameterized by $[0,1]$, with the end points $\gamma(0) = (0,0,0)$ and $\gamma(1) = (0,0,t)$. Using that $|\dot\gamma|$ is constant along the solution, we have the identity in Cauchy's inequality

$$\ell(\gamma) = \int_0^1 |\dot\gamma(s)|\, ds = \Big(\int_0^1 ds\Big)^{1/2} \Big(\int_0^1 |\dot\gamma(s)|^2\Big)^{1/2} = \sqrt{2E},$$

where E is the constant value of the energy along the solution

$$E = \frac{1}{2}|\dot{\gamma}|^2 = \frac{1}{2}(\dot{x}_1^2 + \dot{x}_2^2).$$

1.6.0.5. *Quantization of θ.* θ is constant along the solution and it is an eigenvalue for the following boundary value problem

$$\begin{aligned} \ddot{r} &= -4\theta^2 r \\ r(0) &= 0,\ r(1) = 0. \end{aligned}$$

The solution is $r(s) = A\sin(2\theta s)$. The boundary conditions yield

$$2\theta = m\pi, \tag{1.83}$$

with m integer. Each value of θ will determine a certain length.

1.6.0.6. *The energy and t.* Consider the boundary condition $t(1) = t$. One has $\dot{\phi} = -2\theta$ and then $\phi(s) = \phi_0 - 2\theta s$, where $\phi_0 = \phi(0)$ and $\phi_1 = \phi(1) = \phi_0 - 2\theta = \phi_0 - m\pi$. One may integrate between ϕ_0 and $\phi_1 = \phi_0 - m\pi$ in the equation $\dot{t} = -2r^2\dot{\phi}$.

$$\begin{aligned} t &= -2\int_{\phi_0}^{\phi_0 - m\pi} r^2(\phi)\, d\phi = -2\, r_{max}^2 \int_{\phi_0}^{\phi_0 - m\pi} \sin^2(\phi - \phi_0)\, d\phi \\ &= -2\, r_{max}^2 \int_0^{-m\pi} \sin^2 u\, du = 2\, r_{max}^2 \int_0^{m\pi} \sin^2 u\, du \\ &= 2\, r_{max}^2 (\frac{1}{2}u - \frac{1}{4}\sin 2u)\Big|_0^{m\pi} = r_{max}^2 m\pi = 2\theta r_{max}^2. \end{aligned}$$

From the conservation of energy

$$\frac{1}{2}(\dot{r}^2 + r^2\dot{\phi}^2) = \frac{1}{2}(\dot{r}^2 + 4r^2\theta^2) = E.$$

When $r = r_{max}^2$, then $\dot{r} = 0$. Hence

$$r_{max}^2 = \frac{E}{2\theta^2},$$

and using $t = 2\theta r_{max}^2$, one obtains

$$t = \frac{E}{\theta}.$$

The lengths are

$$(\ell_m)^2 = 2E = 2\theta t = m\pi t\,, \quad m = 1, 2, 3\ldots$$

PROPOSITION 1.56. The equations of the geodesics starting at the origin are

$$\begin{aligned} r_m(\phi) &= \sqrt{\frac{|t|}{m\pi}}\,\sin(\phi - \phi_0) \\ t_m(\phi) &= \frac{|t|}{2m\pi}\Big(\sin(2(\phi - \phi_0)) - 2(\phi - \phi_0)\Big), \quad m \geq 1. \end{aligned}$$

PROOF. Using

$$r_{max} = \sqrt{\frac{E}{2\theta^2}} = \sqrt{\frac{|t|}{2\theta}} = \sqrt{\frac{|t|}{m\pi}}$$

then

$$r_m(\phi) = \sqrt{\frac{|t|}{m\pi}}\sin(\phi - \phi_0).$$

For computing $t_m(\phi)$ one integrates $dt = -2r^2\,d\phi$ between ϕ_0 and ϕ

$$t(\phi) = -2r_{max}^2 \int_{\phi_0}^{\phi} \sin^2(\phi - \phi_0)\,d\phi = -2\frac{|t|}{m\pi}\int_0^{\phi-\phi_0} \sin^2 u\,du.$$

Hence

$$t_m(\phi) = \frac{|t|}{2m\pi}\Big(\sin(2(\phi - \phi_0)) - 2(\phi - \phi_0)\Big).$$

□

COROLLARY 1.57. The projection of the m-th geodesic on the x-plane is a circle with the radius

$$R_m = \frac{1}{2}\sqrt{\frac{|t|}{m\pi}} \tag{1.84}$$

and area

$$\sigma_m = \frac{|t|}{4m}.$$

We note that r_m and t_m depend on ϕ_0 but the corresponding lengths $\ell_m = \sqrt{m\pi|t|}$ do not. This is because of the rotational symmetry around the t-axis.

1.6.0.7. *A Milnor-type property.* In the classical theory of three dimensional curves one defines the total curvature of a curve $c : [0, \tau] \to \mathbb{R}^3$ as $\int_0^\tau k(u)\,du$, where $k(s)$ is the curvature along the curve c. Milnor's theorem states

THEOREM 1.58. *The total curvature of a closed curve in $\mathbb{R}^3$ is $2\pi\chi$, where χ is an integer.*

It is known that χ represents the number of knots of the curve c. In the case of a plane curve, $\chi = 1$. This result was proved independently by Fechnel. For these proofs the reader may consult Millman and Parker [**46**].

We shall show a similar property for the geodesics between the origin and points on the t-axis. We shall prove the following:

PROPOSITION 1.59. The total curvature of a geodesic between the origin and the point $(0, 0, t)$ is $2\pi m$, where $m = 1, 2, \dots$ is an integer which gives the number of rotations of the geodesic.

PROOF. Consider the curve $c : [0, s_f] \to \mathbb{R}^3$ joining $(0, 0, 0)$ and $(0, 0, t)$, parametrized by the arc length. Then $|\dot{c}(u)| = 1$, and by Proposition 1.38, the curvature is $k(u) = |\ddot{c}(u)| = 4|\theta|$. Hence, the total curvature is

$$\int_0^{s_f} k(u)\,du = 4|\theta|s_f.$$

From the boundary value problem

$$\begin{aligned} \ddot{r} &= -4\theta^2 r \\ r(0) &= 0,\ r(s_f) = 0 \end{aligned}$$

it follows that $2\theta s_f = m\pi$. Therefore, the total curvature is equal to $2\pi m$, $m = 1, 2, \dots$ □

Geodesics between any two arbitrary points. Given two points $P_1(x_1, y_1, t_1)$ and $P_2(x_2, y_2, t_2)$, we shall study the connectivity by geodesics in the cases $(x_1, y_1) = (x_2, y_2)$ and $(x_1, y_1) \neq (x_2, y_2)$. Making a left translation by $(-x_1, -y_1, -t_1)$, the point P_1 will become the origin and the point P_2 will have the coordinates

$$\big(x_2 - x_1, y_2 - y_1, t_2 - t_1 - 2(y_1 x_2 - x_1 y_2)\big).$$

By Proposition 1.46 the left translation of a geodesic is a geodesic of the same length. Then theorems 1.52 and 1.53 become:

THEOREM 1.60. *Given the points $P_1(x_1, y_1, t_1)$ and $P_2(x_2, y_2, t_2)$, there are finitely many geodesics between P_1 and P_2 if and only if $(x_1, y_1) \neq (x_2, y_2)$. These geodesics are parametrized by the solutions ζ of*

$$\frac{|t_2 - t_1 - 2(y_1 x_2 - x_1 y_2)|}{(x_2 - x_1)^2 + (y_2 - y_1)^2} = \mu(\zeta). \tag{1.85}$$

There is exactly one such geodesic if and only if

$$|t_2 - t_1 - 2(y_1 x_2 - x_1 y_2)| < \mu(c_1)[(x_2 - x_1)^2 + (y_2 - y_1)^2], \tag{1.86}$$

where c_1 is the first critical point of μ. The number of geodesics increase without bound as

$$\frac{|t_2 - t_1 - 2(y_1 x_2 - x_1 y_2)|}{[(x_2 - x_1)^2 + (y_2 - y_1)^2]} \to \infty.$$

Let $\zeta_1 < \zeta_2 < \cdots < \zeta_N$ be the solutions of (1.85). The square of the length associated with the solution ζ_m is

$$\begin{aligned} s_m^2 &= \Big(\frac{\zeta_m}{\sin \zeta_m}\Big)^2 [(x_2 - x_1)^2 + (y_2 - y_1)^2] \\ &= \nu(\zeta_m)\Big(|t_2 - t_1 - 2(y_1 x_2 - x_1 y_2)| + [(x_2 - x_1)^2 + (y_2 - y_1)^2]\Big), \end{aligned}$$

where

$$\nu(x) = \frac{x^2}{x + \sin^2 x - \sin x \cos x}.$$

Theorem 1.55 becomes:

THEOREM 1.61. *Given the points $P_1(x_1, y_1, t_1)$ and $P_2(x_2, y_2, t_2)$, with $x_1 = x_2$ and $y_1 = y_2$, the geodesics that join P_1 and P_2 have lengths*

$$s_m^2 = m\pi |t_2 - t_1|.$$

1.7. Carnot-Carathéodory distance

Recall that a curve $c(s) = \big(x_1(s), x_2(s), t(s)\big)$ is horizontal if $\dot{c}(s) \in \mathcal{H}_{c(s)}$, *i.e.*,

$$\dot{t} - 2x_2\dot{x}_1 + 2x_1\dot{x}_2 = 0.$$

By Proposition 1.11, any two points P and Q can be joined by a horizontal curve. We have shown that the curve can be considered smooth. Hence the set

$$\{c;\ c(0) = P, c(1) = Q, c \text{ horizontal curve}\} \neq \emptyset.$$

The length of a horizontal curve c is

$$\ell(c) = \int_0^1 \sqrt{g(\dot{c}(s), \dot{c}(s))}\, ds,$$

where $g : \mathcal{H} \times \mathcal{H} \to \mathcal{F}$ is the subRiemannian metric. The Carnot-Carathéodory distance is defined as $d_C : \mathbb{R}^3 \times \mathbb{R}^3 \to [0, \infty)$,

$$d_C(P, Q) = \inf\{\ell(c); c \in S\}. \tag{1.87}$$

One may verify that d_C verifies the distance axioms:
(i) $d_C(P, P) = 0$,
(ii) $d_C(P, Q) = d(Q, P)$,
(iii) $d_C(P, R) \le d_C(P, Q) + d(Q, R)$.

The main result of this section is to show that the Carnot-Carathéodory distance d_C is complete, *i.e.*, any Cauchy sequence with respect to d_C is convergent. We shall prove this using the equivalence between the geodesic completeness and the completeness as a metric space.

DEFINITION 1.62. If for any point P, any geodesic $c(t)$ emanating from P is defined for all $t \in \mathbb{R}$, the geometry is called geodesically complete.

The following theorem can be found in Strichartz [**50**], [**51**]:

THEOREM 1.63. *Let M be a connected step 2 subRiemannian manifold.*
(a) If M is complete, then any two points can be joined by a geodesic.
(b) If there exists a point P such that every geodesic from P can be indefinitely extended, then M is complete.
(c) Every nonconstant geodesic is locally a unique length minimizing curve.
(d) Every length minimizing curve is a geodesic.

It is known that geodesics starting from the origin on the Heisenberg group are infinitely extendable. Using (b) of Theorem 1.63 we obtain:

THEOREM 1.64. *The Carnot-Carathéodory metric d_C is complete.*

Computing the Carnot-Carathéodory distance. In the following we shall describe two ways to obtain the Carnot-Carathéodory distance.

First method:
From (c) and (d) of Theorem 1.63 we obtain a way to compute the Carnot-Carathéodory distance as

$$d_C(P, Q) = \{\ell(c),\ \text{where } c \text{ is the shortest geodesic joinning P and Q}\}.$$

Using Theorem 1.53 with $m = 1$, the Carnot-Carathéodory distance from the origin to (x, t) is

$$d_C(0, (x,t)) = \nu(\zeta_1)\big(|t| + \|x\|^2\big).$$

In particular, if the points are on the same vertical line, the Carnot-Carathéodory distance squared is proportional to the Euclidian distance

$$d_C\big((x,t),(x,t')\big)^2 = \pi|t' - t| = \pi d_E\big((x,t),(x,t')\big).$$

Second method:
Another way to obtain the Carnot-Carathéodory distance is using the complex action. We shall discuss this issue in more detail later in the Complex Hamiltonian Mechanics chapter, (ch 5).
Consider the modified complex action function

$$f(x,t,\tau) = \tau g(x,t,\tau) = -i\tau t + \tau\coth(2\tau)\|x\|^2,$$

where

$$\begin{aligned} g(x,t,\tau) &= -it + \int_0^\tau \{\langle \dot{x}, \xi\rangle - H\}\,ds \\ &= -it + \coth(2\tau)\|x\|^2 \end{aligned}$$

is the complex action. Like the classical action, the complex action g satisfies the Hamilton-Jacobi equation, see [**3**]

$$\frac{\partial g}{\partial \tau} + H\Big(x, \frac{\partial g}{\partial x}\Big) = 0.$$

Using Theorem 1.66 of [**3**], there is a unique critical point with respect to τ of the modified complex action function f in the strip $\{|\mathrm{Im}(\tau)| < \pi/2\}$ given by $\tau_c(x,t) = i\theta_c(x,t)$, where θ_c is the solution of

$$t = \mu(2\tau\theta)\|x\|^2$$

in this interval. At the critical point

$$f(x,t,\tau_c) = \frac{1}{2} d_C(0,(x,t)).$$

This works only in the case $x \neq 0$.

Application to the fundamental solution. We plan to investigate whether we can use the subRiemannian geometry to construct the fundamental solution of the operator

$$\begin{aligned} \Delta_X &= X_1^2 + X_2^2 \\ &= \Big(\partial_{x_1} + 2x_2\partial_t\Big)^2 + \Big(\partial_{x_2} - 2x_1\partial_t\Big)^2 \\ &= \partial_{x_1}^2 + \partial_{x_2}^2 + 4\partial_t\Big(x_2\partial_{x_1} - x_1\partial_{x_2}\Big) + 4(x_1^2 + x_2^2)\partial_t^2. \end{aligned}$$

It is expected that the subRiemannian distance (Carnot-Carathéodory distance) will play a role similar to the one the Euclidian distance plays for the Laplacian, *i.e.*, the fundamental solution is the inverse of the Carnot-Carathéodory distance

$$K(x,t;0,0) = \frac{C}{d_C(x,t)} = \frac{C}{(|x|^4 + t^2)^{1/2}}.$$

We need the following lemma.

LEMMA 1.65. *For any functions $f, h \in C^2(\mathbb{R}^3)$, we have*

$$\Delta_X(fh) = f\Delta_X h + h\Delta f + 2\sum_{i=1}^{2} X_i(f)X_i(h) \tag{1.88}$$

$$\Delta_X(f^2) = 2f\Delta_X f + 2\sum_{i=1}^{2} X_i(f)^2 \tag{1.89}$$

$$\Delta_X\Big(\frac{1}{f}\Big) = \frac{2}{f^3}\sum_{i=1}^{2}(X_i f)^2 - \frac{1}{f^2}\Delta_X f. \tag{1.90}$$

PROOF. For $i = 1, 2$ we have

$$\begin{aligned} X_i^2(fh) &= X_i\Big(X_i(fh)\Big) = X_i\Big(f\, X_i f + h\, X_i f\Big) \\ &= X_i\Big(f\, X_i h\Big) + X_i\Big(h\, X_i f\Big) \\ &= f\, X_i^2 h + h\, X_i^2 f + 2X_i(f)X_i(h). \end{aligned}$$

Summing in the above relation with $i = 1, 2$ we obtain relation (1.88). Making $f = h$ we obtain the second relation. By Lemma 1.65, we have

$$0 = \Delta_X\Big(f\frac{1}{f}\Big) = f\,\Delta_X\Big(\frac{1}{f}\Big) + \frac{1}{f}\Delta_X f + 2\sum_{i=1}^{2} X_i(f)X_i\Big(\frac{1}{f}\Big). \tag{1.91}$$

Using

$$X_i\Big(f\frac{1}{f}\Big) = 0$$

we get

$$X_i\Big(\frac{1}{f}\Big) = -\frac{1}{f^2}X_i(f)$$

and substituting in (1.91) yields

$$\Delta_X\Big(\frac{1}{f}\Big) = \frac{2}{f^3}\sum X_i(f)^2 - \frac{1}{f^2}\Delta_X f. \tag{1.92}$$

□

LEMMA 1.66. *If d is the Carnot-Carathéodory distance from the origin, $d_C(x,t) = \sqrt{|x|^4 + t^2}$, then*

$$\sum X_i(d)^2 = 4|x|^2.$$

PROOF. From $X_1(d^2) = 2dX_1(d)$ we get

$$X_1(d) = \frac{1}{2d}X_1(d^2). \tag{1.93}$$

The right side of (1.93) can be computed explicitly

$$\begin{aligned} X_1(d^2) &= (\partial_{x_2} + 2x_1\partial_t)\big((x_1^2 + x_2^2)^2 + t^2\big) \\ &= 4|x|^2 x_2 + 4x_1 t. \end{aligned}$$

Hence

$$X_1(d) = \frac{2}{d}\Big(|x|^2 x_2 + x_1 t\Big). \tag{1.94}$$

Similarly,

$$X_2(d^2) = 4\Big(|x|^2 x_1 - x_2 t\Big),$$

$$X_2(d) = \frac{2}{d}\Big(|x|^2 x_1 - x_2 t\Big). \tag{1.95}$$

Summing the squares of (1.93) and (1.95) yields

$$\begin{aligned} X_1(d)^2 + X_2(d)^2 &= \frac{4}{d^2}\Big((|x|^2 x_2 + x_1 t)^2 + (|x|^2 x_1 - x_2 t)^2\Big) \\ &= \frac{4}{d^2}\Big(|x|^4(x_1^2 + x_2^2) + t^2(x_1^2 + x_2^2)\Big) \\ &= 4|x|^2. \end{aligned}$$

□

LEMMA 1.67. *We have*

$$\Delta_X(d^2) = 24|x|^2.$$

PROOF. Using the formula for $X_1(d^2)$ from the previous proof

$$\begin{aligned} X_1^2(d^2) &= (\partial_{x_2} + 2x_1\partial_t)(4(x_1^2 + x_2^2)x_2 + 4x_1 t) \\ &= 4((x_1^2 + x_2^2) + 2x_2^2 + 2x_1^2) \\ &= 12|x|^2. \end{aligned}$$

Similarly,

$$X_2^2(d^2) = 12|x|^2.$$

Summing the squares of the last two relations yields the result. □

LEMMA 1.68. *We have*

$$d\Delta_X d = 8|x|^2.$$

PROOF. Substitute the formulas of Lemma 1.66 and Lemma 1.67 into the relation

$$\Delta_X d^2 = 2d\Delta_X d + 2\sum (X_i d)^2.$$

□

PROPOSITION 1.69. For $(x, t) \neq (0, 0)$, we have

$$\Delta_X\Big(\frac{1}{d}\Big) = 0.$$

PROOF. Choosing $f = d$ in the third formula of Lemma 1.65 and using the relations given by Lemma 1.67 and Lemma 1.68 yields

$$\begin{aligned}\Delta_X\Big(\frac{1}{d}\Big) &= \frac{2}{d^3}\sum X_i(d)^2 - \frac{1}{d^2}\Delta_X d \\ &= \frac{2}{d^3}4|x|^2 - \frac{1}{d^2}\frac{8|x|^2}{d} \\ &= 0.\end{aligned}$$

□

One may show that in the sense of distributions

$$\Delta_X\Big(\frac{1}{d}\Big) = \delta_{(0,0)}.$$

For details see Folland [**25**].

1.8. Exercises

EXERCISE 1.1. Let $\delta_\lambda : \mathbb{R}^3 \to \mathbb{R}^3$ given by $\delta_\lambda(x,t) = (\lambda x, \lambda^2 t)$.

i) Show that $(\delta_\lambda)_{\lambda \in \mathbb{R}}$ form a group, called the one-parameter group of dilations.

ii) Show that $(\delta_\lambda)^* X_1 = \lambda^{-1} X_1$, $(\delta_\lambda)^* X_2 = \lambda^{-1} X_2$, where X_1, X_2 are the Heisenberg vector fields.

iii) The length of a horizontal curve is multiplied by $|\lambda|$ under the action of δ_λ.

iv) Show that $d_C(\delta_\lambda p, \delta_\lambda q) = |\lambda| d_C(p,q)$, for any two points $p, q \in \mathbb{R}^3$.

v) Prove the following estimation

$$\frac{1}{3}\Big(|x| + |t|^{1/2}\Big) \le d_C(0,(x,t)) \le 4\Big(|x| + |t|^{1/2}\Big).$$

EXERCISE 1.2. Define the unit ball centered at p as

$$\mathbf{B}(p,1) = \{(x,t) \in \mathbb{R}^3; d_C(x,t) \le 1\},$$

where $d_c(x,t) = (|x|^4 + t^2)^{1/2}$.

i) Draw the unit ball centered at the origin.

ii) Find the volume of the unit ball.

iii) How is the volume of the unit ball changing under left translations?

iv) How is the volume of the unit ball changing under dilations?

v) Draw the unit ball centered at a point (x_0, t_0) in the cases

$$a) \quad x_0 = 0, \qquad\qquad b) \quad x_0 \neq 0.$$

vi) Find two constants $C_1, C_2 > 0$ such that

$$C_1[-\epsilon,\epsilon] \times [-\epsilon,\epsilon] \times [-\epsilon^2,\epsilon^2] \subset \mathbf{B}(0,\epsilon) \subset C_2[-\epsilon,\epsilon] \times [-\epsilon,\epsilon] \times [-\epsilon^2,\epsilon^2].$$

vii) Show that

$$Vol(B(p,\epsilon)) = \epsilon^4\, Vol(B(0,1)), \qquad \forall p \in \mathbb{R}^3,$$

where Vol denotes the usual volume defined by the Lebesgue measure.

EXERCISE 1.3. (see [**20**]) Let $c = (c_1, c_2, c_3) : I \to \mathbf{H}_1$ be a C^1-curve, such that 0 is in the interior of I and $c(0) = g$. Show that if $c_3'(0) \neq 0$, then there exists $\epsilon > 0$ such that

$$\mathbf{B}(g_0, \epsilon) \subset \bigcup_{s \in I} H_{c(s)},$$

where $\mathbf{B}(g_0, \epsilon) = \{g \in \mathbf{H} : d_C(g_0^{-1} g) \leq \epsilon\}$, and H_p denotes the horizontal plane through the point p.

EXERCISE 1.4. Find the expression for the exponential map of the Lie group $\mathbf{H}_1$ as subgroup of $GL(3, \mathbb{R})$.

EXERCISE 1.5. Show that on every vertical line in $\mathbb{R}^3$ the Carnot-Carathéodory metric is locally equivalent to $\sqrt{\text{Euclidean metric}}$.

CHAPTER 2

Geometric Analysis of Step 4 Case

In this chapter we shall study the behavior of subRiemannian geodesics on a certain step 4 subRiemannian manifold defined on $\mathbb{R}^3 = \mathbb{R}^2_x \times \mathbb{R}_t$ by the vector fields

$$X_1 = \partial_{x_1} + 4x_2|x|^2\partial_t \ , \qquad X_2 = \partial_{x_2} - 4x_1|x|^2\partial_t, \tag{2.1}$$

where $|x|^2 = x_1^2 + x_2^2$. Since

$$\begin{aligned} [X_1, X_2] &= -16|x|^2\partial_t, \\ [X_1, [X_1, X_2]] &= -32x_1\partial_t, \\ [X_1, [X_1, [X_1, X_2]]] &= -32\partial_t, \end{aligned}$$

it follows that the hypoelliptic operator $\Delta_X = \frac{1}{2}(X_1^2 + X_2^2)$ is step 4 along the t-axis and step 2 elsewhere. The geometry of this subRiemannian model can be found in Calin's Ph.D. thesis, see [**12**].

We shall compute the length of the subRiemannian geodesics between the origin and any point on the $t-$axis. We can see from the calculation that the t-axis is the conjugate locus of this geometry. We shall also characterize the number of subRiemannian geodesics between the origin and any other point. Finally, we shall state the connectivity and completeness theorems for this model, see [**17**].

The study of the above step 4 model requires the use of elliptic functions. The Heisenberg group needs only elementary trigonometric functions, while steps greater than 4 require hypergeometric functions. We shall provide in the following the definitions of the elliptic functions used in the next sections. For a detailed description the reader may consult Lawden [**42**].

2.1. Elliptic functions

The integral

$$z = \int_0^w \frac{dt}{\sqrt{(1-t^2)(1-k^2t^2)}}, \qquad |k| < 1$$

is called an elliptic integral of the first kind. The integral exists if w is real and $|w| < 1$. Using the substitution $t = \sin\theta$ and $w = \sin\phi$, yields

$$z = \int_0^\phi \frac{d\theta}{\sqrt{1 - k^2\sin^2\theta}}.$$

If $k = 0$, then $z = \sin^{-1} w$ or $w = \sin z$. Similarly, the above integral is denoted by $\mathrm{sn}^{-1}(w; k)$, where $k \neq 0$. The number k is called the modulus. Thus

$$z = \mathrm{sn}^{-1} w = \int_0^w \frac{dt}{\sqrt{(1-t^2)(1-k^2t^2)}}.$$

The function $w = \mathrm{sn}\, z$ is called a Jacobian elliptic function.

Similar to the trigonometric functions, it is convenient to define other elliptic functions

$$\operatorname{cn} z = \sqrt{1 - \operatorname{sn}^2 z}, \qquad \operatorname{dn} z = \sqrt{1 - k^2 \operatorname{sn}^2 z}.$$

A few properties of these functions are

$$\operatorname{sn}(0) = 0, \qquad \operatorname{cn}(0) = 1, \qquad \operatorname{dn}(0) = 1,$$

$$\operatorname{sn}(-z) = \operatorname{sn}(z), \qquad \operatorname{cn}(-z) = \operatorname{cn}(z),$$

$$\frac{d}{dz}\operatorname{sn} z = \operatorname{cn} z \operatorname{dn} z, \; \frac{d}{dz}\operatorname{cn} z = -\operatorname{sn} z \operatorname{dn} z, \; \frac{d}{dz}\operatorname{dn} z = -k^2 \operatorname{sn} z \operatorname{cn} z,$$

$$-1 \le \operatorname{cn} z \le 1, \quad -1 \le \operatorname{sn} z \le 1, \quad 0 \le \operatorname{dn} z \le 1.$$

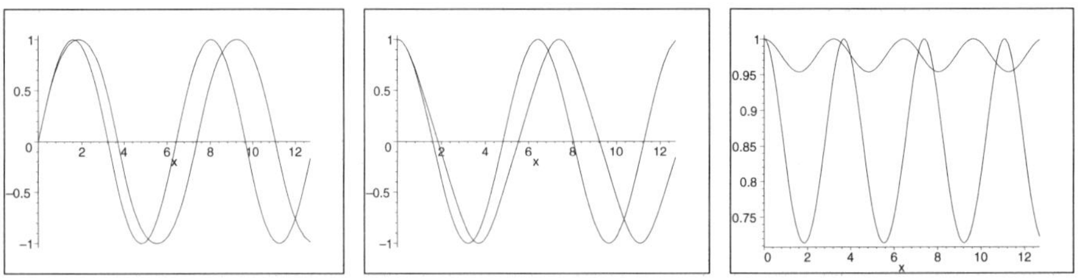

sn(z, k) *cn(z, k)* *dn(z, k)*

Figure 2.1: *The graphs of functions* $sn(z,k)$, $cn(z,k)$ *and* $dn(z,k)$ *for* $k = 0.3$ *and* 0.7

Let

$$K = K(k) = \int_0^1 \frac{dt}{\sqrt{(1-t^2)(1-k^2t^2)}} = \int_0^{\pi/2} \frac{d\theta}{\sqrt{1 - k^2 \sin^2 \theta}}$$

be the complete Jacobi integral. Then, as real functions, the elliptic functions sn and cn are periodic functions of principal period $4K$.

Unlike in the step 2 case, where the vector fields X_1 and X_2 are left invariant with respect to the Heisenberg group law, in this case we have:

PROPOSITION 2.1. There is no Lie group on $\mathbb{R}^3$ underlying the vector fields X_1 and X_2 given by (2.1).

PROOF. Assume there is a Lie group law on $\mathbb{R}^3$ such that X_1 and X_2 are left invariant. Then $[X_1, X_2] = -16|x|^2\partial_t$ is also left invariant. The vector fields $[X_1, X_2]$ and ∂_t are left invariant and proportional. A result of Lie group theory says that the proportionality function has to be a constant, *i.e.*, $|x|^2$ =constant, contradiction. □

This will make the study of the step 4 model more difficult than the Heisenberg group, where any point could be translated to the origin.

2.2. The horizontal distribution

Like in the Heisenberg group case, the horizontal distribution is defined by

$$H : x \to H_x = \text{span}\{X_1, X_2\},$$

where X_1, X_2 are the vector fields given by (2.1).

PROPOSITION 2.2. A curve $c = (x_1, x_2, t)$ is horizontal if and only if

$$\dot{t} = 4|x|^2(\dot{x}_1 x_2 - x_1 \dot{x}_2). \tag{2.2}$$

PROOF. The velocity vector can be written as

$$\begin{aligned} \dot{c} &= \dot{x}_1 \partial_{x_1} + \dot{x}_2 \partial_{x_2} + \dot{t}\partial_t \\ &= \dot{x}_1\big(\partial_{x_1} + 4x_2|x|^2\partial_t\big) - 4\dot{x}_1 x_2|x|^2\partial_t \\ &\quad +\dot{x}_2\big(\partial_{x_2} - 4x_1|x|^2\partial_t\big) + 4x_1\dot{x}_2|x|^2\partial_t + \dot{t}\partial_t \\ &= \dot{x}_1 X_1 + \dot{x}_2 X_2 + \big(\dot{t} + 4x_1|x|^2\dot{x}_2 - 4x_2|x|^2\dot{x}_1\big)\partial_t. \end{aligned}$$

Hence, $\dot{c} \in \mathcal{H}$ iff the coefficient of ∂_t vanishes, *i.e.* (2.2) holds. □

Therefore a horizontal curve $c = (x_1, x_2, t)$ can be written as

$$\dot{c} = \dot{x}_1 X_1 + \dot{x}_2 X_2. \tag{2.3}$$

COROLLARY 2.3. The vector field $[X_1, X_2]$ is not horizontal, and hence H is not an integrable distribution.

PROOF. $[X_1, X_2]$ is proportional to ∂_t and it cannot be written in the form (2.3). □

The horizontality and non-integrability of H can be described also by the 1-connection form

$$\omega = dt - 4|x|^2(x_2 dx_1 - x_1 dx_2). \tag{2.4}$$

A curve $c = (x_1, x_2, t)$ is horizontal if and only if $\omega(\dot{c}) = 0$. Hence the horizontal distribution $H = \ker \omega$. The non-integrability of H can be expressed also by the condition

$$\omega \wedge d\omega \neq 0,$$

called the Frobenius non-integrability condition. Indeed,

$$\begin{aligned} \omega \wedge d\omega &= [dt - 4|x|^2(x_2 dx_1 - x_1 dx_2)] \wedge 16|x|^2 dx_1 \wedge dx_2 \\ &= 16|x|^2 dt \wedge dx_1 \wedge dx_2 \neq 0, \quad \text{for } |x| \neq 0. \end{aligned}$$

Therefore the model is not a contact manifold along the t -axis.

In polar coordinates $x_1 = r\cos\phi$, $x_2 = r\sin\phi$, equation (2.2) yields

$$\begin{aligned} \dot{t} &= 4(\dot{x}_1 x_2 - x_1\dot{x}_2)|x|^2 \\ &= -4r^4\dot{\phi}(\sin^2\phi + \cos^2\phi) \\ &= -4r^4\dot{\phi}. \end{aligned}$$

In differential notation

$$dt = -4r^4\, d\phi, \tag{2.5}$$

and hence the 1-connection form becomes $\omega = dt + 4r^4 d\phi$.

Horizontal connectivity . As the vector fields (2.1) and their iterated brackets span the tangent space at each point, by Chow's theorem (see [5]), any two points can be joined by a piece-wise smooth horizontal curve. However, in this section we shall prove the horizontal connectivity by a direct method.

THEOREM 2.4. *1) Given a point $P(\mathbf{x_1}, \mathbf{x_2}, \mathbf{t}) \in \mathbb{R}^3$, there is a smooth horizontal curve connecting the origin and the point P.*
2) Given two points $P, Q \in \mathbb{R}^3$, there is a piece-wise smooth horizontal curve connecting P and Q.

PROOF. 1) In order to prove the theorem, we shall construct an explicit solution. We shall consider three cases: $\mathbf{x_2} \neq 0$, $\mathbf{x_1} \neq 0$ and $\mathbf{x_1} = \mathbf{x_2} = 0$.
In the case $\mathbf{x_2} \neq 0$, set

$$\begin{aligned} x_1(s) &= as^2 + bs, \\ x_2(s) &= \alpha s^2, \\ t(s) &= -\frac{4}{7}b\alpha(a^2+\alpha^2)s^7 - \frac{4}{3}b^2\alpha a s^6 - \frac{4}{5}b^3\alpha s^5, \quad s \in [0,1] \end{aligned}$$

where a, b and α are constants, which will be found later from the boundary conditions

$$x_1(1) = \mathbf{x_1}, \qquad x_2(1) = \mathbf{x_2}, \qquad t(1) = \mathbf{t}.$$

The curve starts at the origin $x_1(0) = x_2(0) = t(0) = 0$. Using

$$\begin{aligned} \dot{x}_1(s)x_2(s) - x_1(s)\dot{x}_2(s) &= -\alpha b s^2, \\ 4(x_1(s)^2 + x_2(s)^2) &= 4(a^2+\alpha^2)s^4 + 8abs^3 + 4b^2s^2, \end{aligned}$$

one may show by computation

$$\dot{t}(s) = 4\Big(x_1(s)^2 + x_2(s)^2\Big)\Big(\dot{x}_1(s)x_2(s) - x_1(s)\dot{x}_2(s)\Big),$$

i.e., the curve is horizontal.

Next, we shall find the constants a, α, b. For $s = 1$ we have

$$\begin{aligned} a + b &= \mathbf{x_1}, \\ \alpha &= \mathbf{x_2}, \\ \frac{\mathbf{t}}{4} &= -\frac{1}{7}b\alpha(a^2+\alpha^2) - \frac{1}{3}b^2\alpha a - \frac{1}{5}b^3\alpha. \end{aligned}$$

Set $\widetilde{t} = 105\mathbf{t}/4$. The above system yields

$$\begin{aligned} \widetilde{t} &= -15b\alpha(a^2+\alpha^2) - 35b^2\alpha a - 21b^3\alpha \\ &= -15b\mathbf{x_2}(a^2+\mathbf{x_2}^2) - 35b^2\mathbf{x_2}a - 21b^3\mathbf{x_2} \\ &= -15b\mathbf{x_2}((\mathbf{x_1}-b)^2+\mathbf{x_2}^2) - 35b^2\mathbf{x_2}(\mathbf{x_1}-b) - 21b^3\mathbf{x_2} \\ &= -\mathbf{x_2}(b^3 + 5\mathbf{x_1}b^2 + 15|\mathbf{x}|^2 b). \end{aligned}$$

The cubic equation in b

$$b^3 + 5\mathbf{x_1}b^2 + 15|\mathbf{x}|^2 b = \frac{-\widetilde{t}}{\mathbf{x_2}}$$

has at least a real solution $\mathbf{b}$. Choose $a = \mathbf{x_1} - \mathbf{b}$.

In the case $\mathbf{x_1} \neq 0$, we consider the horizontal curve

$$\begin{aligned} x_1(s) &= \alpha s^2, \\ x_2(s) &= as^2 + bs, \\ t(s) &= \frac{4}{7} b\alpha(a^2+\alpha^2)s^7 + \frac{4}{3} b^2 \alpha a s^6 + \frac{4}{5} b^3 \alpha s^5. \end{aligned}$$

In a similar way we arrive at the following cubic equation

$$b^3 + 5\mathbf{x_2} b^2 + 15|\mathbf{x}|^2 b = \frac{\widetilde{t}}{\mathbf{x_1}}.$$

In the case $\mathbf{x_1} = \mathbf{x_2} = 0$, we need to find a horizontal curve joining the origin and the point $(0, 0, \mathbf{t})$. We shall assume $\mathbf{t} < 0$. Set

$$\begin{aligned} x_1(s) &= r\cos(\pi s), \\ x_2(s) &= r\sin(\pi s), \\ t(s) &= \mathbf{t}s, \qquad s \in [0,1], \end{aligned}$$

with the parameter r to be found from the horizontality condition

$$\dot{t}(s) = -4|x(s)|^4 \dot{\phi}(s) = -4r^4\pi,$$

where $\phi = \pi s$. Then $t(s) = -4r^4\pi$, and hence

$$r = \left(\frac{|\mathbf{t}|}{4\pi}\right)^{1/4}.$$

For $\mathbf{t} > 0$, we consider

$$\begin{aligned} x_1(s) &= r\cos(-\pi s), \\ x_2(s) &= r\sin(-\pi s), \\ t(s) &= \mathbf{t}s, \qquad s \in [0,1]. \end{aligned}$$

2) Let c and γ be curves joining the origin with the points P and Q, respectively. Then $\gamma \circ c^{-1}$ is a piece-wise smooth curve joining P and Q. □

Recently Gromov [**29**] proved that Chow's piecewise smooth integral curves may be replaced by smooth integral curves. We shall adapt our direct proof of Chow's theorem to obtain global connectivity. It is interesting to note that if $x_1^0 \neq x_1$ or $x_2^0 \neq x_2$, *i.e.*, the points are not on the same vertical line, the horizontal curve has quadratic x-components.

THEOREM 2.5. *Given two points $P(x_1^0, x_2^0, t^0)$, $Q(x_1, x_2, t) \in \mathbb{R}^3$, there is a smooth horizontal curve connecting P and Q.*

PROOF. We shall consider the cases: $x_1^0 \neq x_1$, $x_2^0 \neq x_2$ and $x_1^0 = x_1, x_2^0 = x_2$. In the case $x_2^0 \neq x_2$, choose

$$\begin{aligned} x_1(s) &= as^2 + bs + x_1^0, \\ x_2(s) &= \alpha s^2 + x_2^0, \end{aligned}$$

where a, b and α are constants, which will be found from the boundary conditions

$$\begin{aligned} x_1(0) = x_1^0, \qquad x_2(0) &= x_2^0, \qquad t(0) = t^0, \\ x_1(1) = x_1, \qquad x_2(1) &= x_2, \qquad t(1) = t. \end{aligned}$$

A computation shows

$$\begin{aligned}
\dot{x}_1(s)x_2(s) - x_1(s)\dot{x}_2(s) &= b(x_2^0 - x_2)s^2 + 2(x_2^0 x_1 - bx_2^0 - x_1^0 x_2)s + bx_2^0, \\
4(x_1(s)^2 + x_2(s)^2) &= \Big(4(x_1 - x_1^0 - b)^2 + 4(x_2 - x_2^0)^2\Big)s^4 \\
&\quad +8b\Big(x_1 - x_1^0 - b)s^3 + (8x_1^0(x_1 - x_1^0 - b) \\
&\quad +4b^2 + 8x_2^0(x_2 - x_2^0)\Big)s^2 \\
&\quad +8bsx_1^0 + 4(x_1^0)^2 + 4(x_2^0)^2.
\end{aligned}$$

Integrating in the horizontality condition (2.2) yields

$$\begin{aligned}
t(s) &= t^0 + \int_0^1 4\Big(x_1(s)^2 + x_2(s)^2\Big)\Big(\dot{x}_1(s)x_2(s) - x_1(s)\dot{x}_2(s)\Big)\,ds \\
&= t^0 + \frac{1}{7}\Big(4(x_1 - x_1^0 - b)^2 + 4(x_2 - x_2^0)^2\Big)(bx_2^0 - bx_2)s^7 \\
&\quad +\Big(\frac{4}{3}b(x_1 - x_1^0 - b)(bx_2^0 - bx_2) + \frac{1}{6}(4(x_1 - x_1^0 - b)^2 \\
&\quad +4(x_2 - x_2^0)^2)(2x_2^0 x_1 - 2bx_2^0 - 2x_1^0 x_2)\Big)s^6 \\
&\quad +\Big(\frac{1}{5}(8x_1^0(x_1 - x_1^0 - b) + 4b^2 + 8x_2^0(x_2 - x_2^0))(bx_2^0 - bx_2) \\
&\quad +\frac{8}{5}b(x_1 - x_1^0 - b)(2x_2^0 x_1 - 2bx_2^0 - 2x_1^0 x_2) \\
&\quad +\frac{1}{5}(4(x_1 - x_1^0 - b)^2 + 4(x_2 - x_2^0)^2)bx_2^0\Big)s^5 \\
&\quad +\Big(2bx_1^0(bx_2^0 - bx_2) + \frac{1}{4}(8x_1^0(x_1 - x_1^0 - b) \\
&\quad +4b^2 + 8x_2^0(x_2 - x_2^0))(2x_2^0 x_1 - 2bx_2^0 - 2x_1^0 x_2) \\
&\quad +2b^2(x_1 - x_1^0 - b)x_2^0\Big)s^4 + \Big(\frac{1}{3}(4(x_1^0)^2 + 4(x_2^0)^2)(bx_2^0 - bx_2) \\
&\quad +\frac{8}{3}x_1^0 b(2x_2^0 x_1 - 2bx_2^0 - 2x_1^0 x_2) \\
&\quad +\frac{1}{3}(8x_1^0(x_1 - x_1^0 - b) + 4b^2 + 8x_2^0(x_2 - x_2^0))bx_2^0\Big)s^3 \\
&\quad +\Big(\frac{1}{2}(4(x_1^0)^2 + 4(x_2^0)^2)(2x_2^0 x_1 - 2bx_2^0 - 2x_1^0 x_2) + 4b^2 x_2^0 x_1^0\Big)s^2 \\
&\quad +\Big(4(x_1^0)^2 + 4(x_2^0)^2\Big)bx_2^0 s.
\end{aligned}$$

For $s = 1$ the boundary conditions for the x-component provide

$$a = x_1 - x_1^0 - b, \tag{2.6}$$

$$\alpha = x_2 - x_2^0, \tag{2.7}$$

$$t = t(1), \tag{2.8}$$

where $t(1)$ can be found substituting $s = 1$ in the above formula. We obtain

$$t = P(b), \tag{2.9}$$

where $P(b)$ is the following polynomial in b

$$
\begin{aligned}
P(b) \quad = \quad & \Big(-\frac{4}{105}x_2 + \frac{4}{105}x_2^0\Big)b^3 \\
& + \Big(-\frac{4}{21}x_1x_2 - \frac{12}{35}x_1^0x_2 + \frac{4}{21}x_2^0x_1 + \frac{12}{35}x_2^0x_1^0\Big)b^2 \\
& + \Big(-\frac{4}{35}x_2(x_2^0)^2 - \frac{116}{105}(x_1^0)^2x_2 - \frac{4}{7}x_2^3 + \frac{4}{7}x_2^0(x_1)^2 + \frac{104}{105}x_1^0x_2^0x_1 \\
& + \frac{116}{105}(x_2^0)^3 - \frac{4}{7}x_1^2x_2 - \frac{104}{105}x_1x_1^0x_2 + \frac{116}{105}x_2^0(x_1^0)^2 - \frac{44}{105}x_2^2x_2^0\Big)b \\
& - \frac{4}{3}(x_2^0)^2x_1^0x_2 - \frac{4}{3}x_2^2x_2^0x_1^0 + \frac{4}{3}x_2(x_2^0)^2x_1 + \frac{4}{3}x_2^2x_2^0x_1 + \frac{4}{3}(x_1^0)^2x_2^0x_1 \\
& - \frac{4}{3}x_1(x_1^0)^2x_2 + \frac{4}{3}x_1^2x_1^0x_2^0 - \frac{4}{3}(x_1)^2x_1^0x_2 - \frac{4}{3}(x_1^0)^3x_2 + \frac{4}{3}x_1^3x_2^0 \\
& - \frac{4}{3}x_2^3x_1^0 + \frac{4}{3}(x_2^0)^3x_1 + t^0.
\end{aligned}
$$

If $x_2 \neq x_2^0$, the above polynomial is of degree 3, and hence the equation (2.9) has at least a real solution b. Using the boundary conditions (2.6) we get a and α. The other two cases are similar and are left to the reader. □

2.3. SubRiemannian geodesics

Consider the Hamiltonian H defined as the principal symbol of the operator $\Delta_X = \frac{1}{2}(X_1^2 + X_2^2)$

$$
H(\xi,\theta,x,t) = \frac{1}{2}\Big(\xi_1 + 4x_2|x|^2\theta\Big)^2 + \frac{1}{2}\Big(\xi_2 - 4x_1|x|^2\theta\Big)^2. \tag{2.10}
$$

DEFINITION 2.6. The *subRiemannian geodesics* between the origin O and the point $P(x,t)$ are the projections on the (x,t)-plane of the solutions of the Hamiltonian system

$$
\begin{cases}
\dot{x} = \partial H/\partial \xi \\
\dot{t} = \partial H/\partial \theta \\
\dot{\xi} = -\partial H/\partial x \\
\dot{\theta} = -\partial H/\partial t
\end{cases} \tag{2.11}
$$

with the boundary conditions

$$
x(0) = t(0) = 0, \;\; x(1) = x, \;\; t(1) = t. \tag{2.12}
$$

PROPOSITION 2.7. Any subRiemannian geodesic is a horizontal curve. There are horizontal curves, which are not subRiemannian geodesics.

PROOF. Let $c(s) = (x_1(s), x_2(s), t(s))$ be a subRiemannian geodesic. Using Hamiton's equations

$$\begin{aligned}
\dot{x}_1 &= \frac{\partial H}{\partial \xi_1} = \xi_1 + 4x_2|x|^2\theta, \\
\dot{x}_2 &= \frac{\partial H}{\partial \xi_2} = \xi_2 - 4x_1|x|^2\theta, \\
\dot{t} &= \frac{\partial H}{\partial \theta} = \left(\xi_1 + 4x_2|x|^2\theta\right) 4x_2|x|^2 - \left(\xi_2 - 4x_1|x|^2\theta\right) 4x_1|x|^2 \\
&= 4|x|^2(\dot{x}_1 x_2 - \dot{x}_2 x_1),
\end{aligned}$$

which is the horizontality condition (2.2).

For the second part of the proof, we need to prove first the following result

LEMMA 2.8. *If $c(s) = (x_1(s), x_2(s), t(s))$ is a subRiemannian geodesic, then*

$$\begin{aligned}
\dot{\xi}_1 &= -8\theta\dot{x}_1 x_1 x_2 + 4\theta|x|^2\dot{x}_2 + 8\theta x_1^2\dot{x}_2, \\
\ddot{x}_1 &= 16\theta|x|^2\dot{x}_2, \\
\theta &= \textit{constant}.
\end{aligned}$$

PROOF. The Hamiltonian equations yield

$$\begin{aligned}
\dot{\theta} &= -\frac{\partial H}{\partial t} = 0 \Longrightarrow \theta = \text{constant}, \\
\dot{\xi}_1 &= -\frac{\partial H}{\partial x_1} = -[\dot{x}_1\, 4\theta\partial_{x_1}(x_2|x|^2) + \dot{x}_2(-4\theta)\partial_{x_1}(x_1|x|^2)] \\
&= -4\theta\dot{x}_1 x_2\, 2x_1 + 4\theta\dot{x}_2(|x|^2 + 2x_1^2) \\
&= -8\theta\dot{x}_1 x_1 x_2 + 4\theta|x|^2\dot{x}_2 + 8\theta x_1^2\dot{x}_2.
\end{aligned}$$

Differentiating $\dot{x}_1 = \xi_1 + 4x_2(x_1^2 + x_2^2)\theta$, we get

$$\begin{aligned}
\ddot{x}_1 &= \dot{\xi}_1 + 4\dot{x}_2|x|^2\theta + 4x_2(2x_1\dot{x}_1 + 2x_2\dot{x}_2)\theta \\
&= -8\theta\dot{x}_1 x_1 x_2 + 4\theta|x|^2\dot{x}_2 + 8\theta x_1^2\dot{x}_2 \\
&\quad +4\dot{x}_2|x|^2\theta + 8x_1x_2\dot{x}_1\theta + 8x_2^2\dot{x}_2\theta \\
&= 8\theta|x|^2\dot{x}_2 + 8\theta(x_1^2 + x_2^2)\dot{x}_2 \\
&= 16|x|^2\dot{x}_2.
\end{aligned}$$

□

We shall show that the horizontal curves constructed in the proof of Theorem 2.4 do not verify the second equation of Lemma 2.8.
First, we note $\theta \neq 0$. If $\theta = 0$, then the Hamiltonian becomes

$$H = \frac{1}{2}(\xi_1^2 + \xi_2^2)$$

with the Hamiltonian system

$$\begin{aligned}
\dot{\xi}_i &= -\frac{\partial H}{\partial x_i} = 0 \Longrightarrow \xi_i = \text{constant}, \\
\dot{x}_i &= \frac{\partial H}{\partial \xi_i} = \xi_i \Longrightarrow x_i(s) = \xi_i s,
\end{aligned}$$

which is in contradiction with the formulas provided by the proof of Theorem 2.4. In the case $x_2(1) \neq 0$, these are

$$x_1(s) = as^2 + bs, \qquad x_2(s) = \alpha s^2, \qquad s \in [0,1]. \tag{2.13}$$

Using

$$\ddot{x}_1(s) = 2a, \qquad \dot{x}_2(s) = 2\alpha s,$$

a computation shows that the relation $\ddot{x}_1 = 16\theta|x|^2\dot{x}_2$ becomes the following polynomial identity

$$\Big(32\theta\alpha(a^2+\alpha^2)\Big)s^5 + 64\theta ab\alpha s^4 + 32\theta\alpha b^2 s^3 - 2a = 0, \quad \forall s \in [0,1].$$

As the coefficients vanish, and $\theta \neq 0$, we have

$$\alpha^3 = 0, \qquad ab\alpha = 0, \qquad a = 0.$$

Hence the equation (2.13) becomes $x_1(s) = bs$, $x_2(s) = 0$, which contradicts the condition $x_2(1) \neq 0$.

In a similar way, one may show that the other horizontal curves provided by the Theorem 2.4 are not satisfying the Hamiltonian system of equations, and hence, they are not subRiemannian geodesics. □

2.4. Lagrangian formalism

We shall associate a Lagrangian with the Hamiltonian (2.10) using a formal Legendre transform. The Hamiltonian is not quadratic, so the Euler-Lagrange and the Hamiltonian equations won't be equivalent. In this section we shall investigate the relationship between the Lagrangian and Hamiltonian formalism.

Using the Hamiltonian equations

$$\dot{x}_1 = \xi_1 + 4x_2|x|^2\theta \Longrightarrow \xi_1 = \dot{x}_1 - 4x_2|x|^2\theta,$$
$$\dot{x}_2 = \xi_2 - 4x_1|x|^2\theta \Longrightarrow \xi_2 = \dot{x}_2 + 4x_1|x|^2\theta,$$

a formal application of the Legendre transform yields

$$\begin{aligned} L &= \xi_1\dot{x}_1 + \xi_2\dot{x}_2 + \theta\dot{t} - H \\ &= (\dot{x}_1 - 4x_2|x|^2\theta)\dot{x}_1 + (\dot{x}_2 + 4x_1|x|^2\theta)\dot{x}_2 + \theta\dot{t} - \frac{1}{2}(\dot{x}_1^2+\dot{x}_2^2) \\ &= \frac{1}{2}(\dot{x}_1^2+\dot{x}_2^2) + \theta\Big(\dot{t} - 4|x|^2(\dot{x}_1x_2 - \dot{x}_2x_1)\Big). \end{aligned}$$

If $c = (x_1, x_2, t)$, the Lagrangian becomes

$$L = \frac{1}{2}(\dot{x}_1^2+\dot{x}_2^2) + \theta\omega(\dot{c}), \tag{2.14}$$

with ω given by (2.4). A computation shows that the Euler-Lagrange system of equations

$$\frac{d}{ds}\Big(\frac{\partial L}{\partial \dot{x}_i}\Big) = \frac{\partial L}{\partial x_i}, \qquad i = 1,2$$

for the Lagrangian (2.14) becomes

$$\begin{cases} \ddot{x}_1 = 16\theta|x|^2\dot{x}_2 \\ \ddot{x}_2 = -16\theta|x|^2\dot{x}_1 \\ \theta = \text{constant}. \end{cases} \tag{2.15}$$

The following proposition will enable us to compute subRiemannian geodesics using the above system. The system (2.15) does not contain information about the t component. This will be recovered from the Hamiltonian, as the horizontality condition.

PROPOSITION 2.9. A horizontal curve, which is a solution of the Euler-Lagrange system (2.15) is a subRiemannian geodesic.

PROOF. Let (x_1, x_2) be a solution of the system (2.15). Let ξ be defined by

$$\begin{aligned} \xi_1 &= \dot{x}_1 - 4x_2|x|^2\theta \\ \xi_2 &= \dot{x}_2 + 4x_1|x|^2\theta. \end{aligned}$$

We shall show that $(x_1, x_2, t, \xi_1, \xi_2, \theta)$ verifies the Hamiltonian system (2.11). From the horizontality condition

$$\dot{t} = 4|x|^2(\dot{x}_1 x_2 - \dot{x}_2 x_1) = \frac{\partial H}{\partial \theta},$$

which is the second equation of (2.11). From the above construction of ξ_1, ξ_2, we have

$$\dot{x} = \frac{\partial H}{\partial \xi}.$$

As $\theta =$ constant,

$$\dot{\theta} = 0 = \frac{\partial H}{\partial t}.$$

We still need to show

$$\dot{\xi}_i = -\frac{\partial H}{\partial x_i}, \qquad i = 1, 2.$$

Let $i = 1$. Taking the derivative with respect to the s variable in the definition of ξ_1, one has

$$\begin{aligned} \dot{\xi}_1 &= \ddot{x}_1 - 4\frac{d}{ds}\left(x_2|x|^2\right)\theta \\ &= \ddot{x}_1 - \left(4\dot{x}_2|x|^2\theta + 8x_1x_2\dot{x}_1\theta + 8x_2^2\dot{x}_2\theta\right) \\ &= 16\theta|x|^2\dot{x}_2 - 4\dot{x}_2|x|^2\theta - 8x_1x_2\dot{x}_1\theta - 8x_2^2\dot{x}_2\theta \\ &= 8\theta|x|^2\dot{x}_2 + 8\theta(x_1^2 + x_2^2)\dot{x}_2 - 4\dot{x}_2|x|^2\theta - 8x_1x_2\dot{x}_1\theta - 8x_2^2\dot{x}_2\theta \\ &= -8x_1x_2\dot{x}_1\theta + 4\theta|x|^2\dot{x}_2 + 8\theta x_1^2\dot{x}_2 \\ &= -4\theta\dot{x}_1x_2\,2x_1 + 4\theta\dot{x}_2(|x|^2 + 2x_1^2) \\ &= -[\dot{x}_1 4\theta\partial_{x_1}(x_2|x|^2) + \dot{x}_2(-4\theta)\partial_{x_1}(x_1|x|^2)] \\ &= -\frac{\partial H}{\partial x_1}. \end{aligned}$$

Hence

$$\dot{\xi}_1 = -\frac{\partial H}{\partial x_1}.$$

A similar computation yields

$$\dot{\xi}_2 = -\frac{\partial H}{\partial x_2}.$$

Hence (x_1, x_2, t) is a subRiemannian geodesic. □

The converse of Proposition 2.9 is also true:

PROPOSITION 2.10. Let $c = (x_1, x_2, t)$ be a subRiemannian geodesic. Then c is a horizontal curve and (x_1, x_2) satisfy the system (2.15).

PROOF. The horizontality part follows from Proposition 2.7. For the second part, Lemma 2.8 provides $\ddot{x}_1 = 16\theta|x|^2\dot{x}_2$, with θ constant. In a similar way $\ddot{x}_2 = -16\theta|x|^2\dot{x}_1$, which yields the system (2.15). □

In the next sections we shall characterize the subRiemannian geodesics using the Propositions 2.9 and 2.10. We shall solve the system (2.15) for different boundary conditions and construct the t-component of the subRiemannian geodesics using the horizontality condition. The system (2.15) can be written as

$$\ddot{x} = 16\theta|x|^2 J\dot{x},$$

where $x = (x_1, x_2)$, and $J^2 = -I$, *i.e.*, J is a $90°$ rotation of $\mathbb{R}^2$. We shall show that for given boundary conditions

$$x(0) = x_0, \qquad x(1) = x_1, \tag{2.16}$$

the constant θ takes only discrete values. θ plays the role of the eigenvalue for the system (2.15) with the boundary conditions (2.16). Because of Lagrangian rotational symmetry, we may rewrite the Lagrangian in terms of polar coordinates $x_1 = r\cos\phi$, $x_2 = r\sin\phi$. We have

$$\dot{x}_1^2 + \dot{x}_2^2 = \dot{r}^2 + r^2\dot{\phi}^2, \tag{2.17}$$

$$x_1\dot{x}_2 - \dot{x}_1 x_2 = r^2\dot{\phi}, \tag{2.18}$$

and the Lagrangian becomes

$$L = \frac{1}{2}(\dot{r}^2 + r^2\dot{\phi}^2) + \theta(\dot{t} + 4r^4\dot{\phi}). \tag{2.19}$$

Since θ is constant along the solutions of the Hamiltonian system, it can be considered as a Lagrange multiplier. The symmetries of the Lagrangian will provide symmetries for the geodesics. The Lagrangian L is invariant under the symmetry

$$S : (r, \phi, t; \theta) \to (r, -\phi, -t; -\theta).$$

If $(r(s), \phi(s), t(s))$ is a geodesic corresponding to θ, then $(r(s), -\phi(s), -t(s))$ is a geodesic corresponding to $-\theta$. Consequently, whatever statement is made for the geodesics joining the origin with the point $(r(s), \phi(s), t(s))$, when $\theta > 0$, it can be also made for the geodesics between the origin and the point $(r(s), -\phi(s), -t(s))$, when $\theta < 0$.

The Euler-Lagrange system of equations can be constructed from the above Lagrangian. A computation shows

$$\frac{d}{ds}\frac{\partial L}{\partial \dot{r}} = \ddot{r} \quad \text{and} \quad \frac{\partial L}{\partial r} = r\dot{\phi}(\dot{\phi} + 16\theta r^2). \tag{2.20}$$

The first Euler-Lagrange equation becomes

$$\ddot{r} = r\dot{\phi}(\dot{\phi} + 16\theta r^2). \tag{2.21}$$

As

$$\frac{\partial L}{\partial \dot{\phi}} = r^2\dot{\phi} + 4\theta r^4 \quad \text{and} \quad \frac{\partial L}{\partial \phi} = 0, \tag{2.22}$$

the second Euler-Lagrange equation is

$$\frac{d}{ds}(r^2\dot{\phi} + 4\theta r^4) = 0,$$

which can be written as

$$r^2(\dot{\phi} + 4\theta r^2) = C, \tag{2.23}$$

where C is a constant.
As

$$\frac{\partial L}{\partial \dot{t}} = \theta \text{ and } \frac{\partial L}{\partial t} = 0, \tag{2.24}$$

then $\theta =$ constant along the solutions. Hence $r(s)$ and $\phi(s)$ will verify the following Euler-Lagrange system

$$\begin{cases} \ddot{r} = r\dot{\phi}(\dot{\phi} + 16\theta r^2) \\ r^2(\dot{\phi} + 4\theta r^2) = C \\ \theta = \text{constant}. \end{cases} \tag{2.25}$$

The system (2.25) is the polar coordinates version of the system (2.15), and hence Propositions 2.9 and 2.10 can be applied.

2.5. Solutions which start from the origin

The solutions which start from the origin will be studied. The system (2.25) will be considered with the following initial conditions

$$r(0) = 0\,, \quad \phi(0) = \phi_0\,, \quad \dot{r}(0) \neq 0\,, \quad \dot{\phi}(0) < \infty. \tag{2.26}$$

From the second equation of (2.25) one obtains $C = 0$ and the system becomes

$$\begin{cases} \ddot{r} = r\dot{\phi}(\dot{\phi} + 16\theta r^2) \\ \dot{\phi} = -4\theta r^2 \\ \theta = \text{constant}. \end{cases} \tag{2.27}$$

In particular, when $\theta = 0$, the system (2.27) becomes

$$\ddot{r} = 0, \qquad \dot{\phi} = 0, \qquad \theta = 0.$$

Hence $r(s) = \dot{r}(0)\,s$ and $\phi(s) = \phi(0)$ =constant. The t-component satisfies

$$\dot{t}(s) = -r^4(s)\dot{\phi}(s) = 0,$$

and hence

$$t(s) = t(0) = 0.$$

We arrive at the following result:

PROPOSITION 2.11. Given a point $P(x, 0)$, there is a unique geodesic between the origin and P. It is a straight line in the plane $\{(x, 0); x \in \mathbb{R}^2\}$ of length $|x|$.

From now on, unless otherwise stated, we shall assume that $\theta < 0$. Substituting the second equation of (2.27) into the first one, we obtain an equation for $r(s)$

$$\ddot{r} = -48\theta^2 r^5. \tag{2.28}$$

Let

$$V(r) = 8\theta^2 r^6 \tag{2.29}$$

be the potential energy. Then the equation (2.28) can be written as the Newton's equation

$$\ddot{r} = -V'(r), \tag{2.30}$$

where $V'(r) = dV(r)/dr$.
Let $\dot{r} = p$ in the equation (2.28). Then

$$\ddot{r} = \frac{dp}{ds} = \frac{dp}{dr}\frac{dr}{ds} = p\frac{dp}{dr} = -48\theta^2 r^5$$

Integrating in

$$p\frac{dp}{dr} = -48\theta^2 r^5 \tag{2.31}$$

yields

$$\frac{p^2}{2} = -8\theta^2 r^6 + E$$

or

$$\frac{1}{2}\dot{r}^2 + 8\theta^2 r^6 = E \tag{2.32}$$

where E is a constant, which denotes the total energy of the system. Using

$$|\dot{x}(s)|^2 = \dot{r}^2(s) + r^2(s)\dot{\phi}(s)$$

and the initial condition $r(0) = 0$ yields

$$E = \frac{1}{2}|\dot{x}(0)|^2 = \dot{r}^2(0)/2.$$

Hence

$$\frac{1}{2}\dot{r}^2 + V(r) = E \tag{2.33}$$

is the law of conservation of energy. The total energy E is equal to the value of the constant Hamiltonian H_0 along the solutions

$$\begin{aligned}\frac{dH}{ds} &= \frac{\partial H}{\partial x}\dot{\xi} + \frac{\partial H}{\partial t}\dot{t} + \frac{\partial H}{\partial \xi}\dot{\xi} + \frac{\partial H}{\partial \theta}\dot{\theta} + \frac{\partial H}{\partial s}\\ &= -\dot{\xi}\dot{x} + \dot{x}\dot{\xi} - \dot{\theta}\dot{t} + \dot{t}\dot{\theta} + \frac{\partial H}{\partial s} = 0 \Longrightarrow H = H_0 = \frac{1}{2}|\dot{x}(0)|^2 = E.\end{aligned}$$

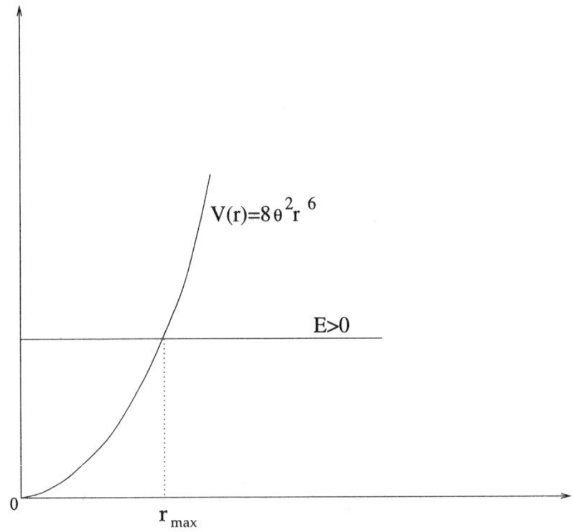

Figure 2.2: The potential $V(r)$ and r_{max}

In the following we shall discuss the components r and ϕ of the geodesics.

The radius $r(s)$

The radius $r(s)$ starts at 0 and increases ($\dot{r} > 0$) until $\dot{r} = 0$, when $r = r_{max}$. After that $r(s)$ decreases ($\dot{r} < 0$) to 0. r_{max} can be found making $\dot{r} = 0$ in (2.33). Hence r_{max} is the positive solution of the following equation (see Figure 2.2):

$$V(r) = E, \tag{2.34}$$

which is

$$r_{max} = \left(\frac{E}{8\theta^2}\right)^{1/6}. \tag{2.35}$$

For each level of energy $E > 0$ there is a corresponding $r_{max} > 0$, which is independent of the initial direction of the solution. The solution $(r(s), \phi(s))$ lies in the disk $D(0, r_{max})$. The greater the energy E is, the larger r_{max}. As the energy $E = |\dot{x}(0)|^2/2$ depends just on the magnitude of the initial speed, r_{max} will do the same.

The angle ϕ

The second equation of (2.27) provides

$$\dot{\phi} = -4\theta r^2. \tag{2.36}$$

As $r(0) = 0$, then $\dot{\phi}(0) = 0$. We know that for $\theta > 0$, $\dot{\phi}(0) < 0$, *i.e.*, ϕ is decreasing, and for $\theta < 0$, $\dot{\phi}(0) > 0$, *i.e.*, ϕ is increasing.

The first two equations of system (2.27) can be written as simultaneous equations

$$\frac{dr}{ds} = \sqrt{2E - 16\theta^2 r^6} \text{ and } \frac{d\phi}{ds} = -4\theta r^2, \tag{2.37}$$

where $r(s)$ was supposed to be increasing. A differential equation connecting r and ϕ but not containing s can be obtained dividing the equations in (2.37)

$$\frac{d\phi}{dr} = \frac{-4\theta r^2}{\sqrt{2E - 16\theta^2 r^6}}. \tag{2.38}$$

Integrating, we obtain the variation of the angle ϕ

$$\phi(s) - \phi(0) = \int_0^{r(s)} \frac{-4\theta x^2}{\sqrt{2E - 16\theta^2 x^6}}\, dx. \tag{2.39}$$

When $r = r_{max}$ the above variation of ϕ is denoted by w and is given by

$$w = \int_0^{r_{max}} \frac{-4\theta x^2}{\sqrt{2E - 16\theta^2 x^6}}\, dx. \tag{2.40}$$

Computing w:
Making the substitution

$$u = \left(\frac{8\theta^2}{E}\right)^{1/6} x,$$

the integral (2.40) yields

$$\begin{aligned} w &= \frac{-4\theta}{\sqrt{2E}} \int_0^1 \frac{\left(\frac{E}{8\theta^2}\right)^{2/6} \left(\frac{E}{8\theta^2}\right)^{1/6} u^2\, du}{\sqrt{1 - u^6}} \\ &= \frac{-\theta}{|\theta|} \int_0^1 \frac{\frac{1}{3} v\, dv}{\sqrt{1 - v^2}} \\ &= -\frac{1}{3} sgn(\theta) \Big(\arcsin(1) - \arcsin(0) \Big) = -sgn(\theta) \frac{\pi}{6}. \end{aligned}$$

PROPOSITION 2.12. The angle w swept by the vectorial radius between $r = 0$ and $r = r_{max}$ is

$$w = \begin{cases} \pi/6, & \text{if } \theta < 0 \\ -\pi/6, & \text{if } \theta > 0. \end{cases} \tag{2.41}$$

This agrees with the fact that the particle is moving clock-wise for $\theta > 0$ and counter-clock-wise for $\theta < 0$, see Figure 2.3.

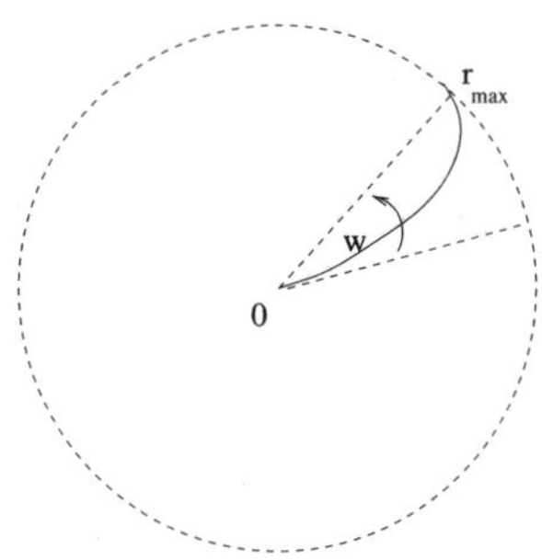

Figure 2.3: A counter clock-wise rotation of angle $w = \pi/6$.

Between $r = r_{max}$ and $r = 0$ the vectorial radius sweeps an angle $\tilde{w}$. As r is decreasing, $\dot{r} < 0$, and

$$\frac{dr}{ds} = -\sqrt{2E - 16\theta^2 r^6}. \tag{2.42}$$

Using again (2.37) and making the division, one obtains

$$\frac{d\phi}{dr} = \frac{4\theta r^2}{\sqrt{2E - 16\theta^2 r^6}}. \tag{2.43}$$

The angle $\tilde{w}$ is given by

$$\tilde{w} = \int_{r_{max}}^{0} \frac{4\theta r^2}{\sqrt{2E - 16\theta^2 r^6}}\, dr = \int_{0}^{r_{max}} \frac{-4\theta r^2}{\sqrt{2E - 16\theta^2 r^6}}\, dr,$$

which is the same integral as that one given in (2.40). Hence

$$\tilde{w} = w = -sgn(\theta)\frac{\pi}{6}. \tag{2.44}$$

The curve $(r(s), \phi(s))$ passes through the origin again after sweeping an angle equal to $w + \tilde{w} = \pi/3$ (clock-wise if $\theta < 0$ and counter-clock-wise if $\theta > 0$). This means that the angle between the tangents at the origin is $\psi = \pi/3$. Since $\psi/2\pi$ is a rational number, the solution will be periodic.

Explicit solution in polar coordinates. In order to avoid using $sgn(\theta)$, we shall assume $\theta < 0$. Then ϕ is increasing and the trajectory is counter-clock-wise. We shall consider a few cases:

• *Assume* $\phi \in [\phi(0), \phi(0) + \pi/6]$ *and* $r \in [0, r_{max}]$
The radius r is increasing and

$$\frac{d\phi}{dr} = \frac{-4\theta r^2}{\sqrt{2E - 16\theta^2 r^6}}, \tag{2.45}$$

which becomes

$$\phi(s) - \phi(0) = -4\theta \int_0^{r(s)} \frac{x^2\, dx}{\sqrt{2E - 16\theta^2 x^6}}, \tag{2.46}$$

where $r(s) \le r_{max}$. Making the substitution

$$u = \left(\frac{16\theta^2}{2E}\right)^{1/6} x, \tag{2.47}$$

and using r_{max} given in (2.35), one may write

$$\begin{aligned} \phi(s) - \phi(0) &= -\frac{4\theta}{\sqrt{2E}}\left(\frac{2E}{16\theta^2}\right)^{2/6}\left(\frac{2E}{16\theta^2}\right)^{1/6} \int_0^{r(s)/r_{max}} \frac{u^2\, du}{\sqrt{1-u^6}} \\ &= -\frac{\theta}{3|\theta|}\Big(\arcsin\Big(\frac{r(s)^3}{r_{max}^3}\Big) - \arcsin(0)\Big). \end{aligned}$$

Let $\phi_0 = \phi(0)$ be the initial argument. Using $\theta < 0$, the relationship between ϕ and r will be given by

$$\phi - \phi_0 = \frac{1}{3}\arcsin\Big(\frac{r^3}{r_{max}^3}\Big), \tag{2.48}$$

which is equivalent to

$$r(\phi) = r_{max} \sin^{1/3}\Big(3(\phi - \phi_0)\Big). \tag{2.49}$$

• *Assume* $\phi \in [\phi(0) + \pi/6, \phi(0) + \pi/3]$
The radius r decreases from r_{max} to 0. Let $M(r_{max}, \phi_0 + \pi/6)$ be the vertex and let $A(r, \phi)$ be a point as in Figure 2.4, *i.e.*, $\phi \in [\phi(0) + \pi/6, \phi(0) + \pi/3]$. On this interval $\dot{r} < 0$, and

$$\frac{dr}{ds} = -\sqrt{2E - 16\theta^2 r^6}. \tag{2.50}$$

Using

$$\frac{d\phi}{ds} = -4\theta r^2,$$

by division of the last two expressions, we obtain

$$\frac{d\phi}{dr} = \frac{4\theta r^2}{\sqrt{2E - 16\theta^2 r^6}}. \tag{2.51}$$

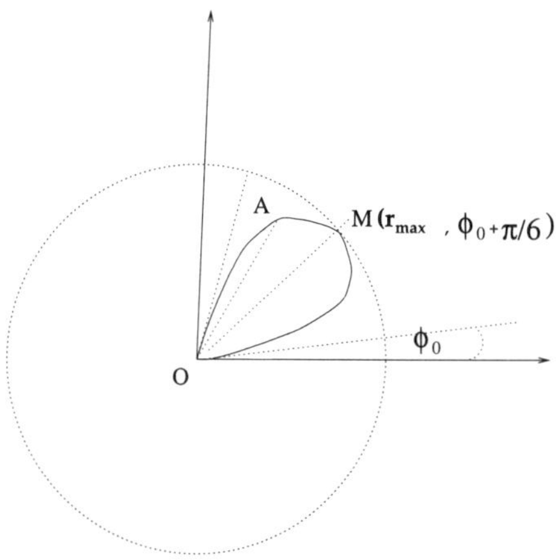

Figure 2.4: After a rotation of $\pi/3$

Integrating between $\phi_0 + \frac{\pi}{6}$ and ϕ, we have

$$\phi - (\phi_0 + \frac{\pi}{6}) = \int_{r_{max}}^{r} \frac{4\theta x^2\,dx}{\sqrt{2E - 16\theta^2 x^6}}. \tag{2.52}$$

The right hand side integral can be decomposed as

$$\int_{r_{max}}^{r} = \int_{r_{max}}^{0} + \int_{0}^{r} = -\int_{0}^{r_{max}} + \int_{0}^{r},$$

and hence

$$\phi - (\phi_0 + \frac{\pi}{6}) = -\int_0^{r_{max}} \frac{4\theta x^2\,dx}{\sqrt{2E - 16\theta^2 x^6}} + \int_0^{r} \frac{4\theta x^2\,dx}{\sqrt{2E - 16\theta^2 x^6}}.$$

The first integral is given in (2.40) and it is equal to $\pi/6$ by Proposition 2.12. The second integral appears in the right hand side of (2.46) with a negative sign and it is given by the negative right hand side of (2.48)

$$\int_0^{r} \frac{4\theta x^2\,dx}{\sqrt{2E - 16\theta^2 x^6}} = -\frac{1}{3}\arcsin(\frac{r^3}{r_{max}^3}). \tag{2.53}$$

Hence (2.52) becomes

$$\begin{aligned} & \phi - (\phi_0 + \frac{\pi}{6}) = \frac{\pi}{6} - \frac{1}{3}\arcsin(\frac{r^3}{r_{max}^3}) \\ \Leftrightarrow \quad & \frac{\pi}{3} - (\phi - \phi_0) = \frac{1}{3}\arcsin(\frac{r^3}{r_{max}^3}) \\ \Leftrightarrow \quad & \sin\Big(\pi - 3(\phi - \phi_0)\Big) = \frac{r^3}{r_{max}^3} \\ \Leftrightarrow \quad & r = r_{max}\sin^{1/3}\Big(3(\phi - \phi_0)\Big), \end{aligned}$$

which agrees with (2.49). We note that

$$r(\phi_0 + \frac{\pi}{3}) = r(\phi_0) = 0, \tag{2.54}$$

which means the solution starts at O with an argument ϕ_0 and bounces back with an argument $\phi_0 + \pi/3$.

• *Assume* $\phi \in [\phi_0 + \pi/3, \phi_0 + 2\pi/3]$

The solution starts from the origin O with an argument $\tilde{\phi}_0 = \phi_0 + \pi/3$, and arrives at O with an argument $\tilde{\phi}_0 + \pi/3$. For $\phi \in [\tilde{\phi}_0, \tilde{\phi}_0 + \pi/3]$ one may write using (2.49)

$$\begin{aligned} r(\phi) &= r_{max} \sin^{1/3}(3(\phi - \tilde{\phi}_0)) \\ &= r_{max} \sin^{1/3}(3\phi - 3\phi_0 + \pi) \\ &= -r_{max} \sin^{1/3}(3(\phi - \phi_0)). \end{aligned}$$

We note that sine function is negative for $3(\phi - \phi_0) \in [\pi, 2\pi]$, and the minus sign in front of the formula makes sense.

In the general case the formula in polar coordinates $r = r(\phi)$ can be written as

$$r(\phi) = r_{max}| \sin^{1/3}(3(\phi - \phi_0))|. \tag{2.55}$$

The solution in the (x_1, x_2)-plane is periodic. It contains 3 loops with an angle of $2\pi/3$ between the symmetry axes, see Figure 2.5.

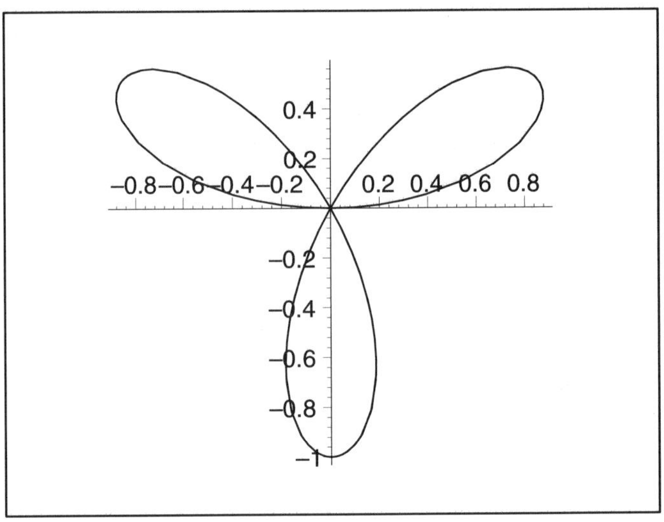

Figure 2.5: The x-projection of the solution

As a conclusion, one may state the following result:

PROPOSITION 2.13. The solution in polar coordinates satisfies

$$r(\phi)^2 = r^2_{max} \sin^{2/3}(3(\phi - \phi_0), \tag{2.56}$$

where

$$r^2_{max} = \frac{1}{2}\Big(\frac{E}{\theta^2}\Big)^{1/3},$$

E being the constant of energy.

2.5.0.8. *The t-component along the solution.* Suppose that $(x_1(s), x_2(s), t(s))$ is a geodesic, *i.e.*, it satisfies the Hamiltonian system of equations given by the Hamiltonian

$$H(\xi, \theta, x, t) = \frac{1}{2}(\xi_1 + 4x_2|x|^2\theta)^2 + \frac{1}{2}(\xi_2 - 4x_1|x|^2\theta)^2. \tag{2.57}$$

One of the Hamiltonian equation is

$$\dot{t} = \frac{\partial H}{\partial \theta} = 4|x|^2(x_2\dot{x}_1 - x_1\dot{x}_2) = -4r^4\dot{\phi}. \tag{2.58}$$

If $\theta < 0$ then ϕ increases, and hence t decreases. Using Proposition 2.13, one obtains

$$dt = -4r^4\, d\phi = -4r^4_{max} \sin^{4/3}(3(\phi - \phi_0))\, d\phi.$$

Integrating, we have

$$\begin{aligned} t(\phi) - t(\phi_0) &= -4r^4_{max} \int_{\phi_0}^{\phi} \sin^{4/3}(3(u - \phi_0))\, du \\ &= -\frac{4}{3}r^4_{max} \int_0^{3(\phi-\phi_0)} \sin^{4/3} v\, dv. \end{aligned}$$

Set

$$\Psi(u) = \int_0^{3u} \sin^{4/3} v\, dv. \tag{2.59}$$

Then

$$t(\phi) - t(\phi_0) = -\frac{4}{3}r^4_{max}\, \Psi(\phi - \phi_0). \tag{2.60}$$

Let

$$T = -\frac{4}{3}r^4_{max}\, \Psi(\pi/3) < 0 \tag{2.61}$$

Then

$$t(\phi_0 + \pi/3) - t(\phi_0) = T \tag{2.62}$$

does not depend on the initial angle ϕ_0.

Conjugate points to the origin. Consider a subRiemannian geodesic $\big(x_1(s), x_2(s), t(s)\big)$ which starts at the origin. In polar coordinates $r(\phi_0) = 0$ and $t(\phi_0) = 0$. When $\phi = \phi_0 + \pi/3$ we have $r(\phi_0 + \pi/3) = 0$, which means that the x-component is zero. The only non-vanishing component is the t-component

$$t(\phi_0 + \pi/3) = T. \tag{2.63}$$

It follows that the point $(0, 0, T)$ belongs to the geodesic and does not depend on the initial condition ϕ_0. A variation of ϕ_0 corresponds to a rotation of the geodesic. The point $(0, 0, T)$ remains fixed during this variation and belongs to all geodesics which are obtained from the initial one by the variation of the initial angle ϕ_0.

The following result states that the conjugate points to the origin along the geodesics are the t-axis.

THEOREM 2.14. *Given the point $P = (0, 0, u)$ on the t-axis, $u \neq 0$, there is a geodesic γ, which starts at the origin, such that P is the first point conjugate to the origin along γ.*

PROOF. The proof is based on the construction of such a geodesic.
(i) case $u < 0$
We choose $\theta < 0$ and require $T = u$. From equation (2.61) one finds

$$r_{max}^4 = -\frac{3w}{4Q}, \tag{2.64}$$

where $Q = \Psi(\pi/3)$. The desired geodesic will be given by

$$\begin{aligned} r(\phi) &= \left(\frac{3|u|}{4Q}\right)^{1/4} \sin^{1/3}(3(\phi - \phi_0), \\ t(\phi) &= \frac{u}{Q}\int_0^{3(\phi-\phi_0)} \sin^{4/3} v\, dv. \end{aligned}$$

It is easy to see that the point $P = (0, 0, u)$ belongs to the above geodesic. Performing a rotation in the x-plane of the initial velocity, all the geodesics obtained will pass through P, and hence P will be the first conjugate point.
(ii) case $w > 0$
In this case $\theta > 0$ and the rest of the proof is similar with the case $\theta < 0$. □

2.6. The length of the geodesics between the origin and $(0, 0, t)$

It is already known from Theorem 2.14 that there are solutions which join the origin to any point on the t-axis. This section answers two questions:
1. How many subRiemannian geodesics of different lengths (parametrized by $[0, 1]$*) join the origin to* $(0, 0, t)$*, for a given* $t \neq 0$*?*
2. What are these lengths?

The length of a horizontal curve is measured with respect to the subRiemannian metric $g : H \times H \to F$ such that $g(X_i, X_j) = \delta_{ij}$, with the vector fields X_i given by (2.1). It is known that if $c = (x_1, x_2, t)$ is a horizontal curve of parameter $s \in [0, 1]$, c is horizontal if and only if

$$\dot{c}(s) = \dot{x}_1(s)X_1 + \dot{x}_2(s)X_2.$$

The length of $\dot{c}$ is

$$|\dot{c}(s)|_g = \sqrt{g(\dot{c}(s), \dot{c}(s))} = \sqrt{\dot{x}_1(s)^2 + \dot{x}_2(s)^2},$$

and the length of c is

$$\ell(c) = \int_0^1 |\dot{c}(s)|_g\, ds = \int_0^1 \sqrt{\dot{x}_1^2 + \dot{x}_2^2}\, ds.$$

PROPOSITION 2.15. *If* $c : [0, 1] \to \mathbb{R}^3$ *is a subRiemannian geodesic, then* $|\dot{c}(s)|_g$ *is constant, and* $\ell(c) = |\dot{c}|_g$.

PROOF. If c is a subRiemannian geodesic, the Hamiltonian is preserved along the geodesic

$$\frac{\dot{x}_1(s)^2 + \dot{x}_2(s)^2}{2} = H_0 \text{ (constant)},$$

and hence

$$|\dot{c}(s)|_g = \sqrt{2H_0}.$$

The second part of the conclusion is obvious. □

In the following we shall find explicit lengths of the subRiemannian geodesics between the origin and points situated on the t-axis, see [**12**]. The following lemma will be useful.

LEMMA 2.16.

$$\int_0^{\sigma(r)^2} \frac{dt}{\sqrt{t(1-t^3)}} = \frac{1}{3^{1/4}} sn^{-1}\Big[\sqrt{1-\frac{g(\sigma)^2}{b^2}}, k\Big],$$

where $b = 2-\sqrt{3}$, $k = \sqrt{b}/2$, *and* $g(\sigma) = \frac{\sqrt{3}}{\sigma^2 + \frac{1+\sqrt{3}}{2}} - 1$.

PROOF. We need to evaluate the integral

$$\int_0^{\sigma^2} \frac{dx}{\sqrt{f(x)}},$$

where $f(x) = x(1-x^3) = -(x^2-x)(1+x+x^2)$.
Making the substitution

$$x = \frac{pt+q}{t+1},$$

with

$$p = \frac{-1-\sqrt{3}}{2}, \quad q = \frac{-1+\sqrt{3}}{2},$$

f becomes

$$\begin{aligned} f &= -\frac{1}{(t+1)^4}\Big((p^2-p)t^2+q^2-q\Big)\Big((p^2+p+1)t^2+(q^2+q+1)\Big) \\ &= -\frac{1}{(t+1)^4}\Big((\frac{3}{2}+\sqrt{3})t^2+(\frac{3}{2}-\sqrt{3})\Big)\Big(\frac{3}{2}t^2+\frac{3}{2}\Big) \\ &= -\frac{1}{(t+1)^4}\frac{3}{2}(\frac{3}{2}+\sqrt{3})(t^2-b^2)(t^2+1), \end{aligned}$$

where $b^2 = 7-4\sqrt{3}$ and $b = 2-\sqrt{3}$. Let

$$g(\sigma) = \frac{\sqrt{3}}{\sigma^2 + \frac{1+\sqrt{3}}{2}} - 1.$$

Then the integral becomes

$$\int_0^{\sigma^2} \frac{dx}{\sqrt{f(x)}} = \frac{2}{\sqrt{3+2\sqrt{3}}} \int_{g(\sigma)}^{b} \frac{1}{\sqrt{(b^2-t^2)(t^2+1)}}\, dt$$

Using the formula (see Lawden, [**9**], chapter 3)

$$\int_x^b \frac{1}{\sqrt{(a^2+t^2)(b^2-t^2)}}\, dt = \frac{1}{\sqrt{a^2+b^2}} sn^{-1}\Big[\frac{\sqrt{b^2-x^2}}{b}, \frac{b}{\sqrt{a^2+b^2}}\Big],$$

one obtains

$$\int_0^{\sigma^2} \frac{dx}{\sqrt{f(x)}} = 3^{-1/4} sn^{-1}\Big[\sqrt{1-\frac{g(\sigma)^2}{b^2}}, \frac{\sqrt{b}}{2}\Big].$$

□

THEOREM 2.17. *The geodesics that join the origin to a point* $(0,0,t)$ *have lengths* d_1, d_2, $d_3 \ldots$, *where*

$$(d_m)^4 = \frac{m^3K^4}{4Q}|t|, \tag{2.65}$$

where

$$K = \int_0^1 \frac{dt}{\sqrt{(1-t^2)(1-k^2t^2)}}, \qquad k = \frac{\sqrt{2-\sqrt{3}}}{2}, \tag{2.66}$$

$$Q = \frac{1}{4}\frac{\Gamma(1/6)}{\Gamma(2/3)}\sqrt{\pi}. \tag{2.67}$$

For each length d_m, *the geodesics of that length are parametrized by the circle* $\mathbb{S}^1$.

PROOF. We need to find the subRiemannian geodesic connecting the origin to $(0,0,t)$. The geodesic is parametrized by $[0,1]$. The boundary conditions are $r(0) = r(1) = 0$, $t(0) = 0$ and $t(1) = t$. Using the Euler-Lagrange equations in polar coordinates, one obtains that the constant θ verifies the following boundary value problem

$$\ddot{r} = -48\theta^2 r^5, \tag{2.68}$$

$$r(0) = r(1) = 0. \tag{2.69}$$

Let $\lambda^2 = 48\theta^2$. Then (2.68) becomes

$$\ddot{r} = -\lambda^2 r^5, \tag{2.70}$$

with the boundary conditions $r(0) = r(1) = 0$. Integrating, we get the conservation law of energy

$$\frac{1}{2}\dot{r}^2 + \frac{1}{6}\lambda^2 r^6 = H_0.$$

H_0 is the constant value of the Hamiltonian along the solutions (denoted also by E). Separate the variables and obtain

$$\frac{dr}{\sqrt{2H_0 - \frac{1}{3}\lambda^2 r^6}} = ds.$$

The radius r is defined by the integral equation

$$\int_0^r \frac{dx}{\sqrt{2H_0 - \frac{1}{3}\lambda^2 x^6}} = s.$$

Making the substitution

$$u = \Big(\frac{\lambda^2}{6H_0}\Big)^{1/6} x,$$

the above expression becomes

$$\int_0^{\sigma(r)} \frac{du}{\sqrt{1-u^6}} = \Big(\frac{4H_0^2\lambda^2}{3}\Big)^{1/6} s, \tag{2.71}$$

where

$$\sigma(r) = \Big(\frac{\lambda^2}{6H_0}\Big)^{1/6} r.$$

Substituting $u^2 = t$, one may write

$$\int_0^{\sigma(r)^2} \frac{dt}{\sqrt{t(1-t^3)}} = \frac{2}{3^{1/6}}(2H_0\lambda)^{1/3}s. \tag{2.72}$$

The left side can be expressed in terms of elliptic integrals. Using Lemma 2.16, relation (2.72) becomes

$$\sqrt{1-\frac{g(\sigma)^2}{b^2}} = sn\,[2^{4/3}3^{1/12}(H_0\lambda)^{1/3}s\;,\; k]. \tag{2.73}$$

For $s=0$, $s=1$, we have $r=0$, and hence $\sigma=0$. Using the expression of g we have $g(0)=\frac{-1+\sqrt{3}}{1+\sqrt{3}}=2-\sqrt{3}=b$, and hence $\sqrt{1-\frac{g(0)^2}{b^2}}=0$. It turns out that for $s=0$ and $s=1$ the left hand side in (2.73) is zero. Then

$$sn\left(2^{4/3}3^{1/12}(H_0\lambda)^{1/3}\right)=0,$$

and hence

$$2^{4/3}\cdot 3^{1/12}(H_0\lambda)^{1/3} = 2mK\;, \qquad m=1,2,3\ldots \tag{2.74}$$

$m\neq 0$ because $H_0\neq 0$ and $\theta\neq 0$. $4K$ is the period of $sn\,(\cdot\,,k)$ with $k=\sqrt{b}/2$. K is defined by the complete Jacobi integral

$$K=\int_0^1 \frac{dt}{\sqrt{(1-t^2)(1-k^2t^2)}}.$$

From (2.74) one has

$$H_0\,\lambda = \frac{m^3K^3}{2\cdot 3^{1/4}}.$$

Using $\lambda = 4\sqrt{3}\,\theta$, the above formula becomes

$$H_0\,\theta = \Big(\frac{mK}{2\cdot 3^{1/4}}\Big)^3.$$

Then θ is given by

$$\theta = \Big(\frac{mK}{2\cdot 3^{1/4}}\Big)^3/H_0,$$

and

$$\frac{H_0}{\theta^2} = (2\cdot 3^{1/4})^6\Big(\frac{H_0}{m^2K^2}\Big)^3.$$

Using (2.35) we have

$$r^4_{max} = \Big(\frac{H_0}{\theta^2}\Big)^{2/3} = 3\frac{2^4H_0^2}{m^4K^4}. \tag{2.75}$$

The left hand side of (2.75) is related to t as follows.
As $dt=-4r^4\,d\phi$, integrating between ϕ_0 and $\phi_0-m\pi/3$, we obtain

$$\begin{aligned} t &= -4\int_{\phi_0}^{\phi_0-m\pi/3} r^4_{max}\sin^{4/3}\Big(3(\phi-\phi_0)\Big)\,d\phi \\ &= -\frac{4}{3}\Big(\frac{H_0}{8\theta^2}\Big)^{2/3}\int_0^{-m\pi}\sin^{4/3}u\,du \\ &= \frac{4m}{3}\Big(\frac{H_0}{8\theta^2}\Big)^{2/3}\int_0^{\pi}\sin^{4/3}u\,du. \end{aligned}$$

Hence

$$t = \frac{m}{3}\left(\frac{H_0}{\theta^2}\right)^{2/3} Q, \tag{2.76}$$

where $Q = \int_0^\pi \sin^{4/3} u\, du$. We used ϕ decreasing because θ was assumed positive, according to the formula $\dot{\phi} = -4\theta r^2$.
Substituting (2.75) in (2.76), we have

$$t = 2^4 Q \, \frac{H_0^2}{m^3 K^4}.$$

Therefore,

$$(2H_0)^2 = \frac{m^3 K^4}{4Q} |t| \,, \qquad m = 1, 2, 3... \tag{2.77}$$

The physical interpretation of the above relation is the fact that the energy is discrete and depends on $|t|$.

The geometrical interpretation is that the geodesics joining the origin O to $P(0,0,t)$ have discrete lengths. Indeed, let $\gamma : [0,1] \to \mathbb{R}^3$ be a geodesic between O and P. As $|\dot{\gamma}|$ is constant in the subRiemannian metric (see Proposition 2.15), we have identity in Cauchy's inequality

$$\ell(\gamma) = \int_0^1 |\dot{\gamma}(s)|\, ds = \left(\int_0^1 ds\right)^{1/2} \left(\int_0^1 |\dot{\gamma}(s)|^2\, ds\right)^{1/2} = \sqrt{2H_0},$$

and hence

$$\ell(\gamma)^4 = (2H_0)^2. \tag{2.78}$$

Using (2.77) we see that the lengths of the solution γ are quantified, namely, $\ell(\gamma) = d_m$ where

$$(d_m)^4 = \frac{m^3 K^4}{4Q} |t| \,, \qquad m = 1, 2, 3, ... \tag{2.79}$$

These lengths do not depend on the angle ϕ_0. When this angle is varied, the solution is rotating around t-axis always having the same length. The constant Q is given in Lemma 2.18. □

The shortest geodesic corresponds to $m = 1$. The Carnot-Carathéodory distance between O and $P(0,0,t)$ is d_1, where

$$(d_1)^4 = \frac{K^4}{4Q}|t|. \tag{2.80}$$

The constant $Q = \int_0^\pi \sin^{4/3} u\, du$ can be expressed in terms of gamma functions as follows.

LEMMA 2.18.

$$Q = \frac{1}{4}\frac{\Gamma(1/6)}{\Gamma(2/3)}\sqrt{\pi}.$$

PROOF. Making the substitution $t = \sin u$ yields

$$Q = \int_0^{\pi} \sin^{4/3} u\, du = 2\int_0^{\frac{\pi}{2}} \sin^{4/3} u\, du = 2\int_0^1 t^{4/3}(1-t^2)^{-1/2}\, dt.$$

With the substitution $t^2 = x$ the integral becomes a beta function, which can be expressed in terms of gamma functions

$$Q = \int_0^1 x^{1/6}(1-x)^{-1/2}\, dx = B(7/6, 1/2) = \frac{\Gamma(7/6)\,\Gamma(1/2)}{\Gamma(7/6+1/2)}.$$

Using $\Gamma(1/2) = \sqrt(\pi)$, $\Gamma(7/6) = \frac{1}{6}\Gamma(1/6)$ and $\Gamma(7/6+1/2) = \frac{2}{3}\Gamma(2/3)$, we obtain

$$Q = \frac{\frac{1}{6}\Gamma(1/6)\Gamma(1/2)}{\frac{2}{3}\Gamma(2/3)} = \frac{1}{4}\frac{\Gamma(1/6)}{\Gamma(2/3)}\sqrt{\pi}.$$

□

2.7. Explicit solutions connecting the origin to $(0,0,t)$

In the previous section we have shown that there are infinitely many geodesics joining the origin with points on the t-axis. In this section we shall find explicit formulae for these geodesics.

From (2.74) and (2.76) we may compute r_{max} as a function of t

$$r_{max}^4 = 3\frac{2^4 H_0^2}{m^4 K^4} = 12\frac{(2H_0)^2}{m^4K^4} = \frac{3|t|}{mQ}.$$

It follows that for a given t, the maximal radius r_{max} is quantized

$$r_{max}^4 = \frac{3|t|}{mQ}, \qquad m = 1, 2, 3\ldots \tag{2.81}$$

As m increases, r_{max} decreases. Using Proposition 2.13 one obtains that the solution in polar coordinates, which joins O to $(0,0,t)$ is given by

$$r_m^4(\phi) = \frac{3|t|}{mQ}\sin^{4/3}\Big(3(\phi-\phi_0)\Big), \qquad m = 1, 2, 3\ldots \tag{2.82}$$

The t-component is computed by integration

$$t_m(\phi) = \int_{\phi_0}^{\phi} -4r_m^4(u)\, du = \frac{-12|t|}{mQ}\int_{\phi_0}^{\phi} \sin^{4/3}\Big(3(u-\phi_0)\Big)\, du.$$

Making the substitution $v = 3(u-\phi_0)$, we obtain

$$t_m(\phi) = \frac{-4|t|}{mQ}\int_0^{3(\phi-\phi_0)} \sin^{4/3}(v)\, dv, \qquad m = 1, 2, 3\ldots \tag{2.83}$$

Formula (2.82) expresses $r = r(\phi)$, but we may compute explicitly the radius $r = r(s)$, which is a solution for the boundary value problem

$$\begin{cases} \ddot{r} = -\lambda^2 r^5 \\ r(0) = r(1) = 0, \end{cases} \tag{2.84}$$

where $\lambda^2 = 48\theta^2$. This will be done in the following.

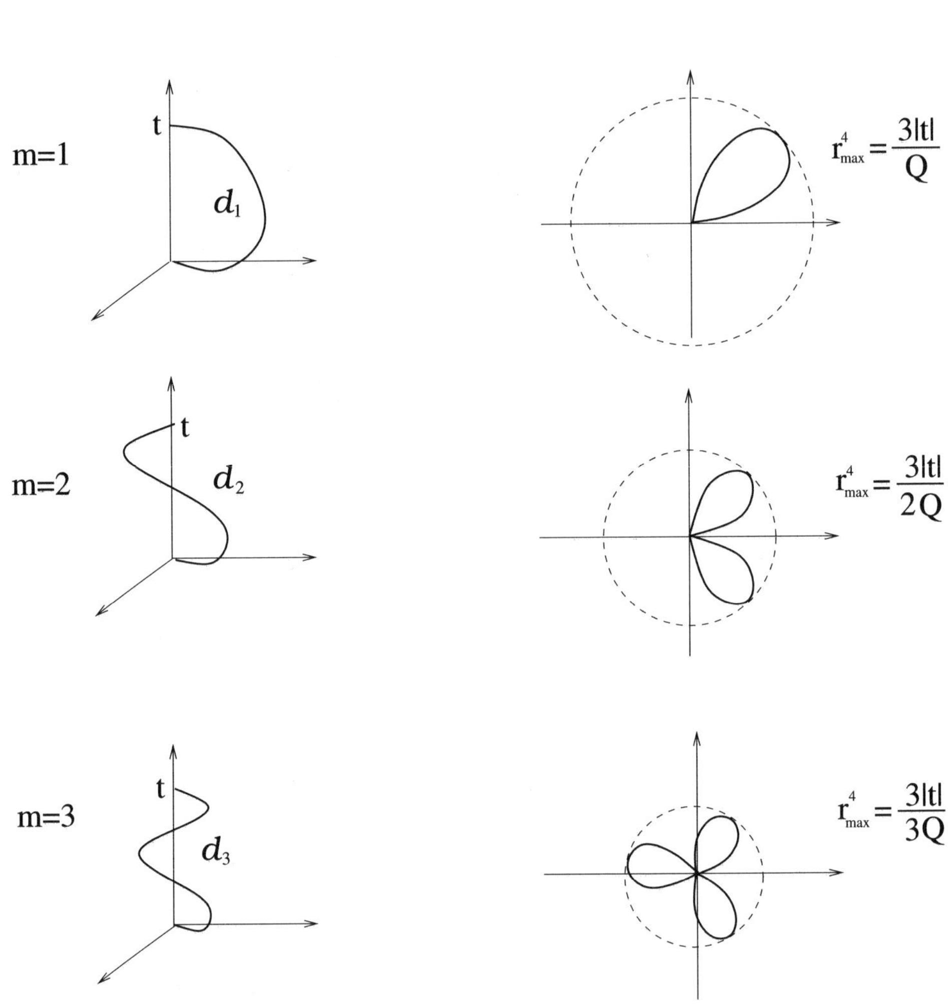

Figure 2.6: The shapes of the solution for $m = 1, 2, 3$.

Using the relation between the elliptic functions $cn = \sqrt{1 - sn^2}$, relation (2.73) becomes

$$\frac{g(\sigma)}{b} = cn\Big[2^{4/3}\, 3^{1/12} (H_0\lambda)^{1/3} s\,, \frac{\sqrt{b}}{2}\Big].$$

Since

$$2^{4/3} \cdot 3^{1/12} (H_0\lambda)^{1/3} = 2mK,$$

then

$$g(\sigma) = b\; cn[2mKs\,, \frac{\sqrt{b}}{2}].$$

Using the definition of $g(\sigma)$, we have

$$\sigma^2 = \frac{\sqrt{3}}{b\; cn\Big(2mKs\,, \frac{\sqrt{b}}{2}\Big) + 1} + p,$$

where $p = \frac{-1-\sqrt{3}}{2}$. Using that

$$\sigma^2 = \Big(\frac{\lambda^2}{6H_0}\Big)^{1/3} r^2 = \Big(\frac{\lambda^2}{6H_0}\Big)^{1/3} r^2 = 2\Big(\frac{\theta^2}{H_0}\Big)^{1/3} r^2 = 2\Big(\frac{mQ}{3|t|}\Big)^{1/2} r^2,$$

the above formula becomes

$$r_m^2(s) = \frac{1}{2}\Big(\frac{3|t|}{mQ}\Big)^{1/2}\Big[p + \frac{\sqrt{3}}{1 + b\,cn[2mKs\,,\,k]}\Big]\,, \quad m = 1,2,3\ldots \tag{2.85}$$

where $k = \sqrt{b}/2$, $b = 2 - \sqrt{3}$, $p = (-1 - \sqrt{3})/2$.

The first three geodesics and their projections on the x-plane are sketched in Figure 2.6. After that the pattern is periodic.
The above results can be stated in the following theorem:

THEOREM 2.19. *The equation of the subRiemannian geodesics which join the origin to a fixed point* $(0, 0, t)$ *in polar coordinates are*

$$r_m^4(\phi) = \frac{3|t|}{mQ}\sin^{4/3}\Big(3(\phi - \phi_0)\Big)\,, \quad m = 1,2,3\ldots \tag{2.86}$$

$$t_m(\phi) = \frac{-4|t|}{mQ}\int_0^{3(\phi-\phi_0)} \sin^{4/3}(v)\,dv. \tag{2.87}$$

The radius $r(s)$ *is given by*

$$r_m^2(s) = \frac{1}{2}\Big(\frac{3|t|}{mQ}\Big)^{1/2}\Big[p + \frac{\sqrt{3}}{1 + b\,cn[2mKs\,,\,k]}\Big], \tag{2.88}$$

where $m = 1, 2, 3\ldots$

2.8. Solutions which start outside the origin

In this case, $r(0) \neq 0$ and the system (2.25) will be considered with the initial conditions

$$r(0) \neq 0\,, \quad \phi(0) = \phi_0. \tag{2.89}$$

The second equation of (2.25) becomes

$$\dot\phi = \frac{C}{r^2} - 4\theta r^2, \tag{2.90}$$

where $C \neq 0$. Substituting in the first equation of (2.25), we obtain

$$\ddot r = r\dot\phi(\dot\phi + 16\theta r^2) = r\Big(\frac{C}{r^2} - 4\theta r^2\Big)\Big(\frac{C}{r^2} + 12\theta r^2\Big) \Leftrightarrow$$

$$\ddot r = \frac{C^2}{r^3} + 8\theta C r - 48\theta^2 r^5. \tag{2.91}$$

Let

$$V(r) = 8\theta^2 r^6 - 4\theta r^2 + \frac{C^2}{2r^2} \tag{2.92}$$

be the potential function. Equation (2.91) can be written as the Newton's equation

$$\ddot r = -V'(r). \tag{2.93}$$

A computation shows that

$$V(r) = \frac{1}{2}\Big(\frac{4\theta r^4 - C}{r}\Big)^2. \tag{2.94}$$

Writing $p = \dot{r}$ the equation (2.91) becomes

$$\begin{aligned} p\frac{dp}{dr} &= -\frac{dV}{dr} \Leftrightarrow \\ \frac{p^2}{2} &= -V(r) + E, \end{aligned}$$

where E is the constant of total energy.

PROPOSITION 2.20. If r is a solution of the Euler-Lagrange system (2.25) with initial conditions (2.89), then the energy is conserved

$$\frac{1}{2}\dot{r}^2 + \frac{1}{2}\Big(\frac{4\theta r^4 - C}{r}\Big)^2 = E. \tag{2.95}$$

We shall give a qualitative description of the solutions of (2.95). There are two cases: $\operatorname{sgn}(C) = \operatorname{sgn}(\theta)$ and $sgn(C) \neq sgn(\theta)$. In each case we shall draw the trajectory in the phase plane.

(1) Case sgn(C) = sgn(θ). In this case $V(r)$ has a root and a global minimum at

$$r_0 = \Big(\frac{C}{4\theta}\Big)^{1/4}, \tag{2.96}$$

and $\lim_{x\to 0^+} V(r) = \lim_{x\to\infty} V(r) = \infty$, see Figure 2.7

Equilibrium points
r_0 corresponds to a stable equilibrium point because

$$V'(r_0) = 0 \ \text{ and } \ V''(r_0) > 0.$$

The corresponding ϕ satisfies

$$\dot{\phi} = \frac{C}{r_0^2} - 4\theta r_0^2 = \frac{C - 4\theta r_0^4}{r_0^2} = 0,$$

and hence ϕ is constant. This means that each point of the circle $C(0, r_0)$ is an equilibrium point (constant solution). At each of these points the energy $E = 0$.

Tangent circles:
When $E > 0$ there are exactly two positive roots r_{min}, r_{max} of the equation

$$V(r) = E, \tag{2.97}$$

such that

$$0 < r_{min} \le r_0 \le r_{max}. \tag{2.98}$$

The phase plane and the potential V are shown in Figure 2.7.

PROPOSITION 2.21. The solution of the Euler-Lagrange system (2.25) in the case $sgn(C) = sgn(\theta)$ lies into the circular ring $W(0, r_{min}, r_{max}) = \{x \in \mathbb{R}^2; r_{min} \le |x| \le r_{max}\}$.

PROOF. As r reaches r_{min} or r_{max} for $\dot{r} = 0$, then the equation of conservation of energy (2.95) shows that r_{min} , r_{max} are solutions of the equation $V(r) = E$, and the solution r has the property

$$0 < r_{min} \leq r \leq r_{max}.$$

□

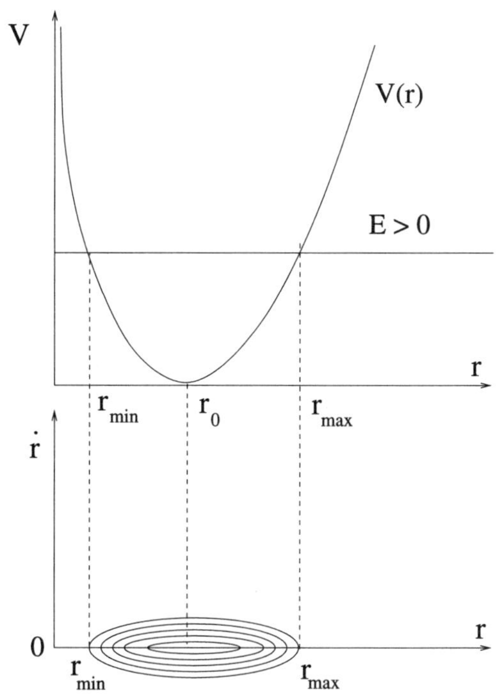

Figure 2.7: The solution in the phase plane

Existence of loops:
The equation $\dot{\phi} = 0$ has the positive root

$$r = \left(\frac{C}{4\theta}\right)^{1/4}. \tag{2.99}$$

Hence $\dot{\phi} = 0$ if and only if $r = r_0$. When the solution intersects the circle of equilibrium points $S(0, r_0)$, the sign of $\dot{\phi}$ changes, *i.e.*, the trajectory is bouncing back, making loops in the ring $W(0, r_{min}, r_{max})$, see Figure 2.8.

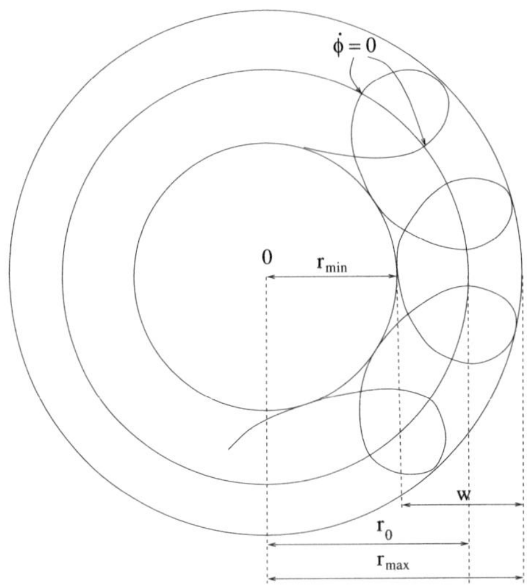

Figure 2.8: The x-projection of the solution

The width of the ring:

The width of the ring $w = r_{max} - r_{min}$ is a function of energy $w = w(E)$. Using the properties of the function $V(r)$, one obtains the following:
(i) $w(E)$ is an increasing function of energy E,
(ii) $w(0) = 0$, case in which $r_{min} = r_0 = r_{max}$,
(iii) $\lim_{E \to \infty} r_{min} = 0$, $\lim_{E \to \infty} r_{max} = \infty$,
(iv) $r_{min} \sim \frac{1}{r_{max}^3}$ as $E \to \infty$.

The asymptotical correspondence (iv) comes from the analysis of the equation

$$4\theta r^3 - \frac{C}{r} = \pm\sqrt{2E},$$

when E is large. We may have $r = r_{min} \to 0$ or $r = r_{max} \to \infty$. When consider $r = r_{min}$, the leading factor is $\frac{C}{r_{min}}$. When $r = r_{max}$, the leading factor is $4\theta r_{max}^3$. See Figure 2.7.

The t-component:

As $\dot{t} = -4r^4\dot{\phi}$ and $\dot{\phi} = 0$ on the circle $S(0, r_0)$, then t will do the same. $\dot{t}(s)$ changes the sign when the solution $x(s)$ crosses the above circle. The component $t(s)$ reaches the maxima and minima on the cylinder $S(0, r_0) \times [0, +\infty)$. See Figure 2.9.

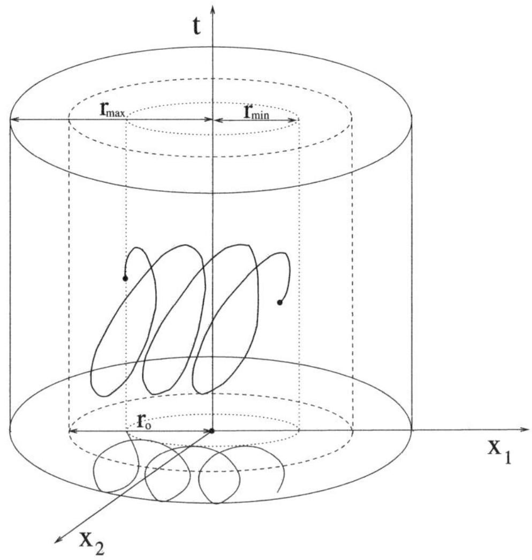

Figure 2.9: The geodesic and its projection in the case sgn(C) ≠ sgn(θ).

(2) Case sgn(C) ≠ sgn(θ), $C \neq 0$. Here

$$\dot{\phi} = \frac{C}{r^2} - 4\theta r^2 \neq 0,$$

which means that ϕ increases (decreases) for $\theta < 0$ ($\theta > 0$), and hence there are no loops. In this case the potential energy is strictly positive

$$V(r) = \frac{1}{2}\Big(\frac{4\theta r^4 - C}{r}\Big)^2 > 0,$$

and it has a minimum at

$$r_1 = \Big(\frac{-C}{12\theta}\Big)^{1/4}. \tag{2.100}$$

Indeed,

$$V'(r) = 48\theta^2 r^5 - 8\theta C r - \frac{C^2}{r^3} = \frac{1}{r^3}(48\theta^2 r^8 - 8\theta C r^4 - C^2),$$

and the equation of critical points

$$V'(r) = 0$$

is equivalent with the following quadratic equation

$$48\theta^2 u^2 - 8\theta C u - C^2 = 0,$$

where $u = r^4$. The equation in u has two roots

$$u_1 = \frac{C}{4\theta} \quad \text{and} \quad u_2 = -\frac{C}{12\theta}.$$

As only the second one is positive, r_1 will be chosen such that

$$(r_1)^4 = -\frac{C}{12\theta}.$$

The equilibrium solution:

The equilibrium solution in this case is the circle of radius r_1 centered at the origin. The corresponding ϕ can be obtained from

$$r_1^2(\dot{\phi} + 4\theta r_1^2) = C,$$

and it is equal to

$$\phi(s) = \Big(\frac{4C}{3}\sqrt{\frac{-12\theta}{C}}\Big)s + \phi_0.$$

The circle is traversed clock-wise if $C > 0$ and counter-clock-wise if $C < 0$.

The bounds for $r(s)$:
There are two positive roots ρ_{min} and ρ_{max} for the equation

$$V(r) = E\ , \quad \text{where } E \geq V(r_1),$$

and

$$0 < \rho_{min} \leq r_1 \leq \rho_{max}.$$

They correspond to the points where $\dot{r} = 0$ in the conservation equation (2.95).

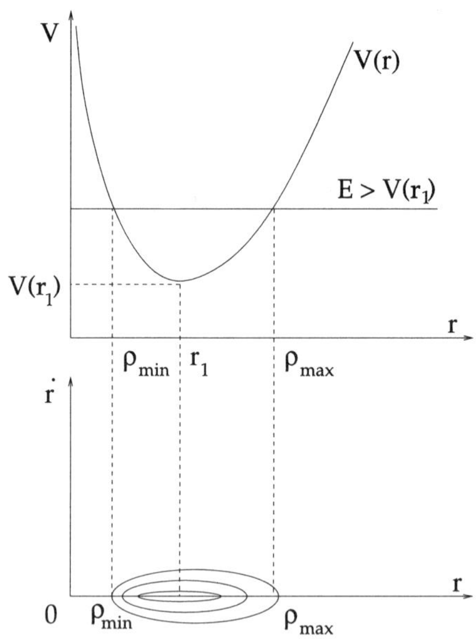

Figure 2.10: The solution in the phase plane

In the phase plane the solution is rotating around the stable equilibrium point $(r_1, 0)$ such that

$$\rho_{min} \leq r(s) \leq \rho_{max},$$

see Figure 2.10.

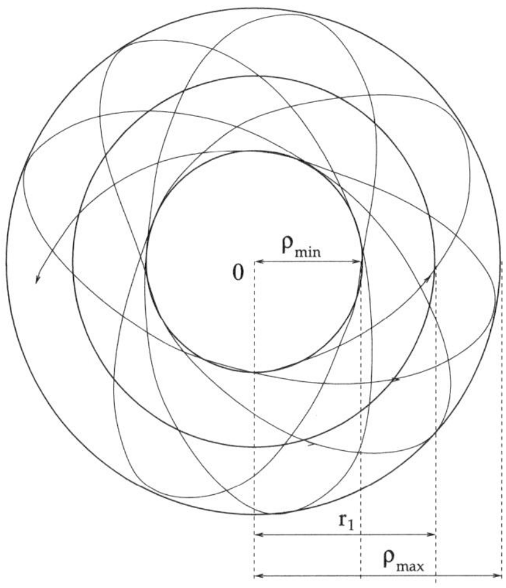

Figure 2.11: The x-projection of the solution

The trajectory in the x-plane belongs to the ring $W(0, \rho_{min}, \rho_{max})$, see Figure 2.11. The width of the ring increases as the total energy E increases, *i.e.*, the initial speed increases, where $E = \frac{1}{2}|\dot{x}(0)|^2$.

The t-component:
In this case $\dot{t} \neq 0$ and $\dot{t} = -4r^4\dot{\phi}$. Hence $t(s)$ only increases (if $\theta > 0$) or only decreases (if $\theta < 0$). See Figure 2.12.

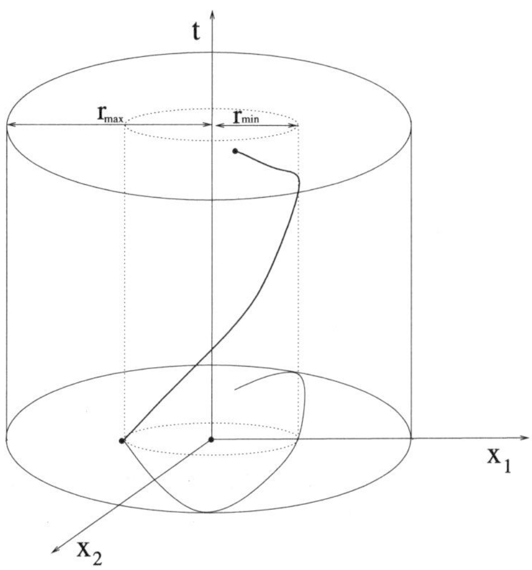

Figure 2.12: The geodesic and its projection in the case $sgn(C) \neq sgn(\theta)$.

2.9. Geodesics between the origin and points away from the t-axis

We have seen that if $|x| = 0$ and $t \neq 0$, there are infinitely many geodesics between the origin and P. In this section we shall study the case $|x| \neq 0$.

THEOREM 2.22. *Let $P(x,t) \in \mathbb{R}^3$ be a point such that $x \neq 0$ and $t \geq 0$. There are finitely many geodesics between the origin and P. This number increases unbounded as $t/|x|^4 \to \infty$.*

PROOF. Let ρ and Φ be the polar coordinates for the endpoint (x_1, x_2). Consider a geodesic which joins the origin and P. The t-component is computed by integration between ϕ_0 and Φ

$$t = -\frac{4}{3} r_{max}^4 \int_0^{3(\Phi-\phi_0)} \sin^{4/3} v \, dv. \tag{2.101}$$

Writing $\rho = r(\Phi)$ in Proposition 2.13, yields

$$\rho^2 = r_{max}^2 \sin^{2/3}\Big(3(\Phi - \phi_0)\Big). \tag{2.102}$$

Eliminating r_{max} from equations (2.101) and (2.102), we obtain

$$\frac{3}{4}\frac{t}{\rho^4} \sin^{4/3}\Big(3(\Phi - \phi_0)\Big) = -\int_0^{3(\Phi-\phi_0)} \sin^{4/3} v \, dv. \tag{2.103}$$

Let

$$F(x) = \int_0^x \sin^{4/3} u \, du. \tag{2.104}$$

It is easy to see that $F(x)$ is an increasing, unbounded and odd function. Let $\psi = 3(\phi_0 - \Phi)$. The relation (2.103) becomes

$$\frac{3}{4}\frac{t}{\rho^4} \sin^{4/3} \psi = F(\psi). \tag{2.105}$$

Since the left hand side is a periodic and bounded function of variable ψ, the above equation has finitely many solutions, see Figure 2.13. This means that for $\rho \neq 0$ there are only a finite number of subRiemannian geodesics joining the origin O and the point P. This number increases unboundedly as the quotient t/ρ^4 increases. For $t = 0$ the equation (2.105) has only the solution $\psi = 0$, *i.e.*, $\phi_0 = \phi = \Phi$. In this case $t(s) = 0$. Hence, the only solution is the segment line OP. □

A detailed analysis of the number of solutions of equation (2.105) is done in the following lemmas, see [**15**].

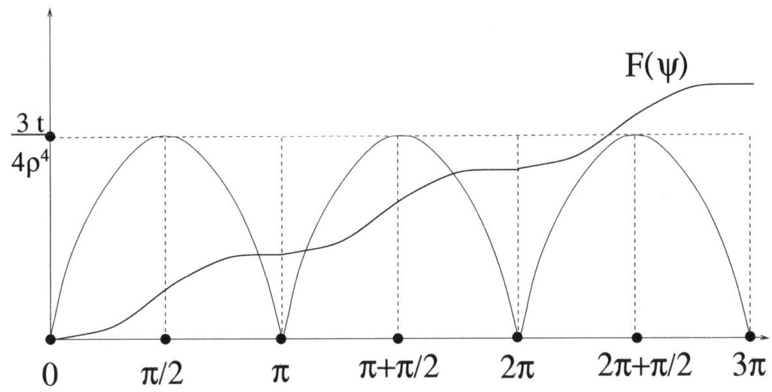

Figure 2.13: The functions $\lambda(\psi) = \frac{3}{4}\frac{t}{\rho^4}\sin^{4/3}\psi$ *and* $F(\psi)$.

Let $\lambda(\psi) = \frac{3}{4}\frac{t}{\rho^4}\sin^{4/3}\psi$. The functions $\lambda(\psi)$ and $F(\psi)$ start from the origin and their graphs are tangent at $\psi = 0$. The following lemma shows that the graph of $F(\psi)$ is below the graph of $\lambda(\psi)$ on a small neighborhood.

LEMMA 2.23. *Let $t > 0$ and $\rho \neq 0$. There is an $\epsilon > 0$ such that*

$$0 < \lambda(\psi) < F(\psi), \quad \textit{for} \quad 0 < \psi < \epsilon.$$

Hence, there are no solutions for the equation (2.105) *on the interval* $(0, \epsilon)$.

PROOF. The derivatives of λ at $\psi = 0$ blow up faster than the derivatives of F, as the following computation shows

$$\begin{aligned} F'(\psi) &= \sin^{4/3}\psi \Longrightarrow F'(0) = 0, \\ F''(\psi) &= \frac{4}{3}\sin^{1/3}\psi\,\cos\psi \Longrightarrow F''(0) = 0, \\ F'''(\psi) &= \frac{4}{9}\cdot\frac{4\cos^2\psi - 3}{\sin^{2/3}\psi} \Longrightarrow F'''(0) = +\infty, \\ \lambda'(0) &= 0, \qquad \lambda''(0) = +\infty. \end{aligned}$$

□

LEMMA 2.24. *Let $t > 0$ and $\rho \neq 0$.*
(i) Let $n \in \mathbb{N}^*$ be an integer such that

$$(n - \frac{1}{2})Q < \frac{3}{4}\frac{t}{\rho^4} < (n + \frac{1}{2})Q, \tag{2.106}$$

where

$$Q = \frac{1}{4}\frac{\Gamma(1/6)}{\Gamma(2/3)}\sqrt{\pi} \approx 1.82.$$

Then the equation (2.105) has not more than $2n + 1$ solutions and not less than $2n - 1$ solutions. All solutions belong to the interval $\Big(0, (n + \frac{1}{2})\pi\Big)$.

(ii) If there is an $n \in \mathbb{N}^*$ such that

$$\frac{3}{4}\frac{t}{\rho^4} = (n+\frac{1}{2})Q,$$

there are exactly $2n+1$ nonzero solutions for the equation (2.105), and they belong to $\Big(0, (n+\frac{1}{2})\pi\Big]$. The largest solution is $(n+\frac{1}{2})\pi$.

(iii) If there is an $n \in \mathbb{N}^*$ such that

$$\frac{3}{4}\frac{t}{\rho^4} > (n+\frac{1}{2})Q,$$

then in the interval $\Big(0, (n+\frac{1}{2})\pi\Big]$ there are $2n$ distinct solutions for the equation (2.105). However, the equation (2.105) may have other solutions outside of the above interval.

PROOF. (*i*) The function $x \to \sin^{4/3} x$ is positive and periodic with the principal period π. Let $n \in \mathbb{N}^*$ be the smallest positive integer such that

$$\frac{3}{4}\frac{t}{\rho^4} < F\Big(n\pi + \frac{\pi}{2}\Big).$$

Then the equation (2.105) has no solutions in the interval $(n\pi + \frac{\pi}{2}, +\infty)$. Hence, if

$$F\Big((n-1)\pi + \frac{\pi}{2}\Big) < \frac{3}{4}\frac{t}{\rho^4} < F\Big(n\pi + \frac{\pi}{2}\Big), \tag{2.107}$$

there are three cases:

(1) If the graph of $F(\psi)$ is above the graph of $\lambda(\psi)$ on the interval $(n\pi, (n+\frac{1}{2})\pi)$, there are exactly $2n-1$ nonzero solutions. See Figure 2.13.

(2) If the graph of $F(\psi)$ is tangent to the graph of $\lambda(\psi)$ on the interval $(n\pi, (n+\frac{1}{2})\pi)$, there are $2n$ nonzero solutions. The largest solution is a double solution. See Figure 2.14(*a*).

(3) If the graph of $F(\psi)$ intersects the graph of $\lambda(\psi)$ twice on the interval $(n\pi, n+\frac{1}{2}\pi)$, there are exactly $2n+1$ nonzero solutions. See Figure 2.14(*b*).

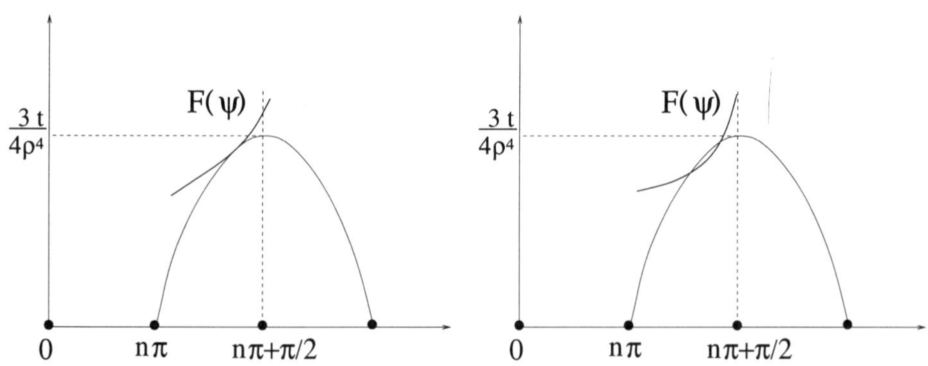

Figure 2.14(*a*) *Figure* 2.14(*b*)

All solutions belong to the interval $\big(0, (n+\frac{1}{2})\pi\big)$. In the following we shall show that

$$F\Big(n\pi + \frac{\pi}{2}\Big) = \Big(n + \frac{1}{2}\Big)Q. \tag{2.108}$$

Using the properties of the integral of a periodic function,

$$\begin{aligned} F\Big(n\pi + \frac{\pi}{2}\Big) &= \int_0^{n\pi+\frac{\pi}{2}} \sin^{\frac{4}{3}} x\,dx \\ &= (2n+1)\int_0^{\frac{\pi}{2}} \sin^{\frac{4}{3}} x\,dx \\ &= (2n+1)\frac{Q}{2} = \Big(n+\frac{1}{2}\Big)Q. \end{aligned}$$

Replacing n by $n-1$, yields

$$F\Big((n-1)\pi + \frac{\pi}{2}\Big) = \Big(n - 1 + \frac{1}{2}\Big)Q = (n - \frac{1}{2})Q. \tag{2.109}$$

Substituting (2.108) in (2.107) yields (2.106).
(*ii*) If there is an integer $n \in \mathbb{N}^*$ such that

$$F\Big(n\pi + \frac{\pi}{2}\Big) = \frac{3}{4}\frac{t}{\rho^4},$$

then

$$(n + \frac{1}{2})Q = \frac{3}{4}\frac{t}{\rho^4},$$

and there is one more solution, equal to $n\pi + \frac{\pi}{2}$.
(*iii*) The condition $\frac{3}{4}\frac{t}{\rho^4} > (n+\frac{1}{2})Q = F\big(n\pi + \frac{\pi}{2}\big)$ implies that the graph of F is below the line $y = \frac{3t}{4\rho^4}$ on the interval $[0, n\pi + \frac{\pi}{2}]$. The graph of F intersects the graph of λ right before and after each turning point such as $\pi, \dots, n\pi$. Hence there are $n+n$ intersections, which correspond to $2n$ solutions in the interval $[0, n\pi+\pi/2]$. See Figure 2.13. □

As a conclusion, one may state the following theorem.

THEOREM 2.25. *Let $P(x,t)$ be a point of $\mathbb{R}^3$, away from the t-axis and x-plane. There are not less than $2n-1$ and not more than $2n+1$ subRiemannian geodesics between the origin and the point P, where the integer n is defined by*

$$(n - \frac{1}{2})Q < \frac{3}{4}\frac{|t|}{|x|^4} \le (n + \frac{1}{2})Q. \tag{2.110}$$

PROOF. The number of subRiemannian geodesics is equal to the solutions of the equation (2.105). Applying Lemma 2.24 leads to the desired conclusion. □

The limit case $|x| \to 0$
One may see from the Figure 2.14 that the solutions of equation (2.105) become in this case

$$0 < \psi_1 = \psi_2 = \pi < \psi_3 = \psi_4 = 2\pi < \psi_5 = \psi_6 = 3\pi < \dots,$$

and hence we have an infinite number of geodesics joining the origin and $P(x,t)$.

2.10. Geodesic completeness in step 4 case

The following definition holds on any subRiemannian manifold.

DEFINITION 2.26. If for any point P, any geodesic $c(t)$ emanating from P is defined for all $t \in \mathbb{R}$, the model is called geodesically complete.

Theorems 2.27 and 2.28 are proved in a more general case in the next chapter. The first theorem states the completeness of the model, while the second one states the global connectivity property.

THEOREM 2.27. *Given a point $P = (x_1^0, x_2^0, t^0) \in \mathbb{R}^3$, any subRiemannian geodesic associated with the model defined by the vector fields (2.1) emanating from P can be extended indefinitely, i.e., the model is geodesically complete.*

THEOREM 2.28. *Given any two points $P, Q \in \mathbb{R}^3$, there is a geodesic joining P and Q.*

2.11. Exercises

EXERCISE 2.1. Let $\operatorname{sn} z = \operatorname{sn}(z,k)$, $\operatorname{cn} z = \operatorname{cn}(z,k)$, $\operatorname{dn} z = \operatorname{dn}(z,k)$. Prove that

$$\begin{aligned}
\frac{d}{dz}\operatorname{sn} z &= \operatorname{cn} z \operatorname{dn} z, \\
\frac{d}{dz}\operatorname{cn} z &= -\operatorname{sn} z \operatorname{dn} z, \\
\frac{d}{dz}\operatorname{dn} z &= -k^2 \operatorname{sn} z \operatorname{cn} z.
\end{aligned}$$

EXERCISE 2.2. Show that the following Maclaurin power series expansions hold about $z = 0$

$$\begin{aligned}
\operatorname{sn} z &= z - \frac{1}{3!}(1+k^2)z^3 + \frac{1}{5!}(1+14k^2+k^4)z^5 - \dots, \\
\operatorname{cn} z &= 1 - \frac{1}{2!}z^2 + \frac{1}{4!}(1+4k^2)z^4 - \dots, \\
\operatorname{dn} z &= 1 - \frac{1}{2!}k^2 z^2 + \frac{1}{4!}(4k^2+k^4)z^4 - \dots.
\end{aligned}$$

EXERCISE 2.3. Show that

$$\lim_{k\to 1} \operatorname{sn}(u,k) = \tanh u, \qquad \lim_{k\to 1} \operatorname{cn}(u,k) = \operatorname{sech} u.$$

EXERCISE 2.4. a) If $0 < u < K$, prove the following inequalities:

$$\operatorname{cn} u < \operatorname{dn} u < 1 < \frac{u}{\operatorname{sn} u} < \frac{1}{\operatorname{cn} u}.$$

b) If $0 < u < \frac{\pi}{2}$ show that

$$\cos u < 1 < \frac{u}{\sin u} < \frac{1}{\cos u}.$$

EXERCISE 2.5. Consider the equation $\ddot{y} = ay^3 - by$, where a and b are constants such that $\frac{a}{2} = b - 1 > 0$. Show that $y(x) = \operatorname{sn}\big(x, \sqrt{b-1}\big)$ is a solution for the above equation.

EXERCISE 2.6. Find the first integral of energy for the equation

$$\ddot{r}(s) = -Kr^{2m+1}(s)$$

where K is a constant and m is an integer.

EXERCISE 2.7. Let $\delta_\lambda : \mathbb{R}^3 \to \mathbb{R}^3$ be given by $\delta_\lambda = (\lambda x, \lambda^4 t)$.
i) Show that

$$\delta_\lambda^* X_1 = \lambda^{-1} X_1, \quad \delta_\lambda^* X_2 = \lambda^{-1} X_2,$$

where X_1 and X_2 are defined in (2.1).
ii) Prove the following estimates

$$C_1(|x|^4 + |t|)^{1/4} \leq d_C\big((0,(x,t)\big) \leq C_2\big(|x|^4 + |t|\big)^{1/4},$$

where C_1 and C_2 are two universal constants and $d_C\big((0,(x,t)\big)$ is the Carnot-Carathéodory distance between the origin and the point (x, t).

CHAPTER 3

The Geometric Analysis of Step $2(k+1)$ Case

In this chapter we study the subRiemannian geometry on the step $2(k+1)$ subRiemannian manifold defined on $\mathbb{R}^3 = \mathbb{R}^2_x \times \mathbb{R}_t$ by the vector fields

$$X_1 = \partial_{x_1} + 2x_2|x|^{2k}\partial_t \ , \qquad X_2 = \partial_{x_2} - 2x_1|x|^{2k}\partial_t, \tag{3.1}$$

where $|x|^{2k} = (x_1^2 + x_2^2)^k$, $k \in \mathbb{N}$.
We shall compute the length of the subRiemannian geodesics between the origin and any point on the $t-$axis, where the conjugate locus is. We shall also characterize the number of subRiemannian geodesics between the origin and any other point. Finally, we shall prove the connectivity and completeness theorems for this model. The subRiemannian lengths are computed in two ways: using polar coordinates and using hypergeometric functions.

With a similar proof as in chapter 2, we have the following result:

PROPOSITION 3.1. There is no Lie group on $\mathbb{R}^3$ underlying the vector fields X_1 and X_2 given by (3.1).

3.1. The horizontal distribution

Let H be the horizontal distribution $x \to H_x = \text{span}\{X_1, X_2\}$. A curve c such that the velocity $\dot{c} \in H$, is called horizontal curve.

PROPOSITION 3.2. A curve $c = (x_1, x_2, t)$ is horizontal if and only if

$$\dot{t} = 2|x|^{2k}(\dot{x}_1 x_2 - x_1 \dot{x}_2). \tag{3.2}$$

PROOF. The velocity vector can be written as

$$\begin{aligned}
\dot{c} &= \dot{x}_1\partial_{x_1} + \dot{x}_2\partial_{x_2} + \dot{t}\partial_t \\
&= \dot{x}_1\big(\partial_{x_1} + 2x_2|x|^{2k}\partial_t\big) - 2\dot{x}_1 x_2|x|^{2k}\partial_t \\
&\quad + \dot{x}_2\big(\partial_{x_2} - 2x_1|x|^{2k}\partial_t\big) + 2x_1\dot{x}_2|x|^{2k}\partial_t + \dot{t}\partial_t \\
&= \dot{x}_1 X_1 + \dot{x}_2 X_2 + \big(\dot{t} + 2x_1|x|^{2k}\dot{x}_2 - 2x_2|x|^{2k}\dot{x}_1\big)\partial_t.
\end{aligned}$$

Hence, $\dot{c} \in \mathcal{H}$ if and only if the coefficient of ∂_t vanishes.

□

Therefore, a horizontal curve $c = (x_1, x_2, t)$ can be written as

$$\dot{c} = \dot{x}_1 X_1 + \dot{x}_2 X_2. \tag{3.3}$$

COROLLARY 3.3. H is not involutive, and hence H is not an integrable distribution.

PROOF. $[X_1, X_2]$ is proportional to ∂_t and it cannot be written in the form (3.3). Therefore H is not involutive. By Frobenius theorem, it is not locally integrable.

□

The 1-connection form

$$\omega = dt - 2|x|^{2k}(x_2 dx_1 - x_1 dx_2)$$

describes the horizontality and non-integrability. A curve $c = (x_1, x_2, t)$ is horizontal if and only if $\omega(\dot{c}) = 0$. The horizontal distribution $H = \ker \omega$.
The Frobenius non-integrability condition can be expressed as

$$\begin{aligned} \omega \wedge d\omega &= [dt - 2|x|^{2k}(x_2 dx_1 - x_1 dx_2)] \wedge 4(k+1)|x|^{2k} dx_1 \wedge dx_2 \\ &= 4(k+1)|x|^{2k} dx_1 \wedge dx_2 \wedge dt \neq 0, \quad \text{for } |x| \neq 0. \end{aligned}$$

Hence the model is not a contact manifold along the t -axis.
In polar coordinates $x_1 = r\cos\phi$, $x_2 = r\sin\phi$, equation (3.2) yields

$$\begin{aligned} \dot{t} &= 2(\dot{x}_1 x_2 - x_1 \dot{x}_2)|x|^{2k} \\ &= -2r^{2k}\dot{\phi}, \end{aligned}$$

and hence $\omega = dt + 2r^{2k} d\phi$.

3.2. Horizontal connectivity

The vector fields (3.1) satisfy Chow's condition, *i.e.*, X_1 , X_2 and their iterated Lie brackets span the tangent space of $\mathbb{R}^3$at each point. The number of brackets plus 1 is called the step of the model, which in this case is $2(k+1)$. By Chow's theorem, (see [**22**]) any two points can be joined by a piece-wise smooth horizontal curve. In the following we shall prove horizontal connectivity by a direct method. The following theorem is the generalization to $2(k+1)$ step of a similar theorem from chapter 2.

THEOREM 3.4. *1) Given a point $P(x_1, x_2, t) \in \mathbb{R}^3$, there is a smooth horizontal curve which joins the origin and the point P.*
2) Given two points $P, Q \in \mathbb{R}^3$, there is a piece-wise smooth horizontal curve which joins P and Q.

PROOF. 1) An explicit solution will be constructed. We shall analyze the following three cases: $x_2 \neq 0$, $x_1 \neq 0$ and $x_1 = x_2 = 0$.
In the case $x_2 \neq 0$, set

$$\begin{aligned} x_1(s) &= as^2 + bs, \\ x_2(s) &= \alpha s^2, \quad s \in [0,1]. \end{aligned}$$

The constants a, b and α will be found later from the boundary conditions

$$x_1(1) = x_1, \qquad x_2(1) = x_2, \qquad t(1) = t.$$

Hence

$$\begin{aligned} a + b &= x_1 \Longrightarrow a = x_1 - b, \\ \alpha &= x_2. \end{aligned}$$

The curve starts at the origin $x_1(0) = x_2(0) = t(0) = 0$. We have

$$\dot{x}_1(s)x_2(s) - x_1(s)\dot{x}_2(s) = -\alpha b s^2.$$

From the horizontality condition

$$\begin{aligned}
\dot{t}(s) &= 2\Big(x_1(s)^2 + x_2(s)^2\Big)^k \Big(\dot{x}_1(s)x_2(s) - x_1(s)\dot{x}_2(s)\Big) \\
&= \Big((as^2+bs)^2 + \alpha^2 s^4\Big)^k (-2\alpha b s^2) \\
&= -2\alpha b s^2 \sum_{r=0}^{k} C_k^r (as^2+bs)^{2r} \alpha^{2(k-r)} s^{4(k-r)} \\
&= -2\alpha b \sum_{r=0}^{k} C_k^r \alpha^{2(k-r)} \Big[\sum_{p=0}^{2r} C_{2r}^p a^p b^{2r-p}\Big] s^{p+4k-2r+2}.
\end{aligned}$$

Integrating,

$$t(s) = \underbrace{t(0)}_{=0} - 2\alpha b \sum_{r=0}^{k} C_k^r \alpha^{2(k-r)} \Big[\sum_{p=0}^{2r} C_{2r}^p a^p b^{2r-p}\Big] s^{p+4k-2r+3}/(p+4k-2r+3).$$

Substituting $a = x_1 - b$, $\alpha = x_2$ and taking $s = 1$ in the above relation, yields

$$\frac{t}{(x_2)^{2(k-r)+1}} = \sum_{r=0}^{k}\sum_{p=0}^{p} C_k^r C_{2r}^p \underbrace{(x_1-b)^p b^{2r-p} b}_{\text{degree}=2r+1} /(p+4k-2r+3). \tag{3.4}$$

The expression on the right side is a polynomial of degree $2k+1$, and hence the equation (3.4) has at least a real root b. One may substitute b back to find a. Then, for any given x_1, $x_2 \neq 0$, t, there are constants a, α and b such that the boundary conditions are satisfied.

The case $x_1 \neq 0$ is similar.

In the case $x_1 = x_2 = 0$, we need to find a horizontal curve joining the origin and the point $(0, 0, t)$. We shall assume $t < 0$. Set

$$\begin{aligned}
x_1(s) &= r\cos(\pi s), \\
x_2(s) &= r\sin(\pi s), \\
t(s) &= ts, \qquad s \in [0,1],
\end{aligned}$$

with the parameter r to be found from the horizontality condition

$$\dot{t}(s) = -2|x(s)|^{2k}\dot{\phi}(s) = -2r^{2k}\pi,$$

where $\phi = \pi s$. Then $t(s) = -2r^{2k}\pi s$, and hence making $t = 1$ yields

$$r = \Big(\frac{|t|}{2\pi}\Big)^{1/2k}.$$

For $t > 0$, we consider

$$\begin{aligned}
x_1(s) &= r\cos(-\pi s), \\
x_2(s) &= r\sin(-\pi s), \\
t(s) &= ts, \qquad s \in [0,1].
\end{aligned}$$

2) Any two points can be connected by a horizontal curve which passes through the origin.

□

It is interesting to note that if the point $P(x_1, x_2, t)$ is not on the t-axis, the x-components of the horizontal curve are quadratic in s.

3.3. SubRiemannian geodesics

The principal symbol of the sub-elliptic operator $\Delta_X = \frac{1}{2}\left(X_1^2 + X_2^2\right)$ is given by

$$H(\xi, \theta, x, t) = \frac{1}{2}\left(\xi_1 + 2x_2|x|^{2k}\theta\right)^2 + \frac{1}{2}\left(\xi_2 - 2x_1|x|^{2k}\theta\right)^2 \tag{3.5}$$

and it will be called Hamiltonian.

DEFINITION 3.5. A subRiemannian geodesic connecting the points $P(x_0, t_0)$ and $Q(x_f, t_f)$ is the projection on the (x, t)-space of a solution of the Hamiltonian system

$$\dot{x} = H_\xi, \qquad \dot{t} = H_\theta,$$

$$\dot{\xi} = -H_x, \qquad \dot{\theta} = -H_t$$

with the boundary conditions $x(0) = x_0$, $t(0) = t_0$, $x(s_f) = x_f$, $t(s_f) = t_f$.

As $dH/ds = \partial H/\partial s = 0$, the Hamiltonian is conserved along the geodesics. In the following sections we cannot assume $s_f = 1$ and $H = \frac{1}{2}$ simultaneously. If we use the arc length parametrization, we take $H = \frac{1}{2}$, but in this case s_f is left free. If the parametrization used is not by arc length, we may fix $s_f = 1$ and consider the constant of energy H as a parameter.

Euler-Lagrange system. The Lagrangian associated with the Hamiltonian (3.5) is given formally by the Legendre transform

$$\begin{aligned} L(x, y, t, \dot{x}, \dot{y}, \dot{t}) &= \xi_1\dot{x}_1 + \xi_2\dot{x}_2 + \theta\dot{t} - H \\ &= \frac{1}{2}(\dot{x}_1^2 + \dot{x}_2^2) + \theta\big(\dot{t} - 2|x|^{2k}(x_2\dot{x}_1 - x_1\dot{x}_2)\big). \end{aligned}$$

The Euler-Lagrange system of equations

$$\frac{d}{ds}\frac{\partial L}{\partial \dot{x}_1} = \frac{\partial L}{\partial x_1}, \quad \frac{d}{ds}\frac{\partial L}{\partial \dot{x}_2} = \frac{\partial L}{\partial x_2}, \quad \frac{d}{ds}\frac{\partial L}{\partial \dot{t}} = \frac{\partial L}{\partial t}$$

becomes

$$\begin{cases} \ddot{x}_1 = 4\theta(k+1)|x|^{2k}\dot{x}_2 \\ \ddot{x}_2 = -4\theta(k+1)|x|^{2k}\dot{x}_1 \\ \theta = \text{constant.} \end{cases} \tag{3.6}$$

PROPOSITION 3.6. Any subRiemannian geodesic is a horizontal curve. There are horizontal curves which are not subRiemannian geodesics.

PROOF. Let $c(s) = (x_1(s), x_2(s), t(s))$ be a subRiemannian geodesic. Then, using the Hamiltonian equations, we have

$$\begin{aligned}
\dot{x}_1 &= \frac{\partial H}{\partial \xi_1} = \xi_1 + 2x_2|x|^{2k}\theta, \\
\dot{x}_2 &= \frac{\partial H}{\partial \xi_2} = \xi_2 - 2x_1|x|^{2k}\theta, \\
\dot{t} &= \frac{\partial H}{\partial \theta} = \left(\xi_1 + 2x_2|x|^{2k}\theta\right) 2x_2|x|^{2k} - \left(\xi_2 - 2x_1|x|^{2k}\theta\right) 2x_1|x|^{2k} \\
&= 2|x|^{2k}(\dot{x}_1 x_2 - \dot{x}_2 x_1),
\end{aligned}$$

which is the horizontality condition (3.2).
For the second part of the proof, we may check that the horizontal curves constructed in the proof of Theorem 3.4 do not satisfy the system (3.6), and hence they are not subRiemannian geodesics.

□

PROPOSITION 3.7. Any subRiemannian geodesic satisfies the system (3.6).

PROOF. From the Hamiltonian system

$$\dot{\theta} = -\frac{\partial H}{\partial t} = 0 \Longrightarrow \theta = \text{constant}.$$

$$\begin{aligned}
\dot{\xi}_1 &= -\frac{\partial H}{\partial x_1} = -\left[\dot{x}_1\, 2\theta \partial_{x_1}(x_2|x|^{2k}) + \dot{x}_2(-2\theta)\partial_{x_1}(x_1|x|^{2k})\right] \\
&= -2\theta k \dot{x}_1 x_2 2x_1|x|^{2k-2} + 2\theta\dot{x}_2\left(|x|^{2k} + 2kx_1^2|x|^{2k-2}\right) \\
&= -4\theta k x_1 x_2 \dot{x}_1|x|^{2k-2} + 2\theta\dot{x}_2|x|^{2k} + 4\theta k x_1^2\dot{x}_2|x|^{2k-2}.
\end{aligned}$$

$$\begin{aligned}
\ddot{x}_1 &= \frac{d}{ds}\left(\xi_1 + 2x_2|x|^{2k}\theta\right) \\
&= \dot{\xi}_1 + 2\dot{x}_2|x|^{2k}\theta + 2x_2\theta k|x|^{2k-2}(2x_1\dot{x}_1 + 2x_2\dot{x}_2) \\
&= -4\theta k x_1 x_2 \dot{x}_1|x|^{2k-2} + 2\theta\dot{x}_2|x|^{2k} + 4\theta k x_1^2\dot{x}_2|x|^{2k-2} \\
&\quad + 2\dot{x}_2|x|^{2k}\theta + 2x_2\theta k|x|^{2k-2}(2x_1\dot{x}_1 + 2x_2\dot{x}_2) \\
&= 4\theta|x|^{2k}\dot{x}_2 + 4\theta k\dot{x}_2|x|^{2k-2}(x_1^2 + x_2^2) \\
&= 4\theta(k+1)|x|^{2k}\dot{x}_2.
\end{aligned}$$

Similar calculations will lead to

$$\ddot{x}_2 = -4\theta(k+1)|x|^{2k}\dot{x}_1.$$

□

THEOREM 3.8. *A horizontal curve which satisfies the system* (3.6) *is a subRiemannian geodesic.*

PROOF. Let $s \to (x_1(s), x_2(s), t(s))$ be a horizontal curve. Consider

$$\begin{aligned}
\xi_1 &= \dot{x}_1 - 2\theta x_2|x|^{2k} \\
\xi_2 &= \dot{x}_2 + 2\theta x_1|x|^{2k} \\
\theta &= \text{constant}.
\end{aligned}$$

We shall show that $(x_1, x_2, t, \xi_1, \xi_2, \theta)$ satisfies the Hamiltonian system. Indeed,

$$\begin{aligned} \dot\theta &= 0 = \frac{\partial H}{\partial \theta}, \\ \dot x_1 &= \xi_1 + 2\theta x_2 |x|^{2k} = \frac{\partial H}{\partial \xi_1}, \\ \dot x_2 &= \xi_2 - 2\theta x_1 |x|^{2k} = \frac{\partial H}{\partial \xi_2}. \end{aligned}$$

From the horizontality condition

$$\dot t = 2|x|^{2k}(\dot x_1 x_2 - \dot x_2 x_1) = \frac{\partial H}{\partial \theta}.$$

We shall show in the following that

$$\dot\xi_1 = -\frac{\partial H}{\partial x_1}.$$

Using a computation from the previous proof,

$$-\frac{\partial H}{\partial x_1} = -4\theta k x_1 x_2 \dot x_1 |x|^{2k-2} + 2\theta \dot x_2 |x|^{2k} + 4\theta k x_1^2 \dot x_2 |x|^{2k-2}. \tag{3.7}$$

On the other hand, the definition of ξ_1 yields

$$\begin{aligned} \dot\xi_1 &= \frac{d}{ds}\Big(\dot x_1 - 2\theta x_2 |x|^{2k}\Big) \\ &= 4\theta(k+1)\dot x_2 |x|^{2k} - 2\theta \frac{d}{ds}\Big(x_2|x|^{2k}\Big) \\ &= 4\theta k \dot x_2 |x|^{2k} + 4\theta \dot x_2 |x|^{2k} - 2\theta \dot x_2 |x|^{2k} - 2\theta k x_2 |x|^{2k-2}(2x_1\dot x_1 + 2x_2 \dot x_2) \\ &= 4\theta k \dot x_2 |x|^{2k} + 4\theta |x|^{2k}\dot x_2 - 2\theta \dot x_2 |x|^{2k} - 4\theta k x_1 x_2 \dot x_1 |x|^{2k-2} - 4\theta k x_2^2 |x|^{2k-2}\dot x_2 \\ &= -4\theta k x_1 x_2 \dot x_1 |x|^{2k-2} + 2\theta \dot x_2 |x|^{2k} + 4\theta k \dot x_2 \Big(|x|^{2k-2}(x_1^2 + x_2^2) - |x|^{2k-2}x_2^2\Big) \\ &= -4\theta k x_1 x_2 \dot x_1 |x|^{2k-2} + 2\theta \dot x_2 |x|^{2k} + 4\theta k x_1^2 \dot x_2 |x|^{2k-2}. \end{aligned}$$

Using (3.7) we get

$$\dot\xi_1 = -\frac{\partial H}{\partial x_1},$$

and hence ξ_1 satisfies the Hamiltonian system.
By direct computation,

$$\frac{\partial H}{\partial x_2} = -4\theta k x_1 x_2 \dot x_2 |x|^{2k-2} + 2\theta \dot x_1 |x|^{2k} + 4\theta k x_2^2 \dot x_1 |x|^{2k-2}.$$

Differentiating in the definition of ξ_2, and using $\ddot x_2 = -4\theta(k+1)|x|^{2k}\dot x_1$, yields

$$\begin{aligned} \dot\xi_2 &= \frac{d}{ds}\Big(\dot x_2 + 2\theta x_1 |x|^{2k}\Big) \\ &= -4\theta(k+1)|x|^{2k}\dot x_1 + 2\theta \frac{d}{ds}(x_1|x|^{2k}) \\ &= -4\theta k |x|^{2k}\dot x_1 - 4\theta |x|^{2k}\dot x_1 + 2\theta \dot x_1 |x|^{2k} + 2\theta k x_1 |x|^{2k-2}(2x_1\dot x_1 + 2x_2\dot x_2) \\ &= -4\theta k |x|^{2k-2}\dot x_1(x_1^2 + x_2^2) + 4\theta k x_1^2 \dot x_1 |x|^{2k-2} - 2\theta |x|^{2k}\dot x_1 + 4\theta k x_1 x_2 \dot x_2 |x|^{2k-2} \\ &= 4\theta k x_1 x_2 \dot x_2 |x|^{2k-2} - 2\theta |x|^{2k}\dot x_1 - 4\theta k x_2^2 \dot x_1 |x|^{2k-2} \\ &= -\frac{\partial H}{\partial x_2}, \end{aligned}$$

and hence ξ_2 satisfies the Hamiltonian equation.

□

3.4. Euler-Lagrange system in polar coordinates

In polar coordinates $x_1 = r\cos\phi$, $x_2 = r\sin\phi$, the Lagrangian becomes

$$L = \frac{1}{2}(\dot{r}^2 + r^2\dot{\phi}^2) + \theta\big(\dot{t} + 2r^{2k+2}\dot{\phi}\big). \tag{3.8}$$

The momentum θ is constant because $\dot{\theta} = -\partial H/\partial t = 0$, and can be considered as a Lagrange multiplier.

The symmetries for the Lagrangian (3.8) will provide symmetries for the geodesics. The Lagrangian (3.8) is invariant under the symmetry

$$S : (r, \phi, t; \theta) \to (r, -\phi, -t; -\theta). \tag{3.9}$$

If $(r(s), \phi(s), t(s))$ is a geodesic corresponding to θ, then $(r(s), -\phi(s), -t(s))$ is a geodesic corresponding to $-\theta$.

Consequently, whatever statement is made for the geodesics joining the origin and the point $(r(s), \phi(s), t(s))$, when $\theta > 0$, can be also made for the geodesics between the origin and $(r(s), -\phi(s), -t(s))$, when $\theta < 0$. This allows us to do the analysis only for the case $\theta > 0$.

Euler-Lagrange equations . A computation shows

$$\frac{d}{ds}\frac{\partial L}{\partial \dot{r}} = \ddot{r}, \qquad \frac{\partial L}{\partial r} = r\dot{\phi}\big(\dot{\phi} + 4(k+1)\theta r^{2k}\big), \tag{3.10}$$

$$\frac{\partial L}{\partial \dot{\phi}} = r^2\dot{\phi} + 2\theta r^{2(k+1)}, \qquad \frac{\partial L}{\partial \phi} = 0, \tag{3.11}$$

$$\frac{\partial L}{\partial \dot{t}} = \theta, \qquad \frac{\partial L}{\partial t} = 0, \tag{3.12}$$

and hence $r(s)$, $\phi(s)$ and θ satisfy the Euler-Lagrange system

$$\begin{cases} \ddot{r} = r\dot{\phi}\big(\dot{\phi} + 4(k+1)\theta r^{2k}\big) \\ r^2(\dot{\phi} + 2\theta r^{2k}) = C(\text{constant}) \\ \theta = \theta_0(\text{constant}). \end{cases} \tag{3.13}$$

3.5. Geodesics starting at the origin

If the geodesic starts at the origin, then $r(0) = 0$ and hence $C = 0$. The second equation of (3.13) yields $\dot{\phi} = -2\theta r^{2k}$. With the assumption $\theta > 0$, the argument angle ϕ rotates clock-wise.

When $\theta = 0$, the system (3.13) becomes

$$\begin{cases} \ddot{r} = 0 \\ \dot{\phi} = 0 \\ \theta = 0. \end{cases} \tag{3.14}$$

PROPOSITION 3.9. Given a point $P(x,0)$, there is a unique geodesic between the origin and P. It is a straight line in the plane $\{t=0\}$ of length $|x|$, and it is obtained for $\theta = 0$.

PROOF. In the case $\theta = 0$, the Hamiltonian is $H(\xi) = \frac{1}{2}\xi_1^2 + \frac{1}{2}\xi_2^2$. From the Hamiltonian equation $\dot{t} = \partial H/\partial \theta = 0$, $t(s)$ is constant, and as $t(0) = 0$, it follows that $t(s) = 0$ and the solution belongs to the x-plane. Using the equation $\ddot{r} = 0$, it follows the solution is a straight line. □

From now on, unless otherwise stated, we shall assume $\theta > 0$.

Solving the Hamiltonian system for $\theta \neq 0$. As $dH/ds = (\partial H/\partial \xi)\dot{\xi} + (\partial H/\partial \theta)\dot{\theta} + (\partial H/\partial x)\dot{x} + (\partial H/\partial t)\dot{t} + \partial H/\partial s = 0$, the Hamiltonian H is preserved along the solutions. From the Hamiltonian equations

$$\dot{x}_1 = \partial H/\partial \xi_1 = \xi_1 + 2x_2|x|^{2k}\theta,$$
$$\dot{x}_2 = \partial H/\partial \xi_2 = \xi_2 - 2x_1|x|^{2k}\theta,$$

and the Hamiltonian (3.5) becomes

$$H = \frac{1}{2}(\dot{x}_1^2 + \dot{x}_2^2) = \frac{1}{2}, \tag{3.15}$$

where we considered s the arc length parameter along the solution. In polar coordinates, $\dot{x}_1^2 + \dot{x}_2^2 = \dot{r}^2 + r^2\dot{\phi}^2$. Using $\dot{\phi} = -2\theta r^{2k}$, yields

$$\dot{r}^2 + 4\theta^2 r^{2(2k+1)} = 1. \tag{3.16}$$

When $\dot{r} = 0$ we get $r = r_{max}$, with

$$r_{max} = \left(\frac{1}{2\theta}\right)^{\frac{1}{2k+1}}. \tag{3.17}$$

We have the following simultaneous system of equations

$$\begin{cases} dr/ds = \pm\sqrt{1 - 4\theta^2 r^{2(2k+1)}} \\ d\phi/ds = -2\theta r^{2k}. \end{cases} \tag{3.18}$$

The " + " sign corresponds to the case when the radius $r(s)$ increases between 0 and r_{max}. Similarly, the " − " sign is considered for the case when $r(s)$ decreases between r_{max} and 0. In the following we denote by ϕ_0 the initial argument.

Taking the " + " sign, dividing the equations in (3.18) and integrating, we obtain the angle swept by the vectorial radius between $r(0) = 0$ and $r = r_{max}$

$$w = \int_0^{r_{max}} \frac{-2\theta r^{2k}}{\sqrt{1 - 4\theta^2 r^{2(2k+1)}}}\, dr.$$

The substitution $v = 2\theta r^{2k+1}$ yields

$$\begin{aligned} w &= -\frac{1}{2k+1}\int_0^1 \frac{dv}{\sqrt{1-v^2}} \\ &= -\frac{1}{2k+1}(\arcsin(1) - \arcsin(0)) = \frac{-\pi}{2(2k+1)}. \end{aligned}$$

It is interesting to note that the angle w is independent of θ. The negative sign shows the rotation is done clock-wise.

Taking the " $-$ " sign, dividing the equations in (3.18) and integrating, we obtain the angle swept by the vectorial radius between $r = r_{max}$ and $r(0) = 0$

$$\begin{aligned}\widetilde{w} &= \int_{r_{max}}^{0} \frac{2\theta r^{2k}}{\sqrt{1-4\theta^2 r^{2(2k+1)}}}\,dr = \int_{0}^{r_{max}} \frac{-2\theta r^{2k}}{\sqrt{1-4\theta^2 r^{2(2k+1)}}} \\ &= w = \frac{-\pi}{2(2k+1)}.\end{aligned}$$

Hence we have the following proposition:

PROPOSITION 3.10. The projection of the solution on the x-space passes again through the origin after it sweeps an angle equal to $w + \widetilde{w} = \frac{-\pi}{2k+1}$.

THEOREM 3.11. *For $\phi_0 - \phi \in [0, \frac{\pi}{2k+1}]$, the solution in polar coordinates is given by*

$$r(\phi) = \Big(\frac{1}{2\theta}\Big)^{\frac{1}{2k+1}} \sin^{\frac{1}{2k+1}}\big((2k+1)(\phi_0 - \phi)\big). \tag{3.19}$$

PROOF. We have two cases to investigate:

Case $\phi_0 - \phi \in [0, \frac{\pi}{2(2k+1)}]$:

The radius r increases between $r(0) = 0$ and r_{max}. Considering the positive sign and dividing equations (3.18) we obtain

$$\frac{d\phi}{dr} = \frac{-2\theta r^{2k}}{\sqrt{1-4\theta^2 r^{2(2k+1)}}}.$$

Integrating between 0 and r

$$\phi - \phi_0 = -\int_0^r \frac{2\theta r^{2k}}{\sqrt{1-4\theta^2 r^{2(2k+1)}}} = -\frac{1}{2k+1}\arcsin(2\theta r^{2k+1}).$$

Taking the sine on both sides yields (3.19).

Case $\phi_0 - \phi \in [\frac{\pi}{2(2k+1)}, \frac{\pi}{(2k+1)}]$:

The radius r decreases between r_{max} and 0. Considering the negative sign and dividing equations (3.18)

$$\frac{d\phi}{dr} = \frac{2\theta r^{2k}}{\sqrt{1-4\theta^2 r^{2(2k+1)}}}.$$

Integrating between $\phi_0 - \frac{\pi}{2(2k+1)}$ and ϕ on the left side and between r_{max} and r on the right side, yields

$$\phi - \phi_0 + \frac{\pi}{2(2k+1)} = \int_{r_{max}}^{r} \frac{2\theta r^{2k}}{\sqrt{1-4\theta^2 r^{2(2k+1)}}}. \tag{3.20}$$

Using the decomposition

$$\int_{r_{max}}^{r} = \int_{r_{max}}^{0} + \int_0^r,$$

the right side of (3.20) becomes

$$\begin{aligned}&- \frac{\pi}{2(2k+1)} + \int_0^r \frac{2\theta r^{2k}}{\sqrt{1-4\theta^2 r^{2(2k+1)}}} \\ &= -\frac{\pi}{2(2k+1)} + \frac{1}{2k+1}\arcsin(2\theta r^{2k+1}).\end{aligned}$$

Then (3.20) is equivalent with the following sequence of identities:

$$\begin{aligned} \phi - \phi_0 + \frac{\pi}{2k+1} &= \frac{1}{2k+1} \arcsin(2\theta r^{2k+1}) \\ \sin\big((2k+1)(\phi - \phi_0) + \pi\big) &= 2\theta r^{2k+1} \\ -\sin\big((2k+1)(\phi - \phi_0)\big) &= 2\theta r^{2k+1}, \end{aligned}$$

which leads to (3.19). □

COROLLARY 3.12. For any angle ϕ and number $p \in \mathbb{N}$,

$$r(\phi)^{2p} = \Big(\frac{1}{2\theta}\Big)^{\frac{2p}{2k+1}} \sin^{\frac{2p}{2k+1}}\big((2k+1)(\phi_0 - \phi)\big).$$

PROOF. Similarly, one may show that if $\phi_0 - \phi \in \left[\frac{m\pi}{2k+1}, \frac{(m+1)\pi}{2k+1}\right]$, then

$$r(\phi) = (-1)^m \Big(\frac{1}{2\theta}\Big)^{\frac{1}{2k+1}} \sin^{\frac{1}{2k+1}}\big((2k+1)(\phi_0 - \phi)\big).$$

Raising to an even power eliminates the minus sign. □

Consider the boundary conditions

$$\begin{aligned} r(0) = 0, \ \ \phi(0) = \phi_0, \ \ t(0) = 0, \\ r(s) = R, \ \ \phi(s) = \phi_1, \ \ t(s) = t. \end{aligned}$$

Using the rotational symmetry, we may assume $\phi_0 = 0$. In the following theorem the parameter s represents the arc length.

THEOREM 3.13. *The following relations take place among the boundary conditions:*

$$\int_0^{(2k+1)\phi_1} \frac{dv}{\sin^{\frac{2k}{2k+1}}(v)} = -(2k+1)(2\theta)^{\frac{1}{2k+1}} s, \tag{3.21}$$

$$R^{2(k+1)} = \Big(\frac{1}{2\theta}\Big)^{\frac{2(k+1)}{2k+1}} \sin^{\frac{2(k+1)}{2k+1}}\big((2k+1)(\phi_1)\big), \tag{3.22}$$

$$t = -\frac{2}{2k+1}\Big(\frac{1}{2\theta}\Big)^{\frac{2(k+1)}{2k+1}} \int_0^{(2k+1)\phi_1} \sin^{\frac{2(k+1)}{2k+1}}(v)\, dv, \quad t > 0, \tag{3.23}$$

$$\frac{t}{R^{2(k+1)}} = \mu_k(-\phi_1), \qquad \text{for } t > 0,\ R \neq 0, \tag{3.24}$$

with

$$\mu_k(x) = \frac{2}{2k+1} \frac{\int_0^{(2k+1)x} \sin^{\frac{2(k+1)}{2k+1}}(v)\, dv}{\sin^{\frac{2(k+1)}{2k+1}}(x)}. \tag{3.25}$$

PROOF. Using $\frac{d\phi}{ds} = -2\theta r^{2k}$ and Corollary 3.12 with $p = k$, yields

$$\frac{d\phi}{ds} = -2\theta\Big(\frac{1}{2\theta}\Big)^{\frac{2k}{2k+1}} \sin^{\frac{2k}{2k+1}}\big(-(2k+1)\phi\big).$$

Separating

$$\frac{d\phi}{\sin^{\frac{2k}{2k+1}}\big((2k+1)\phi\big)} = -(2\theta)^{\frac{1}{2k+1}}\, ds,$$

and integrating

$$\int_0^{\phi_1} \frac{d\phi}{\sin^{\frac{2k}{2k+1}}\big((2k+1)\phi\big)} = -(2\theta)^{\frac{1}{2k+1}} s.$$

Changing the variable,

$$\int_0^{(2k+1)\phi_1} \frac{dv}{\sin^{\frac{2k}{2k+1}}(v)} = -(2k+1)(2\theta)^{\frac{1}{2k+1}} s,$$

which is equation (3.21).

Replacing ϕ by ϕ_1, r by R, and p by $k+1$ in Corollary 3.12, yields equation (3.22).

From the Hamiltonian equation

$$\dot{t}(s) = \frac{\partial H}{\partial \theta} = 2|x|^{2k}(x_2\dot{x}_1 - x_1\dot{x}_2) = -2r^{2(k+1)}\dot{\phi}.$$

As $\dot{\phi} = -2\theta r^{2k} < 0$, $t(s)$ is increasing from $t(0) = 0$ to $t > 0$. Making $p = k+1$ in Corollary 3.12 and integrating between $\phi_0 = 0$ and ϕ_1, yields

$$t = -2\Big(\frac{1}{2\theta}\Big)^{\frac{2(k+1)}{2k+1}} \frac{1}{2k+1} \int_0^{(2k+1)\phi_1} \sin^{\frac{2(k+1)}{2k+1}}(v)\, dv,$$

which is equivalent to (3.23).
Dividing equations (3.23) and (3.22) yields equation (3.24).

□

The function μ_k. The behavior of the function μ_k given by (3.25) is very important in understanding the subRiemannian geometry defined by the vector fields (3.1). For the sake of simplicity we shall denote in the following μ_k by μ.

The function $\mu(x)$ is odd. It suffices to study it just for $x > 0$. Furthermore, since $\lim_{x\to 0}\mu(x) = 0$, we shall define $\mu(0) = 0$. Indeed, applying l'Hospital rule

$$\begin{aligned}\lim_{x\to 0}\mu(x) &= \lim_{x\to 0}\frac{2}{2k+1}\cdot\frac{(2k+1)\sin^{\frac{2(k+1)}{2k+1}}((2k+1)x)}{\frac{2(k+1)}{2k+1}\sin^{\frac{1}{2k+1}}((2k+1)x)\cos((2k+1)x)\,(2k+1)}\\ &= \lim_{x\to 0}\frac{2\sin^{\frac{1}{2k+1}}((2k+1)x)}{\cos((2k+1)x)} = 0.\end{aligned}$$

LEMMA 3.14. *(i) μ satisfies the ordinary differential equation*

$$\mu'(x) = 2 - 2(k+1)\cot((2k+1)x)\mu(x).$$

(ii) If x_c is a critical point for $\mu(x)$, then

$$\mu(x_c) = \frac{1}{k+1}\tan((2k+1)x_c).$$

PROOF. (i) Differentiating in equation (3.25)

$$\begin{aligned}
\mu'(x) &= \frac{2}{2k+1}\cdot\frac{(2k+1)\sin^{\frac{4(k+1)}{2k+1}}((2k+1)x)}{\sin^{\frac{4(k+1)}{2k+1}}((2k+1)x)} \\
&- \frac{2}{2k+1}\frac{2(k+1)\sin^{\frac{1}{2k+1}}((2k+1)x)\cos((2k+1)x)\int_0^{(2k+1)x}\sin^{\frac{2(k+1)}{2k+1}}(v)\,dv}{\sin^{\frac{4(k+1)}{2k+1}}((2k+1)x)} \\
&= 2\big[1-(k+1)\cot((2k+1)x)\mu(x)\big]
\end{aligned}$$

(ii) It is a simple consequence of (i).

□

The next result deals with the second derivative of μ.

LEMMA 3.15. *If $x_c > 0$ is a critical point for μ, then*

$$\mu''(x_c) > 4(2k+1) > 0,$$

and hence all the positive critical points are local minima.

PROOF. Differentiating in relation (i) of Lemma 3.14

$$\begin{aligned}
\mu''(x) = &- 2(k+1)\big[-(2k+1)\big(1+\cot^2((2k+1)x)\big)\mu(x)\big] \\
&+ \cot((2k+1)x)\big(2-2(k+1)\cot((2k+1)x)\mu(x)\big) \\
= &- 2(k+1)\big[-(2k+1)\mu(x)+2\cot((2k+1)x) \\
&- (4k+3)\cot((2k+1)x)\mu(x)\big].
\end{aligned}$$

The relation (ii) of Lemma 3.14 yields

$$\begin{aligned}
\mu''(x_c) = &- 2(k+1)\Big[-\frac{2k+1}{k+1}\tan((2k+1)x_c)+2\cot((2k+1)x_c) \\
&- \frac{(4k+3)}{k+1}\cot^2((2k+1)x_c)\tan((2k+1)x_c)\Big] \\
= &\quad 2(2k+1)\big[\tan((2k+1)x_c)+\cot((2k+1)x_c)\big] \ge 4(2k+1),
\end{aligned}$$

where we used the inequality $a + \frac{1}{a} \ge 2$, for $a > 0$. In the above case

$$a = \tan((2k+1)x_c) = (k+1)\mu(x_c) > 0.$$

□

LEMMA 3.16. *μ_k is a monotone increasing diffeomorphism of the interval $\big(-\frac{\pi}{2k+1}, \frac{\pi}{2k+1}\big)$ onto $\mathbb{R}$. On each interval $\big(\frac{m\pi}{2k+1}, \frac{(m+1)\pi}{2k+1}\big)$, $m = 1, 2, \dots$, μ_k has a unique critical point x_m. On this interval μ_k decreases strictly from $+\infty$ to $\mu_k(x_m)$ and then increases strictly from $\mu_k(x_m)$ to $+\infty$. Moreover*

$$\mu_k(x_m) + q < \mu_k(x_{m+1}), \tag{3.26}$$

with $q > 0$ independent of m.

PROOF. As μ is odd and $\mu(\frac{\pi}{2k+1}, -) = +\infty$, it suffices to show that μ is increasing on $\big(0, \frac{\pi}{2k+1}\big)$. As $\mu(0) = 0$ and $\mu(x) > 0$ for $x > 0$, μ is increasing on a right neighborhood $(0, \delta)$, with $\delta > 0$ small enough. Let

$$I = \{x \in \big(0, \frac{\pi}{2k+1}\big); \mu \text{ increasing on } I\}.$$

$I \neq \emptyset$ because $(0, \delta) \subset I$. The set I is an interval, because otherwise, there are $a < c < b$ with $a, b \in I$ and $c \notin I$. μ is increasing at a and decreasing at c and hence it has a maximum point in the interval (a, c), which is in contradiction with Lemma 3.15.
Let $\gamma = \sup I$. Suppose $\gamma \neq \frac{\pi}{2k+1}$. There is an α such that $\gamma < \alpha < \frac{\pi}{2k+1}$ with $\mu_{|(0,\gamma)}$ increasing and $\mu_{|(\gamma,\alpha)}$ decreasing. Then there is a local maximum for μ on (γ, α), contradiction. It follows that $\gamma = \frac{\pi}{2k+1}$ and $I = (0, \frac{\pi}{2k+1})$. Hence μ is a monotone increasing diffeomorphism of the interval $\big(-\frac{\pi}{2k+1}, \frac{\pi}{2k+1}\big)$ onto $\mathbb{R}$.

As $\mu(\frac{m\pi}{2k+1}, +) = \mu(\frac{(m+1)\pi}{2k+1}, -) = +\infty$, there is at least a critical point in the interval $\big(\frac{m\pi}{2k+1}, \frac{(m+1)\pi}{2k+1}\big)$. As all the critical points are local minima, the critical point has to be unique (more than one critical point will imply the existence of a local maximum, contradiction).

We shall compute a lower bound for the difference between two successive minima.

$$\begin{aligned}
\mu(x + \frac{\pi}{2k+1}) &= \frac{2}{2k+1} \cdot \frac{\int_0^{(2k+1)x+\pi} \sin^{\frac{2(k+1)}{2k+1}}(v)\, dv}{\sin^{\frac{2(k+1)}{2k+1}}((2k+1)x + \pi)} \\
&= \frac{2}{2k+1} \cdot \frac{\int_0^{\pi} + \int_{\pi}^{(2k+1)x+\pi}}{\sin^{\frac{2(k+1)}{2k+1}}((2k+1)x)} \\
&= \frac{q}{\sin^{\frac{2(k+1)}{2k+1}}((2k+1)x)} + \mu(x),
\end{aligned}$$

and hence

$$\mu(x + \frac{\pi}{2k+1}) - \mu(x) > q > 0.$$

The constant q is independent of m and it is evaluated below:

$$\begin{aligned}
q &= \frac{2}{2k+1} \int_0^{\pi} \sin^{\frac{2(k+1)}{2k+1}}(v)\, dv = \frac{4}{2k+1} \int_0^{\pi/2} \sin^{\frac{2(k+1)}{2k+1}}(v)\, dv \\
&= \frac{4}{2k+1} \int_0^1 t^{\frac{2(k+1)}{2k+1}} (1 - t^2)^{-1/2}\, dt = \frac{2}{2k+1} \int_0^1 x^{\frac{1}{2(2k+1)}} (1 - x)^{-1/2}\, dx,
\end{aligned}$$

and hence q can be expressed in terms of beta functions as

$$q = \frac{2}{2k+1} \cdot B\Big(\frac{4k+3}{4k+2}, \frac{1}{2}\Big) > 0.$$

The values of μ are shifted up by at least q when μ changes from $\Big(\frac{m\pi}{2k+1}, \frac{(m+1)\pi}{2k+1}\Big)$ to $\Big(\frac{(m+1)\pi}{2k+1}, \frac{(m+2)\pi}{2k+1}\Big)$. Hence the minimum will be shifted up by at least q, *i.e.*, $\mu(x_{m+1}) - \mu(x_m) > q$. Note that $\lim_{m\to\infty} \mu(x_m) = +\infty$.

□

The function μ has a similar behavior for all k. The graph of the function μ in the step 2 case ($k = 0$) is given in Figure 1.2.

3.6. Carnot-Carathéodory distances from the origin

We shall study the connectivity and the length of the geodesics starting from the origin. The complete picture is given in theorems 3.17 and 3.18.

THEOREM 3.17. *There are finitely many geodesics that join the origin to (x,t) if and only if $x \neq 0$. These geodesics are parametrized by the solutions ψ of:*

$$\frac{|t|}{\|x\|^{2(k+1)}} = \mu(\psi), \tag{3.27}$$

There is exactly one such geodesic if and only if:

$$|t| < \mu(x_1)\|x\|^{2(k+1)},$$

where x_1 is the first critical point for μ. The number of geodesics increases without bound as $\frac{|t|}{\|x\|^{2(k+1)}} \to \infty$.

If $0 \leq \psi_1 < \cdots < \psi_N$ are the solutions of equation (3.27), there are exactly N geodesics, and their lengths are given by

$$s_m^{2(k+1)} = \nu(\psi_m)\Big(|t| + \|x\|^{2(k+1)}\Big), \tag{3.28}$$

where

$$\nu(\psi) = \frac{\left[\int_0^{(2k+1)\psi} \sin^{-\frac{2k}{2k+1}}(v)\,dv\right]^{2(k+1)}}{(k+1)^{2(k+1)}\big(1+\mu(\psi)\big)\sin^{\frac{2(k+1)}{2k+1}}((2k+1)\psi)}. \tag{3.29}$$

PROOF. The enumeration of geodesics follows from Lemma 3.16. In order to compute the lengths, multiply equation (3.22) with equation (3.21) raised to power $2(k+1)$ to eliminate θ

$$(s_m)^{2(k+1)} = \Big(\frac{1}{2k+1}\Big)^{2(k+1)} R^{2(k+1)} \frac{\left[\int_0^{(2k+1)\psi_m} \sin^{-\frac{2k}{2k+1}}(v)\,dv\right]^{2(k+1)}}{\sin^{\frac{2(k+1)}{2k+1}}((2k+1)\psi_m)}. \tag{3.30}$$

On the other hand, (3.24) yields

$$|t| + R^{2(k+1)} = (1+\mu(\psi_m))R^{2(k+1)}.$$

Then

$$(s_m)^{2(k+1)} = \nu(\psi_m)\Big(|t| + R^{2(k+1)}\Big) = \nu(\psi_m)(1+\mu(\psi_m))R^{2(k+1)}. \tag{3.31}$$

Choosing ν as in formula (3.29), the right side of the equations (3.30) and (3.31) are the same.

□

The graph of the function ν in the step 2 case ($k=0$) is given in Figure 1.4. We note that the function ν is not increasing. But the values at $\frac{m\pi}{2k+1}$ are increasing as m increases.

THEOREM 3.18. *The geodesics that join the origin to a point $(0,t)$ have lengths $s_1, s_2, \ldots,$ where*

$$(s_m)^{2(k+1)} = \Big(\frac{m}{2k+1}\Big)^{2k+1} \frac{M^{2(k+1)}}{Q}|t|, \tag{3.32}$$

with the constants M and Q expressed in terms of beta functions

$$\begin{aligned} M &= B\Big(\frac{1}{4k+2}, \frac{1}{2}\Big), \\ Q &= (2k+1)q = 2B\Big(\frac{4k+3}{4k+2}, \frac{1}{2}\Big). \end{aligned}$$

For each length s_m, the geodesics of that length are parametrized by the circle $\mathbb{S}^1$.

PROOF. When $\|x\| = 0$, the equation

$$\frac{|t|}{\|x\|^{2(k+1)}} = \mu(\psi)$$

is replaced by $\mu(\psi) = +\infty$. This equation has solutions $\psi_m = \frac{m\pi}{2k+1}$, $m = 1, 2, \ldots$. The function μ has vertical asymptotes at these points. We obtain infinitely many geodesics joining the origin with $(0, t)$ with lengths given by formula (3.28)

$$(s_m)^{2(k+1)} = \nu\Big(\frac{m\pi}{2k+1}\Big)|t|. \tag{3.33}$$

In the following we shall compute $\nu\big(\frac{m\pi}{2k+1}\big)$ in terms of beta functions. Using formula (3.29) and the definition of the μ function, after cancellations we get

$$\begin{aligned}
\nu\Big(\frac{m\pi}{2k+1}\Big) &= \frac{\big[\int_0^{m\pi} \sin^{\frac{-2k}{2k+1}}(v)\,dv\big]^{2(k+1)}}{(2k+1)^{2(k+1)}\cdot\frac{2}{2k+1}\int_0^{m\pi}\sin^{\frac{2(k+1)}{2k+1}}(v)\,dv}\\
&= \frac{(2m)^{2(k+1)}\big[\int_0^{\pi/2}\sin^{\frac{-2k}{2k+1}}(v)\,dv\big]^{2(k+1)}}{2(2k+1)^{2k+1}\cdot m\int_0^{\pi}\sin^{\frac{2(k+1)}{2k+1}}(v)\,dv}
\end{aligned}$$

$$= \Big(\frac{2m}{2k+1}\Big)^{2k+1}\cdot\frac{\big[\int_0^{\pi/2}\sin^{\frac{-2k}{2k+1}}(v)\,dv\big]^{2(k+1)}}{\int_0^{\pi}\sin^{\frac{2(k+1)}{2k+1}}(v)\,dv}. \tag{3.34}$$

The bottom integral was found as a part of computation of q in Lemma 3.16

$$\int_0^{\pi}\sin^{\frac{2(k+1)}{2k+1}}(v)\,dv = B\Big(\frac{4k+3}{4k+2},\frac{1}{2}\Big).$$

With the substitutions $v = \arcsin u$ and $t = u^2$, the top integral becomes

$$\begin{aligned}
\int_0^{\pi/2}\sin^{\frac{-2k}{2k+1}}(v)\,dv &= \int_0^1 \frac{du}{u^{\frac{2k}{2k+1}}(1-u^2)^{\frac{1}{2}}}\\
&= \frac{1}{2}\int_0^1 t^{-\frac{4k+1}{4k+2}}(1-t)^{-\frac{1}{2}}\,dt = \frac{1}{2}B\Big(\frac{1}{4k+2},\frac{1}{2}\Big).
\end{aligned}$$

Hence (3.34) yields

$$\begin{aligned}
\nu\Big(\frac{m\pi}{2k+1}\Big) &= \Big(\frac{2m}{2k+1}\Big)^{2k+1}\frac{1}{2^{2k+2}}\frac{B\Big(\frac{1}{4k+2},\frac{1}{2}\Big)^{2(k+1)}}{B\Big(\frac{4k+3}{4k+2},\frac{1}{2}\Big)}\\
&= \Big(\frac{m}{2k+1}\Big)^{2k+1}\frac{M^{2(k+1)}}{Q}.
\end{aligned}$$

Substituting in formula (3.33) we get the distances (3.32).

□

3.7. Particular cases

Step 4 case. When $k=1$ the vector fields

$$X_1 = \frac{\partial}{\partial x_1} + 2x_2|x|^2\frac{\partial}{\partial_t}, \quad X_2 = \frac{\partial}{\partial x_2} - 2x_1|x|^2\frac{\partial}{\partial_t}, \tag{3.35}$$

define a step 4 model. There is no Lie group law on $\mathbb{R}^3$ underlying these vector fields. Theorems 3.17 and 3.18 in this case are:

THEOREM 3.19. *There are finitely many geodesics that join the origin to (x,t) if and only if $x \neq 0$. These geodesics are parametrized by the solutions ψ of:*

$$\frac{|t|}{\|x\|^4} = \mu(\psi), \tag{3.36}$$

There is exactly one such geodesic if and only if:

$$|t| < \mu(x_1)\|x\|^4.$$

The number of geodesics increases without bound as $\frac{|t|}{\|x\|^4} \to \infty$.
If $0 \le \psi_1 < \cdots < \psi_N$ are the solutions of equation (3.36), there are exactly N geodesics, and their lengths are given by

$$s_m^4 = \nu(\psi_m)\Big(|t| + \|x\|^4\Big), \tag{3.37}$$

where

$$\nu(x) = \frac{\left[\int_0^{3x} \sin^{-\frac{2}{3}}(v)\,dv\right]^4}{16(1+\mu(x))\sin^{\frac{4}{3}}(3x)}, \tag{3.38}$$

and

$$\mu(x) = \mu_1(x) = \frac{2}{3}\frac{\int_0^{3x} \sin^{\frac{4}{3}}(v)\,dv}{\sin^{\frac{4}{3}} x}. \tag{3.39}$$

THEOREM 3.20. *The geodesics that join the origin to a point $(0,t)$ have lengths $s_1, s_2, \ldots$, where*

$$(s_m)^4 = 2\Big(\frac{mM}{3}\Big)^3|t|, \tag{3.40}$$

with $M = B\Big(\frac{1}{6},\frac{1}{2}\Big)$. For each length s_m, the geodesics of that length are parametrized by the circle $\mathbb{S}^1$.

PROOF. Using the properties of beta functions and taking $k=1$, yields

$$\begin{aligned} M &= B\Big(\frac{1}{6},\frac{1}{2}\Big) = \frac{\Gamma(\frac{1}{6})\Gamma(\frac{1}{2})}{\Gamma(\frac{2}{3})}. \\ Q &= 2B\Big(\frac{7}{6},\frac{1}{2}\Big) = 2\frac{\Gamma(\frac{7}{6})\Gamma(\frac{1}{2})}{\Gamma(\frac{5}{3})} = \frac{1}{2}\frac{\Gamma(\frac{1}{6})\Gamma(\frac{1}{2})}{\Gamma(\frac{2}{3})} \\ &= \frac{M}{2}. \end{aligned}$$

Then

$$\frac{M^4}{Q} = 2M^3.$$

Substituting $k = 1$ in equation (3.32), yields

$$(s_m)^4 = \left(\frac{m}{3}\right)^3 \frac{M^4}{Q}|t| = \left(\frac{m}{3}\right)^3 \cdot 2M^3|t|,$$

which leads to (3.40).

□

The geodesic of length s_m intersects the t-axis between $(0,0,0)$ and $(0,0,t)$ $m-1$ times. The projection of the geodesics on the x-plane consists of loops which return back to the origin after a clock-wise rotation of $\frac{\pi}{3}$. The projection is tangent to a circle which narrows down as m increases. The radius of the circle is given by the following proposition:

PROPOSITION 3.21. The m-th geodesic between the origin and $(0,t)$ is tangent to a cylinder centered on the t-axis with the radius $\left(\frac{3|t|}{2mQ}\right)^{1/4}$, where $Q = 2B(\frac{7}{6}, \frac{1}{2})$.

PROOF. The solution in polar coordinates satisfies

$$r(\phi)^4 = r_{max}^4 \sin^{\frac{4}{3}}(3(\phi - \phi_0)).$$

After m clock-wise loops, the angle is

$$\phi_1 = \phi_0 - \frac{m\pi}{3}.$$

Integrating between ϕ_0 and ϕ_1 in the equation

$$\dot{t} = -4r^4\dot{\phi},$$

and using the boundary conditions $t(\phi_0) = 0$, $t(\phi_1) = t$

$$\begin{aligned}
t &= -4r_{max}^4 \int_{\phi_0}^{\phi_0 - \frac{m\pi}{3}} \sin^{\frac{4}{3}}(3(\phi - \phi_0))\, d\phi \\
&= \frac{4}{3} r_{max}^4 \int_0^{m\pi} \sin^{\frac{4}{3}} v\, dv = \frac{4m}{3} r_{max}^4 \int_0^{\pi} \sin^{\frac{4}{3}} v\, dv \\
&= \frac{4m}{3} r_{max}^4\, 2\int_0^{\pi/2} \sin^{\frac{4}{3}} v\, dv = \frac{4m}{3} r_{max}^4 \int_0^1 x^{\frac{1}{6}}(1-x)^{-\frac{1}{2}}\, dx \\
&= \frac{4m}{3} r_{max}^4 B\left(\frac{7}{6}, \frac{1}{2}\right) = \frac{2m}{3} r_{max}^4 Q.
\end{aligned}$$

Hence

$$r_{max}^4 = \frac{3|t|}{2mQ}.$$

□

Heisenberg group case. When $k = 0$ the vector fields

$$X_1 = \frac{\partial}{\partial x_1} + 2x_2\frac{\partial}{\partial_t}, \quad X_2 = \frac{\partial}{\partial x_2} - 2x_1\frac{\partial}{\partial_t}, \tag{3.41}$$

define a step 2 model. They are left invariant vector fields with respect to the 1-dimensional Heisenberg group $\mathbf{H}_1 = (\mathbb{R}^3, \circ)$, with the group law

$$(x,t) \circ (x',t') = (x + x', t + t' + 2x_2x_1' - 2x_1x_2').$$

In this case, we may recover Theorems 1.52 and 1.53 and hence Theorems 1.36 and 1.41 in [**1**]. We state the result as the following theorem:

THEOREM 3.22. *There are finitely many geodesics that join the origin to (x,t) if and only if $x \neq 0$. These geodesics are parametrized by the solutions ψ of:*

$$\frac{|t|}{\|x\|^2} = \mu(\psi). \tag{3.42}$$

There is exactly one such geodesic if and only if:

$$|t| < \mu(x_1)\|x\|^2.$$

The number of geodesics increases without bound as $\frac{|t|}{\|x\|^2} \to \infty$.
If $0 \leq \psi_1 < \cdots < \psi_N$ are the solutions of equation (3.42), there are N geodesics joining the origin and their lengths are given by

$$s_m^2 = \nu(\psi_m)\Big(|t| + \|x\|^2\Big), \tag{3.43}$$

where

$$\nu(x) = \frac{x^2}{x + \sin^2 x - \sin x \cos x}, \tag{3.44}$$

and

$$\mu(x) = \frac{x}{\sin^2 x} - \cot x. \tag{3.45}$$

PROOF. It comes from Theorem 3.17. With the substitution $k = 0$ in equation (3.77)

$$\mu_0(\psi) = 2\frac{\int_0^\psi \sin^2 v\, dv}{\sin^2 \psi} = \frac{\psi - \sin\psi\cos\psi}{\sin^2\psi} = \frac{\psi}{\sin^2\psi} - \cot\psi,$$

which is equation (1.26) of [**1**].
Substituting $k = 0$ in (3.29) yields

$$\begin{aligned}\nu(\psi) = \frac{\psi^2}{(1+\mu_0(\psi))\sin^2\psi} &= \frac{\psi^2}{\sin^2\psi + \frac{\phi - \sin\psi\cos\psi}{\sin^2\psi}\sin^2\psi} \\ &= \frac{\psi^2}{\sin^2\psi + \psi - \sin\psi\cos\psi},\end{aligned}$$

which is (3.44). □

When $k = 0$, the constants M and Q can be computed explicitly. Hence we recover Theorem 1.55 if chapter 1. Since the proof is different, we state it as a new theorem as follows.

THEOREM 3.23. *The geodesics that join the origin to a point $(0,t)$ have lengths $s_1, s_2, \ldots$, where*

$$(s_m)^2 = m\pi|t|. \tag{3.46}$$

For each length s_m, the geodesics of that length are parametrized by the circle $\mathbb{S}^1$.

PROOF. Substituting $k = 0$ in Theorem 3.18 yields

$$\begin{aligned} M &= B\Big(\frac{1}{2}, \frac{1}{2}\Big) = \frac{\Gamma(\frac{1}{2})\Gamma(\frac{1}{2})}{\Gamma(1)} = \frac{(\sqrt{\pi})^2}{1} = \pi, \\ Q &= 2B\Big(\frac{3}{2}, \frac{1}{2}\Big) = 2\frac{\Gamma(\frac{3}{2})\Gamma(\frac{1}{2})}{\Gamma(\frac{3}{2}+\frac{1}{2})} = 2\frac{\frac{1}{2}\Gamma(\frac{1}{2})^2}{\Gamma(2)} = \pi. \end{aligned}$$

Then equation (3.32) becomes

$$(s_m)^2 = m\frac{M^2}{Q}|t| = m\pi|t|.$$

□

Corollary 3.12 in this case becomes

$$r(\phi)^2 = r_{max}^2 \sin^2(\phi - \phi_0),$$

with $r_{max} = \frac{1}{2\theta}$. The projection on the x-plane is a circle of radius $1/(2\theta)$. In this case the solution reaches the t-axis again after a clock-wise rotation (for $\theta > 0$) of angle $2w = \pi$. The m-th geodesic between the origin and $(0, t)$ intersects the t-axis $m - 1$ times (without counting endpoints). Because of the rotational symmetry, we may always choose the initial argument $\phi_0 = 0$. After m loops the angle is $\phi_1 = m\pi$. The equation (3.23) yields

$$t = -2\Big(\frac{1}{2\theta}\Big)^2 \int_0^{-m\pi} \sin^2 v\, dv = 2m\Big(\frac{1}{2\theta}\Big)^2 \int_0^{\pi} \sin^2 v\, dv = \frac{m\pi}{(2\theta)^2}.$$

We arrive at:

PROPOSITION 3.24. The m-th geodesic between the origin and $(0, t)$ is tangent to a cylinder which contains the t-axis and has the base diameter $\left(\frac{|t|}{m\pi}\right)^{1/2}$.

Note that in the step 2 case, the cylinder contains the t-axis, while in step $2(k+1)$, $k \geq 1$, the cylinder is centered at the t-axis. The projections on the x-space for three different values of $m = 1, 2, 3$ in the case of steps $2, 4, 6$ and 8 are given in Figures 3.1 - 3.4.

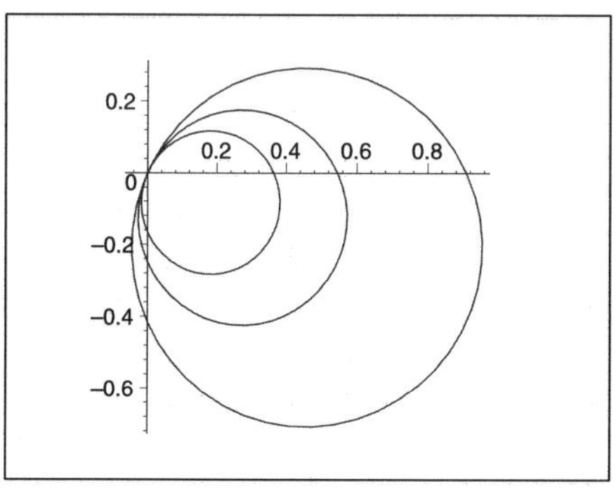

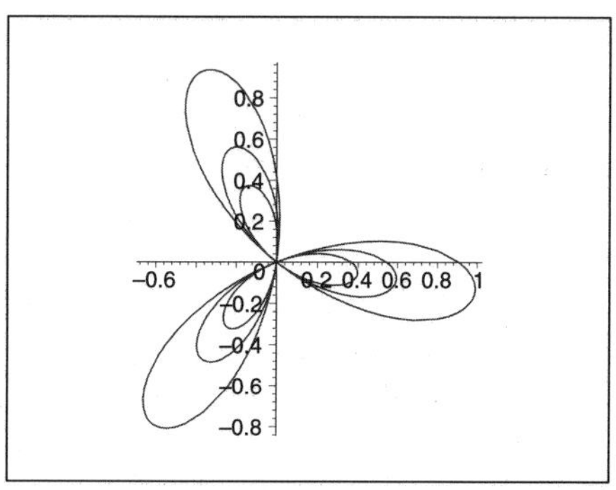

Figure 3.1: x-projection in Heisenberg case. *Figure 3.2: x-projection in step 4 case.*

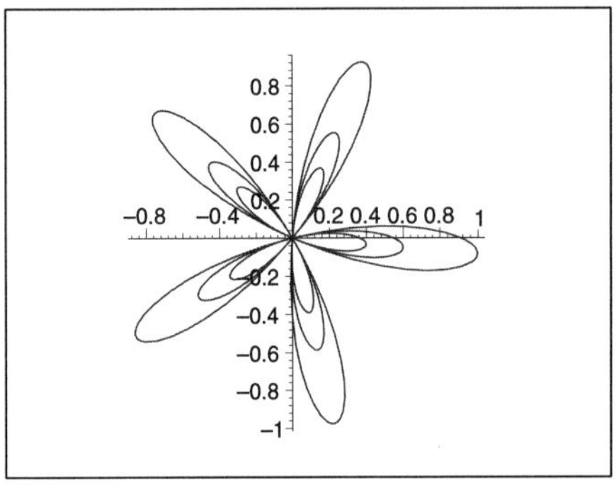

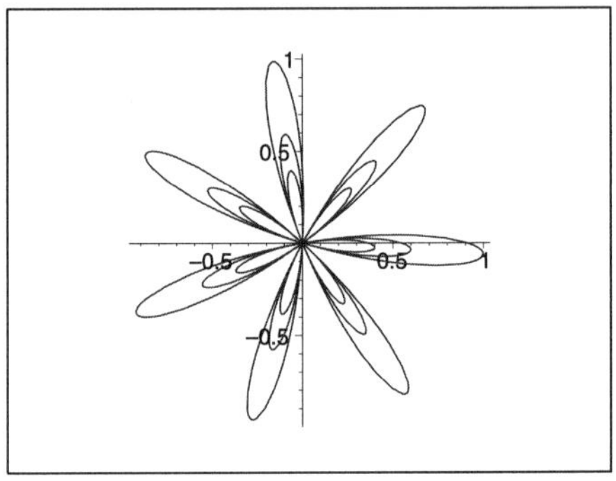

Figure 3.3: *x-projection in step* 6 *case. Figure* 3.4: *x-projection in step* 8 *case.*

The number of loops . The number of total distinct loops depends on the integer k and is denoted by $N(k)$. We have the following result.

PROPOSITION 3.25. The number of loops is given by

$$N(k) = 2k + 1.$$

PROOF. The angle of a loop is $\frac{\pi}{2k+1} = \frac{2\pi}{4k+2}$, which means there are $4k+2$ such angles around the origin. The angles divide the plane in $4k+2$ sector areas as in Figure 3.5.

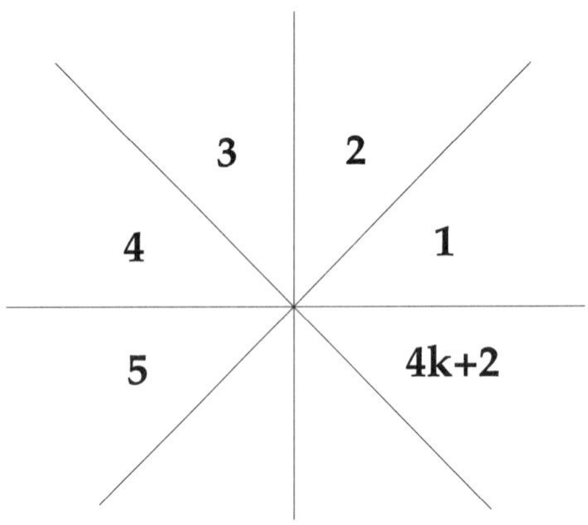

Figure 3.5: $4k+2$ *sector areas.*

All the sector areas between $\phi_0 + 2n\pi$ and $\phi_0 + \frac{\pi}{2k+1} + 2n\pi$ are denoted by **1** (the class of equivalence of the first sector area). The sector areas between $\phi_0 + \frac{\pi}{2k+1} + 2n\pi$ and $\phi_0 + \frac{2\pi}{2k+1} + 2n\pi$ are denoted by **2** (the class of equivalence of the second sector area). The equivalence classes are modulo $4k+2$ (*i.e.*, $\mathbf{j} + (\mathbf{4k}+\mathbf{2})\mathbf{m} = \mathbf{j}$, where $\mathbf{m} \in \mathbb{Z}_+$). Each loop belongs to a certain sector area, for instance the first loop (which starts with ϕ_0 argument) lies into the sector area **1**.

The second loop continues the first loop in such that they have a common tangent at 0. As $\theta < 0$ (our initial assumption), the particle is moving counter clock-wise in the sector area $\mathbf{2k+1+2} = \mathbf{2k+3}$. Indeed, an increment with $\mathbf{2k+1}$ means to go to the class which corresponds to the sector situated with π forward. This happens for any loop.

Fact 1: If a loop belongs to the sector area $\mathbf{m}$ then the following loop will belong to the sector area $\mathbf{m+2k+2}$. As the second loop belongs to the sector area $\mathbf{2k+3}$, using *Fact 1*, the third will belong to $(\mathbf{2k+3})+(\mathbf{2k+2})=\mathbf{4k+2+3}=\mathbf{3}$ sector area.

Fact 2: In general, the $2\ell+1$ loop lies into the $\mathbf{2\ell+1}$ sector area. Indeed, using *Fact 1* for 2ℓ times, the loop $2\ell+1$ belongs to the class of

$$1+(2k+2)+\cdots+(2k+2)=1+2\ell(2k+2)=1+2\ell+(4k+2),$$

which is the class $\mathbf{2\ell+1}$. From *Fact 2* each odd loop lies in the corresponding odd number class area sector. Hence each odd area sector contains a loop. We will show that the even sector areas remain empty. This will come from the fact that the even loops lie also in the odd sector areas. Indeed, the second loop lies into $\mathbf{2k+3}$ sector area, which is an odd number. In general, using *Fact 1*, the 2ℓ loop belongs to the class of

$$1+(2\ell-1)(2k+2)=1+2(2\ell-1)(k+1),$$

which is odd. As there are $4k+2=2(2k+1)$ sectors only half are odd. No loops belong to even class sector areas. Hence there only $2k+1$ loops. This completes the proof of the proposition.

Using Proposition 3.25, $N(0)=1$. In this case, the single loop is in fact a circle which starts and closes at the origin. The equation in polar coordinates is $r(\phi)=r_{max}\sin(\phi-\phi_0)$, where $\phi-\phi_0\in[0,\pi]$. If $k=1$, then $N(1)=3$. There are 3 loops with an angle of 120 degrees between the symmetry axes. This corresponds to the step 4 case. The equation of the first loop is $r(\phi)=r_{max}\sin^{1/3}(3(\phi-\phi_0))$, where $\phi-\phi_0\in[0,\pi/3]$. If $k=2$, then $N(2)=5$. The equation of the first loop is $r(\phi)=r_{max}\sin^{1/5}(5(\phi-\phi_0))$, where $\phi-\phi_0\in[0,\pi/5]$. See Figures 3.1 - 3.4. $\square$

3.8. Conjugate points to the origin

Consider a geodesic (x_1,x_2,t) which starts at the origin. In polar coordinates, one has $r(\phi_0)=0$ and $t(\phi_0)=0$. When $\phi_1=\phi_0-\pi/(2k+1)$, then $r(\phi_1)=0$, which means that the x-component is zero. The t-component is given by Theorem 3.13, equation (3.23)

$$\begin{aligned} t\Big(\phi_0-\pi/(2k+1)\Big) &= -\frac{2}{2k+1}\Big(\frac{1}{2\theta}\Big)^{\frac{2(k+1)}{2k+1}}\int_0^{(2k+1)\phi_1}\sin^{\frac{2(k+1)}{2k+1}}(v)\,dv \\ &= -\frac{2}{2k+1}\Big(\frac{1}{2\theta}\Big)^{\frac{2(k+1)}{2k+1}}\int_0^{\pi}\sin^{\frac{2(k+1)}{2k+1}}(v)\,dv=T \end{aligned}$$

does not depend on the initial argument ϕ_0. On the other hand, for each ϕ_0 there is a geodesic which starts with the initial argument $\phi(0)=\phi_0$. A variation of ϕ_0 corresponds to a variation of the geodesic. The point $(0,0,T)$ remains fixed during the variation, and hence it is a conjugate point to the origin. The following theorem shows that the converse is also true.

THEOREM 3.26. *Given a point $P(0,0,\tau)$ on the t-axis, $\tau\neq 0$, there is a geodesic γ starting at the origin, such that P is the first conjugate point to the origin along γ.*

PROOF. The proof is based on an explicit construction of the geodesic. Consider $\tau < 0$ and hence, $\theta < 0$. From the equation $\tau = T$, we have

$$\left(\frac{1}{2\theta}\right)^{\frac{2(k+1)}{2k+1}} = \frac{-\tau(2k+1)}{2}\left(\int_0^{\pi} \sin^{\frac{2(k+1)}{2k+1}}(v)\, dv\right)^{-1} = \frac{-\tau(2k+1)}{Q}. \tag{3.47}$$

Given ϕ_0, using Theorem 3.11, one may construct the projection on the x-plane of the geodesic

$$r(\phi) = \left(\frac{1}{2\theta}\right)^{\frac{1}{2k+1}} \sin^{\frac{1}{2k+1}}\left((2k+1)(\phi_0 - \phi)\right). \tag{3.48}$$

The t-component can be found by integrating the horizontal condition

$$t(\phi) = -2\int_{\phi_0}^{\phi} r^{2k}(\phi)\, d\phi.$$

□

3.9. The use of the hypergeometric function

In this section we shall compute the length of geodesics between $(0,0,0)$ and $(0,0,t)$, using the arc length parametrization, *i.e.*, $\dot{x}_1^2 + \dot{x}_2^2 = 1$, and hypergeometric functions. We recover the formula for the lengths obtained in the previous section.

The Hamiltonian

$$H(\xi,\theta,x,t) = \frac{1}{2}\left(\xi_1 + 2x_2|x|^{2k}\theta\right)^2 + \frac{1}{2}\left(\xi_2 - 2x_1|x|^{2k}\theta\right)^2,$$

is preserved along the geodesics. Using the Hamiltonian equations

$$\dot{x}_1 = \partial H/\partial \xi_1 = \xi_1 + 2x_2|x|^{2k}\theta, \qquad \dot{x}_2 = \partial H/\partial \xi_2 = \xi_2 - 2x_1|x|^{2k}\theta,$$

and considering the arc length parametrization, the Hamiltonian becomes

$$H = \frac{1}{2}(\dot{x}_1^2 + \dot{x}_2^2) = \frac{1}{2}.$$

In polar coordinates $\dot{x}_1^2 + \dot{x}_2^2 = \dot{r}^2 + r^2\dot{\phi}^2 = \dot{r}^2 + 4\theta^2 r^{2(2k+1)}$, where we used $\dot{\phi} = -2\theta r^{2k}$. The radius $r(s)$ verifies the equation

$$\dot{r}^2 + 4\theta^2 r^{2(2k+1)} = 1.$$

This can be written as

$$\frac{dr}{\sqrt{1 - 4\theta^2 r^{2(2k+1)}}} = \pm ds.$$

As the geodesic starts at the origin, $r(0) = 0$. Then we shall consider the positive sign in the right side. Integrating

$$\int_0^{r(s)} \frac{dx}{\sqrt{1 - 4\theta^2 x^{2(2k+1)}}} = s.$$

The substitution $u = x^2$ yields

$$\int_0^{r^2(s)} \frac{du}{\sqrt{u(1 - 4\theta^2 u^{2k+1})}} = 2s.$$

Let $v = u(4\theta^2)^{\frac{1}{2k+1}}$. Then $du = (4\theta^2)^{-\frac{1}{2k+1}}\, dv$ and the above integral equation becomes

$$\int_0^{\sigma(s)} \frac{dv}{\sqrt{v(1-v^{2k+1})}} = 2(2\theta)^{\frac{1}{2k+1}} s, \tag{3.49}$$

where $\sigma(s) = (4\theta^2)^{\frac{1}{2k+1}}$ and we considered $\theta > 0$.
The above integral can be written in terms of Barnes's extended hypergeometric function (see Erdélyi [**23**])

$$\int_0^{\sigma(s)} \frac{dv}{\sqrt{v(1-v^{2k+1})}} = 2\sqrt{\sigma(s)}\, F\Big(\Big[\frac{1}{2}, \frac{1}{2(2k+1)}\Big], \Big[1 + \frac{1}{2(2k+1)}\Big], \sigma(s)^{2k+1}\Big). \tag{3.50}$$

Using $2\sqrt{\sigma(s)} = 2\,(2\theta)^{\frac{1}{2k+1}}\, r(s)$, the equation (3.50) becomes

$$r(s)\, F\Big(\Big[\frac{1}{2}, \frac{1}{2(2k+1)}\Big], \Big[1 + \frac{1}{2(2k+1)}\Big], \sigma(s)^{2k+1}\Big) = s. \tag{3.51}$$

The function

$$z \to F\Big(\Big[\frac{1}{2}, \frac{1}{2(2k+1)}\Big], \Big[1 + \frac{1}{2(2k+1)}\Big], z\Big)$$

is increasing on the domain $[0, 1]$. See Figure 3.6.

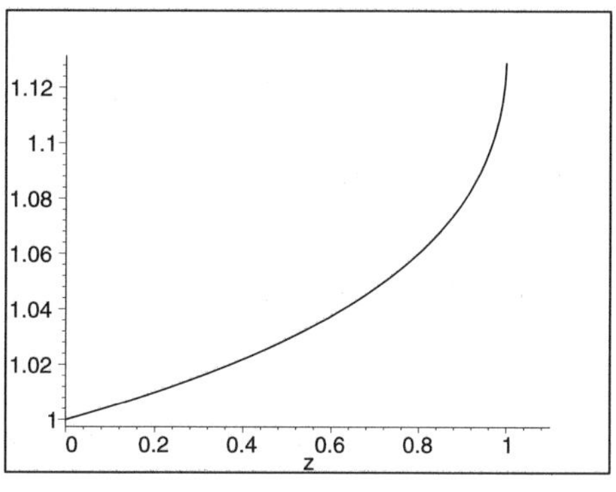

Figure 3.6: *The function $z \to F(z)$.*

The minimum and maximum values are

$$F\Big(\Big[\frac{1}{2}, \frac{1}{2(2k+1)}\Big], \Big[1 + \frac{1}{2(2k+1)}\Big], 0\Big) = 1,$$

$$F\Big(\Big[\frac{1}{2}, \frac{1}{2(2k+1)}\Big], \Big[1 + \frac{1}{2(2k+1)}\Big], 1\Big) = \mathrm{F}.$$

Choose $r(s)$ such that $\sigma(s) = 1$, and hence the hypergeometric function reaches the maximum in relation (3.51)

$$r(s) = \Big(\frac{1}{2\theta}\Big)^{\frac{1}{2k+1}},$$

which is the same as r_{max}, see formula (3.17).
Let s_1 be the arc length parameter for which $r(s_1) = r_{max}$. Then (3.51) yields

$$r_{max}\, \mathrm{F} = s_1. \tag{3.52}$$

At $r(s_1)$ the trajectory starts to bounce back, and at $s = 2s_1$ it reaches the t-axis again, *i.e.*, $r(2s_1) = 0$. Hence, the shortest geodesic will have the length

$$\ell_1 = 2s_1 = 2r_{max}\,\mathrm{F}. \tag{3.53}$$

If the geodesic winds m times around the t-axis, joining $(0,0,0)$ and $(0,0,t)$, the length is

$$\ell_m = 2ms_1 = 2m\,r_{max}\,\mathrm{F}. \tag{3.54}$$

The momentum θ depends on the boundary condition t. This relationship will be found below.

A Hamiltonian equation yields

$$\dot{t} = H_\theta = -2r^{2k+2}\dot{\phi}.$$

We shall consider the following boundary conditions for the argument angles

$$\phi_0 = \phi(0), \qquad \phi_1 = \phi(2ms_1).$$

Let $\theta > 0$. Then the angle ϕ moves clock-wise, *i.e.*, decreasing. After one complete loop we have $\phi(2s_1) = -\frac{\pi}{2k+1}$, and hence after m loops

$$\phi_1 = m\phi(2s_1) = -\frac{m\pi}{2k+1}.$$

Integrating between ϕ_0 and ϕ_1, yields

$$t(\phi_1) - t(\phi_0) = -2\int_{\phi_0}^{\phi_1} r^{2k+2}(\phi)\,d\phi.$$

Using the boundary conditions for $t(s)$

$$t(\phi_0) = 0, \quad t(\phi_1) = t,$$

and the formula in polar coordinates for the solution given by Corollary 3.12, we have

$$t = -2r_{max}^{2k+2}\int_{\phi_0}^{\phi_1} \sin((2k+1)(\phi-\phi_0))^{\frac{2k+2}{2k+1}}\,d\phi.$$

Using the substitution $v = (2k+1)(\phi - \phi_0)$ yields

$$t = \frac{2r_{max}^{2k+2}}{2k+1}\int_0^{m\pi} \sin(v)^{\frac{2k+2}{2k+1}}\,dv = \frac{2mr_{max}^{2k+2}}{2k+1}\int_0^{\pi} \sin(v)^{\frac{2k+2}{2k+1}}\,dv = \frac{mQr_{max}^{2k+2}}{2k+1},$$

and hence

$$r_{max}^{2k+2} = \frac{(2k+1)t}{mQ},$$

where $Q = 2\int_0^\pi \sin(v)^{\frac{2k+2}{2k+1}}\,dv$. Substituting in the equation (3.54) we obtain the following result.

THEOREM 3.27. *The lengths of the geodesics joining the origin and the point $(0,0,t)$ are*

$$\ell_m^{2k+2} = (2k+1)\frac{m^{2k+1}\,(2\mathrm{F})^{2k+2}}{Q}|t|. \tag{3.55}$$

The Carnot-Carathéodory distance between $(0,0,0)$ *and* $(0,0,t)$ *is* ℓ_1 *with*

$$\ell_1^{2k+2} = (2k+1)\frac{(2\mathrm{F})^{2k+2}}{Q}|t|. \tag{3.56}$$

The above formulas are consistent with the formulas obtained in Theorem 3.18 if we take $M = 2(2k+1)\mathrm{F}$. This provides a formula for F in terms of beta functions.

The constant F. In the case $k = 0$ the constant F is related to the period of the sine function while in the case $k = 1$, it is related to the period of the elliptic function sn. Indeed, making $\sigma(s) = 1$ in equation (3.50), yields

$$2\mathrm{F} = \int_0^1 \frac{dv}{\sqrt{v(1-v^{2k+1})}}. \tag{3.57}$$

For $k = 0$, in the step 2 case

$$2\mathrm{F} = \int_0^1 \frac{dv}{\sqrt{v(1-v)}} = \arcsin(2v-1)\Big|_{v=0}^{v=1} = \pi,$$

i.e., half of the period of the sine function.
For $k = 1$, in the step 4 case

$$2\mathrm{F} = \int_0^1 \frac{dv}{\sqrt{v(1-v^3)}} = \frac{2}{3^{1/4}} K\left(\frac{\sqrt{2}}{2(1+\sqrt{3})}\right),$$

where $K(\kappa)$ is the complete elliptic integral defined by

$$K(\kappa) = \int_0^1 \frac{dt}{\sqrt{(1-t^2)(1-\kappa^2 t^2)}}.$$

In the above case $\kappa = \frac{\sqrt{2}}{2(1+\sqrt{3})}$, and

$$K\left(\frac{\sqrt{2}}{2(1+\sqrt{3})}\right) \approx 1.598.$$

2F is proportional to $2K(\kappa)$ which is half of the period of the elliptic function $\mathrm{sn}(\,\cdot\,,\kappa)$.

For the general step case, we shall write F as an infinite numerical series.

PROPOSITION 3.28. We have

$$\mathrm{F} = \sum_{n=0}^{\infty} \frac{a_n}{2^n\, n!}$$

where

$$a_n = \frac{\prod_{p=0}^{k-1}(1+2p)}{1+2n(n+1)}.$$

PROOF. The Barnes' extended hypergeometric function can be written as a series

$$F([n],[d],z) = \sum_{k=0}^{\infty} \frac{\prod_{i=0}^{p} \frac{\Gamma(n_i+k)}{\Gamma(n_i)}}{k! \prod_{i=0}^{q} \frac{\Gamma(d_i+k)}{\Gamma(d_i)}} z^k$$

where $[n] = [n_1, \ldots, n_p]$ and $[d] = [d_1, \ldots, d_q]$.
In our case $p = 2$ and $q = 1$. Then

$$n_1 = \frac{1}{2}, \quad n_2 = \frac{1}{2(k+1)}, \quad d_1 = 1 + \frac{1}{2(k+1)}.$$

Using $\Gamma(z+1) = z\Gamma(z)$, we have

$$\begin{aligned} \mathrm{F} &= F\Big(\Big[\frac{1}{2}, \frac{1}{2(k+1)}\Big], \Big[1 + \frac{1}{2(k+1)}\Big], 1\Big) \\ &= \sum_{k=0}^{\infty} \frac{1}{k!} \frac{\Gamma(\frac{1}{2}+k)\,\Gamma\big(\frac{1}{2(k+1)}+k\big)}{\Gamma(\frac{1}{2})\,\Gamma\big(\frac{1}{2(k+1)}\big)} \frac{\Gamma\big(1+\frac{1}{2(k+1)}\big)}{\Gamma\big(1+\frac{1}{2(k+1)}+k\big)} \\ &= \sum_{k=0}^{\infty} \frac{1}{k!} \frac{\Gamma(\frac{1}{2}+k)}{\Gamma(\frac{1}{2})} \cdot \frac{1}{1+2k(k+1)}. \end{aligned}$$

Using

$$\begin{aligned} \Gamma\Big(\frac{1}{2}+k\Big) &= \Big(\frac{1}{2}+k-1\Big)\Big(\frac{1}{2}+k-2\Big)\ldots\Big(\frac{1}{2}+k-k\Big)\Gamma\Big(\frac{1}{2}\Big) \\ &= \frac{\prod_{p=0}^{k-1}(1+2p)}{2^k}\Gamma\Big(\frac{1}{2}\Big), \end{aligned}$$

yields the formula for a_k.

□

3.10. Geodesics starting outside the origin

Recall Euler-Lagrange system of equations

$$(3.58) \qquad \begin{cases} \ddot{r} = r\dot{\phi}(\dot{\phi} + 4(k+1)\theta r^{2k}) \\ r^2(\dot{\phi} + 2\theta r^{2k}) = C(\text{constant}) \\ \theta = \theta_0(\text{constant}). \end{cases}$$

The solutions with $r(0) \neq 0$ will be discussed. The system (3.58) is considered with the initial conditions

$$(3.59) \qquad r(0) \neq 0\,, \quad \phi(0) = \phi_0.$$

The second equation of (3.58) becomes

$$(3.60) \qquad \dot{\phi} = \frac{C}{r^2} - 2\theta r^{2k}.$$

Substituting in the first equation of (3.58), one obtains

$$(3.61) \qquad \ddot{r} = \frac{C^2}{r^3} + 4k\theta C r^{2k-1} - 4\theta^2(2k+1)r^{4k+1}.$$

Let

$$(3.62) \qquad V(r) = \frac{1}{2}\Big(\frac{C^2}{r^2} - 4\theta C r^{2k} + 4\theta^2 r^{4k+2}\Big) = \frac{1}{2}\left(\frac{2\theta r^{2k+2} - C}{r}\right)^2.$$

Then, one may rewrite equation (3.61) as Newton's equation

$$(3.63) \qquad \ddot{r} = -V'(r).$$

Substituting $p = \dot{r}$, equation (3.61) becomes

$$p\frac{dp}{dr} = -\frac{dV}{dr},$$

which yields

$$\frac{p^2}{2} = -V(r) + E,$$

where E is the constant of the total energy.

PROPOSITION 3.29. If r is a solution of the Euler-Lagrange system with initial conditions (3.59), then the energy is conserved:

$$\frac{1}{2}\dot{r}^2 + \frac{1}{2}\left(\frac{2\theta r^{2k+2} - C}{r}\right)^2 = E. \tag{3.64}$$

We shall give a qualitative description of the solutions of (3.64). There are two cases: $sgn(C) = sgn(\theta)$ and $sgn(C) \neq sgn(\theta)$. In each case we shall draw the graph of the potential energy V and the trajectory of the solution in the phase plane.

Case $\mathbf{sgn}(C) = \mathbf{sgn}(\theta)$. In this case $V(r)$ has a zero and a global minimum at

$$r_0 = \left(\frac{C}{2\theta}\right)^{1/2(k+1)}, \tag{3.65}$$

and $\lim_{x\to 0^+} V(r) = \lim_{x\to\infty} V(r) = \infty$.

Equilibrium points: r_0 corresponds to a stable equilibrium point,

$$V'(r_0) = 0 \text{ and } V''(r_0) > 0.$$

The corresponding ϕ satisfies

$$\dot{\phi} = \frac{C}{r_0^2} - 2\theta r_0^{2k} = \frac{C - 2\theta r_0^{2(k+1)}}{r_0^2} = 0$$

and hence ϕ is constant. This means that each point of the circle $\mathcal{S}(0, r_0)$ is an equilibrium point (constant solution). At each of these points the energy $E = 0$.

Tangent circles: When $E > 0$, there are exactly two positive roots r_{min} and r_{max} of the equation

$$V(r) = E, \tag{3.66}$$

such that

$$0 < r_{min} \leq r_0 \leq r_{max}. \tag{3.67}$$

PROPOSITION 3.30. The solution of the Euler-Lagrange system (3.58) in the case $sgn(C) = sgn(\theta)$ lies in the circular ring $\mathcal{W}(0, r_{min}, r_{max}) = \{x \in \mathbb{R}^2; r_{min} \leq |x| \leq r_{max}\}$.

PROOF. r is equal to r_{min} or r_{max} when $\dot{r} = 0$. The equation of conservation of energy (3.64) shows that r_{min} and r_{max} are solutions of the equation $V(r) = E$ and the solution r has the property

$$0 < r_{min} \leq r \leq r_{max}.$$

□

Existence of loops: The equation $\dot{\phi} = 0$ has the positive root

$$r = \Big(\frac{C}{2\theta}\Big)^{\frac{1}{2(k+1)}}. \tag{3.68}$$

Hence $\dot{\phi} = 0$ if and only if $r = r_0$. When the solution intersects the circle of equilibrium points $\mathcal{S}(0, r_0)$, the sign of $\dot{\phi}$ changes, *i.e.*, the trajectory is bouncing back, making loops in the ring $\mathcal{W}(0, r_{min}, r_{max})$, see Figure 2.8.

The ring width: $w = r_{max} - r_{min}$ is a function of energy $w = w(E)$. Using properties of the function $V(r)$, we find
(i) $w(E)$ is an increasing function of energy E;
(ii) $w(0) = 0$ if and only if $r_{min} = r_0 = r_{max}$;
(iii) $\lim_{E\to\infty} r_{min} = 0$, $\lim_{E\to\infty} r_{max} = \infty$.

The t-component: From the Hamiltonian equation

$$\dot{t} = \frac{\partial H}{\partial \theta} = 2|x|^{2k}(x_2\dot{x}_1 - x_1\dot{x}_2) = -2r^{2(k+1)}\dot{\phi},$$

$\dot{\phi} = 0$ on the circle $\mathcal{S}(0, r_0)$ implies that $\dot{t}(s)$ changes sign and reaches the local maxima and minima when the geodesic crosses the cylinder $\mathcal{S}(0, r_0) \times \mathbb{R}$.

Case sgn(C) $\neq$ sgn(θ), $C \neq 0$. Here

$$\dot{\phi} = \frac{C - 2\theta r^{2(k+1)}}{r^2} \neq 0,$$

which means that ϕ increases (decreases) for $\theta < 0$ ($\theta > 0$) and hence there are no loops. In this case the potential energy is strictly positive

$$V(r) = \frac{1}{2}\Big(\frac{2\theta r^{2k+2} - C}{r}\Big)^2 > 0,$$

and has a minimum at

$$r_1 = \Big(\frac{-C}{2\theta(2k+1)}\Big)^{\frac{1}{2(k+1)}}. \tag{3.69}$$

Indeed, $V'(r) = 0$ if and only if

$$4\theta^2(2k+1)r^{4k+4} - 4k\theta C r^{2k+2} - C^2 = 0.$$

Substituting $u = r^{2k+2}$ we obtain a quadratic equation in u

$$4\theta^2(2k+1)u^2 - 4k\theta C u - C^2 = 0,$$

with the roots

$$u_1 = \frac{C}{2\theta} \quad \text{and} \quad u_2 = -\frac{C}{2\theta(2k+1)}.$$

As r_1 is positive, we choose

$$(r_1)^{2(k+1)} = -\frac{C}{2\theta(2k+1)}.$$

The equilibrium solution. The equilibrium solution is the circle of radius r_1 centered at the origin. The corresponding ϕ can be obtained from

$$r_1^2(\dot{\phi} + 2\theta r_1^{2k}) = C,$$

and it is equal to

$$\phi(s) = \Big(\frac{C}{r_1^2} - 2\theta r_1^{2k}\Big) + \phi_0.$$

Bounds for $r(s)$: There are two positive roots ρ_{min} and ρ_{max} for the equation

$$V(r) = E\ , \qquad \text{where } E \geq V(r_1)$$

and $0 < \rho_{min} \leq r_1 \leq \rho_{max}$.
In the phase plane the solution is rotating around the stable equilibrium point $(r_1, 0)$ such that

$$\rho_{min} \leq r(s) \leq \rho_{max}.$$

The projection of the geodesics in the x-plane is contained in the ring $\mathcal{W}(0, \rho_{min}, \rho_{max})$. The width of the ring increases as the total energy E increases, where $E = \frac{1}{2}|\dot{x}(0)|^2$. See Figure 2.11.

The t-component: In this case $\dot{t} = -2r^{2(k+1)}\dot{\phi} \neq 0$. Hence $t(s)$ increases if $\theta > 0$, decreases if $\theta < 0$, it is constant if $\theta = 0$.

3.11. Geodesic completeness

DEFINITION 3.31. If for any point P, any geodesic $c(t)$ emanating from P is defined for all $t \in \mathbb{R}$, the geometry is called geodesically complete.

The main result of this section is the following theorem.

THEOREM 3.32. *The geometry induced by the vector fields (3.1) is geodesically complete.*

We note that geodesic completeness does not hold in general. A counter-example in the step 2 case is provided in Calin [**4**]. To prove Theorem 3.32 we need a few results regarding extendability of solutions of differential equations. The following result can be found for instance in Hartman [**14**].

LEMMA 3.33. *Let $f(t, y)$ be a continuous function on a strip $t_0 \leq t \leq t_0 + a < \infty$, $y \in \mathbb{R}^d$ arbitrary. Let $y = y(t)$ denote a solution of*

$$y' = f(t, y), \quad y(t_0) = y_0 \tag{3.70}$$

on a right maximal interval J. Then either $J = [t_0, t_0 + a]$ or $J = [t_0, \delta)$, $\delta \leq t_0 + a$, and $|y(t)| \to \infty$ as $t \to \delta$.

The Hamiltonian system of equations can be written in the form (3.70). Consider $y \in \mathbb{R}^{2n}$, $y = (x, p)$, with $x, p \in \mathbb{R}^n$. Let $H(y)$ be the Hamiltonian function, and set

$$\nabla H(y) = \Big(\frac{\partial H}{\partial x}, \frac{\partial H}{\partial p}\Big),$$

and

$$\mathcal{J} = \begin{pmatrix} O_n & I_n \\ -I_n & O_n \end{pmatrix} \in \mathcal{M}_{2n}(\mathbb{R}).$$

Then

$$\mathcal{J}\,\nabla H(y) = \Big(\frac{\partial H}{\partial p}, -\frac{\partial H}{\partial x}\Big).$$

Hamiltonian system

$$\begin{cases} \dot{x}(s) = \partial H/\partial p \\ \dot{p}(s) = -\partial H/\partial x \end{cases} \tag{3.71}$$

can be written as

$$\dot{y}(s) = f(y(s)), \tag{3.72}$$

where $f(y) = \mathcal{J}\nabla H(y)$. Since f is independent of s, we may rewrite Lemma 3.33 as

COROLLARY 3.34. Let $f(y)$ be an arbitrary continuous function for $y \in \mathbb{R}^{2n}$ and let $y(s)$ be a solution for

$$\dot{y}(s) = f(y(s)), \quad y(0) = y_0$$

on a right maximal interval I. Then either $I = [0, \infty)$ or $I = [0, \delta)$ with $|y(s)| \to \infty$ as $s \to \delta$.

In other words, if the solution $y(s)$ doesn't blow up for a finite value of the parameter s, then the maximal right interval is $[0, \infty)$.
We are working in coordinates r, ϕ, t with the corresponding momenta p, η, θ. Consider a solution for the Hamiltonian system

$$y(s) = \big(r(s), \phi(s), t(s); p(s), \eta(s), \theta(s)\big).$$

We shall show that $|y(s)|$ cannot blow up for finite values of s. This is equivalent to showing that each component of $y(s)$ has this property. The momentum θ is bounded because it is constant. Recall the Lagrangian in polar coordinates is

$$L = \frac{1}{2}(\dot{r}^2 + r^2\dot{\phi}^2) + \theta\dot{t} + 2\theta r^{2k+2}\dot{\phi}. \tag{3.73}$$

Equation (3.73) yields $\dot{\eta} = \partial L/\partial\phi = 0$. This implies that $\eta(s) = \eta$ is a constant and hence bounded. The momentum p is given by $p = \partial L/\partial\dot{r} = \dot{r}$. Using the conservation of energy, yields

$$|p(s)| = |\dot{r}(s)| \leq \sqrt{2E}.$$

As E is constant along the solutions, $|p(s)|$ is bounded. Here we used the standard definition of the momentum

$$p = \frac{\partial L}{\partial\dot{q}},$$

and the Euler-Lagrange equation

$$\dot{p} = \frac{\partial L}{\partial q}.$$

Next, we deal with the bounds for $r(s)$. We have seen that for $E > 0$ there are exactly two positive roots r_{min}, r_{max} of the equation

$$V(r) = E,$$

and

$$0 < r_{min} \leq r(s) \leq r_{max},$$

so $r(s)$ is bounded. The projection of the trajectory in the x-plane is contained in the ring defined by the circles of radii r_{min} and r_{max}. See the Figures 2.8 and 2.11. In order to deal with the bounds for the ϕ component, we use

$$\dot{\phi}(s) = \begin{cases} \frac{C}{r^2(s)} - 2\theta r^{2k}(s), & \text{if } r(0) \neq 0, \\ -2\theta r^{2k}, & \text{if } r(0) = 0. \end{cases} \tag{3.74}$$

The inequality

$$0 \le r_{min} \le r(s) \le r_{max}$$

yields the upper bound

$$|\dot{\phi}(s)| \le C_0, \ \ \forall s,$$

with

$$C_0 = \begin{cases} |C|/r^2_{min} + 2|\theta| r^{2k}_{max}, & \text{if } r(0) \ne 0, \\ 2|\theta| r^{2k}_{max}, & \text{if } r(0) = 0. \end{cases} \tag{3.75}$$

Thus for all s

$$|\phi(s)| \le |\phi(s) - \phi(0)| + |\phi(0)| \le C_0|s| + |\phi(0)|,$$

so

$$\lim_{s\to\delta} |\phi(s)| \le C\delta + |\phi_0|,$$

and $|\phi(s)|$ cannot blow up for finite δ.

We still need to show that $t(s)$ doesn't blow up for finite s. From the Hamiltonian equation

$$\dot{t}(s) = \frac{\partial H}{\partial \theta} = -2r^{2(k+1)}(s)\dot{\phi}(s).$$

Using the boundedness of $|\dot{\phi}|$,

$$\begin{aligned} |t(s)| &\le |t(s) - t(0)| + |t(0)| \le \max_{u\in[0,s]} |\dot{t}(u)||s| + |t(0)| \\ &\le 2r^{2(k+1)}_{max}|\dot{\phi}(s)| + |t(0)| \le 2C_0 r^{2(k+1)}_{max}|s| + |t(0)| \\ &= C_1|s| + |t(0)|. \end{aligned}$$

Hence $t(s)$ cannot blow up for a finite value of s. Using Corollary 3.34 the bicharacteristic solution $y(s) = (r(s), \phi(s), t(s), p(s), \eta(s), \theta(s))$ is defined on $[0, +\infty)$. Using a similar argument to the left we get $y(s)$ defined on the entire real line. In particular, the projection on the (r, ϕ)-plane has the same property.

3.12. Global connectivity by geodesics

Consider the subRiemannian geometry induced by the vector fields (3.1). In this section we shall prove the following global connectivity result:

THEOREM 3.35. *Given any two points $A, B \in \mathbb{R}^3$, there is a geodesic joining A and B.*

It is known that geodesics are locally length minimizing, see Hamenstädt [**30**], Lee and Markus [**43**], Strichartz [**50**], [**51**], Belaïche [**7**]. However, there are examples of length minimizing curves in steps greater than 2, which are not geodesics, see Liu and Sussmann [**44**], Sussmann [**52**].

DEFINITION 3.36. The distribution $\mathcal{H}$ generated by the vector fields X_1 and X_2 given by (3.1) is called the horizontal distribution. A curve c is called horizontal if $\dot{c} \in \mathcal{H}$. The Riemannian metric g defined on $\mathcal{H}$ with respect to which X_1 and X_2 are orthonormal is called the subRiemannian metric.

By Chow's theorem, [**22**], arbitrary points P and Q can be joined by a piece-wise horizontal curve. Using the subRiemannian metric, one may define the distance

$$d(P,Q) = \inf\{\text{length}(c); c \text{ horizontal, joining P and Q}\},$$

where $\text{length}(c) = \int_I \sqrt{g(\dot{c},\dot{c})}$.

DEFINITION 3.37. A subRiemannian manifold is called complete if it is complete as a metric space, *i.e.*, the distance function d given by the subRiemannian metric is complete.

The following theorem can be found in Strichartz [**50**], [**51**]:

THEOREM 3.38. *Let M be a connected step 2 subRiemannian manifold.*
(a) If M is complete, then any two points can be joined by a geodesic.
(b) If there exists a point P such that every geodesic from P can be indefinitely extended, then M is complete.
(c) Every nonconstant geodesic is locally a unique length minimizing curve.
(d) Every length minimizing curve is a geodesic.

Proof of Theorem 3.35
Let $\mathcal{M} = \mathbb{R}^3 \backslash \{(0,t); t \in \mathbb{R}\}$. The subRiemannian model defined by the vector fields (3.1) is step 2 on $\mathcal{M}$ and step $2(k+1)$ on $\{(0,t); t \in \mathbb{R}\}$. Let A, $B \in \mathbb{R}^3$.
(i) $A, B \in \mathcal{M}$:
From Theorem 3.32, any point of $\mathcal{M}$ has the property (b) of Theorem 3.38. Therefore $\mathcal{M}$ is complete. Using Theorem 3.38 (a), any two points in M can be joined by a minimizing geodesic. By property (d) of Theorem 3.38, the minimizing geodesic is a geodesic. Hence A and B can be connected by a normal geodesic.
(ii) $A \in \{(0,t); t \in \mathbb{R}\}$ and $B \in \mathcal{M}$:
After a translation along the t-axis, Theorem 3.17 has the following consequence:

COROLLARY 3.39. There are finitely many geodesics that join the point $(0, t_0)$ to (x,t) if and only if $x \neq 0$. These geodesics are parametrized by the solutions ψ of:

$$\frac{|t - t_0|}{|x|^{2(k+1)}} = \mu(\psi), \tag{3.76}$$

There is exactly one such geodesic if and only if:

$$|t - t_0| < \mu(c_1)|x|^{2(k+1)},$$

where c_1 is the first critical point for μ and

$$\mu(x) = \frac{2}{2k+1} \frac{\int_0^{(2k+1)x} \sin^{\frac{2(k+1)}{2k+1}}(v)\, dv}{\sin^{\frac{2(k+1)}{2k+1}}(x)}. \tag{3.77}$$

Clearly, (ii) is an immediate consequence.
(iii) $A, B \in \{(0,t); t \in \mathbb{R}\}$:
This is the limit case, $|x| \to 0$, of Corollary 3.39. Namely, (3.76) becomes $\mu(\psi) = \infty$. Hence $\psi \in \frac{\pi}{2k+1}\mathbb{Z}$. There are infinitely many geodesics between A and B.

3.13. Kaehler metrics

With the definitions given in Chapter 1, we have the following generalization of Proposition 1.32

PROPOSITION 3.40. The subRiemannian metric g in which $\{X_1, X_2\}$ are orthonormal is a Kaehler metric on $\mathcal{H}_x$, for any $x \in \mathbb{R}^3$. The fundamental 2-form Φ satisfies $4(k+1)|x|^{2k}\Phi = \Omega$. Hence

$$\Omega(U,V) = 4(k+1)|x|^{2k} g(U, JV), \quad \text{for all horizontal vectors } U \text{ and } V.$$

PROOF. Consider $V = \mathcal{H}_x$. We shall show first that g is a Hermitian metric. Let $U = U^1X_1 + U^2X_2$ and $V = V^1X_1 + V^2X_2$ be two horizontal vector fields. Using $JX_1 = -X_2$ and $JX_2 = X_1$, yields

$$JU = -U^1X_2 + U^2X_1, \qquad JV = -V^1X_2 + V^2X_1.$$

Using the orthonormality of X_1 and X_2

$$\begin{aligned} g(JU, JV) &= g(U^2X_1 - U^1X_2, V^2X_1 - V^1X_2) \\ &= U^1V^1 + U^2V^2 \\ &= g(U^1X_1 + U^2X_2, V^1X_1 + V^2X_2) \\ &= g(U,V), \end{aligned}$$

and hence g is invariant by J. The 2-form Ω is closed because it is exact $\Omega = d\omega$.

$$\begin{aligned} \Omega(U,V) &= \Omega(U^1X_1 + U^2X_2, V^1X_1 + V^2X_2) \\ &= (U^1V^2 - U^2V^1)\Omega(X_1, X_2) \\ &= (U^1V^2 - U^2V^1)\big(X_1(\omega(X_2)) - X_2(\omega(X_1)) - \omega([X_1, X_2])\big) \\ &= (U^1V^2 - U^2V^1)\omega(4(k+1)|x|^{2k}\partial_t) \\ &= 4(k+1)|x|^{2k}(U^1V^2 - U^2V^1) \\ &= 4(k+1)|x|^{2k} g(U^1X_1 + U^2X_2, V^2X_1 - V^1X_2) \\ &= 4(k+1)|x|^{2k} g(U, JV). \end{aligned}$$

□

The skew-symmetry of Ω yields:

COROLLARY 3.41.

$$g(U, JU) = 0, \quad \text{for any horizontal vector } U.$$

The following result was proved in Chapter 1 and it is true in general step:

PROPOSITION 3.42. For any horizontal vector field U we have

$$J(U) = \frac{1}{4}\Big(\Omega(X_1, U)X_1 + \Omega(X_2, U)X_2\Big).$$

PROOF. As both sides are linear, it suffices to check the relation only for the basic vector fields. Using $\Omega(X_1, X_2) = -\Omega(X_2, X_1) = 4$, yields

$$JX_1 = -X_2 = \frac{1}{4}\Omega(X_2, X_1)X_2 = \frac{1}{4}\Big(\Omega(X_1, X_1)X_1 + \Omega(X_2, X_1)X_2\Big)$$

$$JX_2 = X_1 = \frac{1}{4}\Omega(X_1, X_2)X_1 = \frac{1}{4}\Big(\Omega(X_1, X_2)X_1 + \Omega(X_2, X_2)X_2\Big).$$

□

3.14. The classical action

The case of step $2(k+1)$. The Lagrangian associated with the vector fields

$$X_1 = \partial_{x_1} + 2x_2|x|^{2k}\partial_t, \quad X_2 = \partial_{x_2} - 2x_1|x|^{2k}\partial_t \tag{3.78}$$

is given in polar coordinates by

$$\begin{aligned} L &= \frac{1}{2}(\dot{x}_1^2 + \dot{x}_2^2) + \theta(\dot{t} - 4\theta r^{4k+2}) \\ &= E + \theta\dot{t} - 4\theta^2 r^{4k+2}. \end{aligned}$$

Consider the action along the geodesics with the classical boundary conditions

$$x(0) = 0, \qquad t(0) = 0, \qquad t(\tau) = t, \qquad x(\tau) = x,$$

where $r = |x|$. The classical action is given by

$$\begin{aligned} S(r,t,\theta,\tau) &= \int_0^\tau L\,ds = \int_0^\tau (E + \theta\dot{t} - 4\theta^2 r^{4k+2})\,ds \\ &= E\tau + \theta t - 4\theta^2 \int_0^\tau r^{4k+2}(s)\,ds \\ &= E\tau + \theta t - 4\theta^2 J, \end{aligned}$$

Hence

$$S = E\tau + \theta t - 4\theta^2 J, \tag{3.79}$$

where

$$J = \int_0^\tau r^{4k+2}(s)\,ds. \tag{3.80}$$

We shall compute in the following the integral J.
The energy conservation law $\dot{x}_1^2 + \dot{x}_2^2 = 2E$, written in polar coordinates is

$$\dot{r}^2 + 4\theta^2 r^{4k+2} = 2E.$$

This is equivalent to

$$ds = \frac{dr}{\sqrt{2E - 4\theta^2 r^{4k+2}}}, \tag{3.81}$$

where $r(0) = 0$ and $\dot{r}(0) > 0$. Substituting (3.81) in relation (3.80), yields

$$4J = 4\int_0^\tau \frac{r^{4k+2}}{\sqrt{2E - 4\theta^2 r^{4k+2}}}\,dr. \tag{3.82}$$

Substituting

$$v = u\Big(\frac{2\theta^2}{E}\Big)^{\frac{1}{4k+2}}, \qquad \sigma = r\Big(\frac{2\theta^2}{E}\Big)^{\frac{1}{4k+2}}$$

in relation (3.82), yields

$$\begin{aligned} 4J &= \int_0^r \frac{4u^{4k+2}\,du}{\sqrt{2E - 4\theta^2 u^{4k+2}}} = \frac{4}{\sqrt{2E}}\int_0^r \frac{u^{4k+2}\,du}{\sqrt{1 - \frac{2\theta^2}{E}u^{4k+2}}} \\ &= \frac{4}{\sqrt{2E}}\cdot\frac{E}{2\theta^2}\cdot\Big(\frac{E}{2\theta^2}\Big)^{\frac{1}{4k+2}}\int_0^\sigma \frac{v^{4k+2}}{\sqrt{1 - v^{4k+2}}}\,dv. \end{aligned}$$

Then

$$4J = 2^{\frac{k}{2k+1}} E^{\frac{k+1}{2k+1}} \theta^{-\frac{4k+3}{2k+1}} I, \tag{3.83}$$

where

$$I = \int_0^\sigma \frac{v^{4k+2}\,dv}{\sqrt{1-v^{4k+2}}}. \tag{3.84}$$

In the following we shall compute the integral I. Integrating in (3.81) yields

$$\int_0^r \frac{du}{\sqrt{2E-4\theta^2 u^{4k+2}}} = \tau \iff \int_0^r \frac{du}{\sqrt{1-\frac{4\theta^2}{2E}u^{4k+2}}} = \sqrt{2E}\tau$$

$$\int_0^\sigma \frac{\left(\frac{E}{2\theta^2}\right)^{\frac{1}{4k+2}} dv}{\sqrt{1-v^{4k+2}}} = \sqrt{2E}\tau \iff \int_0^\sigma \frac{dv}{\sqrt{1-v^{4k+2}}} = \tau\left(2^{k+1}E^k\theta\right)^{\frac{1}{2k+1}}.$$

Integrating by parts in (3.84) and using the above relation, one obtains an equation for I

$$\begin{aligned}
I &= \int_0^\sigma \frac{v^{4k+2}\,dv}{\sqrt{1-v^{4k+2}}} = \frac{1}{2k+1}\int_0^\sigma v\cdot\frac{(2k+1)v^{4k+1}}{\sqrt{1-v^{4k+2}}}\,dv \\
&= \frac{1}{2k+1}\int_0^\sigma v\,[-\sqrt{1-v^{4k+2}}]'\,dv \\
&= -\frac{1}{2k+1}\Big[\sigma\sqrt{1-\sigma^{4k+2}} - \int_0^\sigma \sqrt{1-v^{4k+2}}\,dv\Big] \\
&= -\frac{1}{2k+1}\Big[\sigma\sqrt{1-\sigma^{4k+2}} - \int_0^\sigma \frac{1-v^{4k+2}}{\sqrt{1-v^{4k+2}}}\,dv\Big] \\
&= \frac{1}{2k+1}\Big[-\sigma\sqrt{1-\sigma^{4k+2}} + \int_0^\sigma \frac{1}{\sqrt{1-v^{4k+2}}}\,dv - \int_0^\sigma \frac{v^{4k+2}}{\sqrt{1-v^{4k+2}}}\,dv\Big] \\
&= \frac{1}{2k+1}\Big[-\sigma\sqrt{1-\sigma^{4k+2}} + \tau\big(2^{k+1}E^k\theta\big)^{\frac{1}{2k+1}} - I\Big].
\end{aligned}$$

Then

$$(2k+1)I = -\sigma\sqrt{1-\sigma^{4k+2}} + \tau\big(2^{k+1}E^k\theta\big)^{\frac{1}{2k+1}} - I$$

and solving for I, yields

$$I = -\frac{\sigma\sqrt{1-\sigma^{4k+2}}}{2(k+1)} + \frac{\tau}{k+1}\cdot\left(\frac{\theta E^k}{2^k}\right)^{\frac{1}{2k+1}}. \tag{3.85}$$

In the following we shall compute the first term in (3.85). Using $\sigma = r\left(\frac{2\theta^2}{E}\right)^{\frac{1}{4k+2}}$, we have

$$\begin{aligned}
\sigma\sqrt{1-\sigma^{4k+2}} &= r\left(\frac{2\theta^2}{E}\right)^{\frac{1}{4k+2}}\sqrt{1-\frac{2\theta^2}{E}r^{4k+2}} \\
&= \left(\frac{2}{E}\right)^{\frac{k+1}{2k+1}}\theta^{\frac{1}{2k+1}}\,r\sqrt{\frac{E}{2}-\theta^2 r^{4k+2}}.
\end{aligned}$$

Substituting in (3.85), relation (3.83) becomes

$$\begin{aligned} 4J &= 2^{\frac{k}{2k+1}} \cdot E^{\frac{k+1}{2k+1}} \cdot \theta^{-\frac{4k+3}{2k+1}} \cdot \theta^{\frac{1}{2k+1}} \cdot E^{\frac{k}{2k+1}} \cdot 2^{-\frac{k}{2k+1}} \cdot \frac{\tau}{k+1} \\ &\quad -2^{\frac{k}{2k+1}} \cdot E^{\frac{k+1}{2k+1}} \cdot \theta^{-\frac{4k+3}{2k+1}} \cdot \frac{1}{2(k+1)} \cdot \frac{2^{\frac{k+1}{2k+1}}}{E^{\frac{k+1}{2k+1}}} \cdot \theta^{\frac{1}{2k+1}} \cdot r\sqrt{\frac{E}{2} - \theta^2 r^{4k+2}} \\ &= \frac{E\tau}{\theta^2(k+1)} - \frac{r}{\theta^2(k+1)}\sqrt{\frac{E}{2} - \theta^2 r^{4k+2}}. \end{aligned}$$

Hence

$$4J = \frac{E\tau}{\theta^2(k+1)} - \frac{r}{\theta^2(k+1)}\sqrt{\frac{E}{2} - \theta^2 r^{4k+2}}. \tag{3.86}$$

Substituting (3.86) in equation (3.79) yields

$$\begin{aligned} S &= E\tau + \theta t - \theta^2\left[\frac{E\tau}{\theta^2(k+1)} - \frac{r}{\theta^2(k+1)}\sqrt{\frac{E}{2} - \theta^2 r^{4k+2}}\right] \\ &= E\tau + \theta t - \frac{E\tau}{k+1} + \frac{r}{k+1}\sqrt{\frac{E}{2} - \theta^2 r^{4k+2}}. \end{aligned}$$

Hence the classical action is given by

$$S = \theta t + \frac{1}{k+1}\left[kE\tau + r\sqrt{\frac{E}{2} - \theta^2 r^{4k+2}}\right]. \tag{3.87}$$

The action $S = S(r, t, \theta, \tau, E)$ depends on the boundary conditions

$$r = r(\tau) = |x(\tau)|, \quad t = t(\tau),$$

the energy constant E, the time τ and the momentum constant θ.

The Heisenberg group case. In the case $k = 0$, formula (3.87) becomes

$$S = \theta t + r\sqrt{\frac{E}{2} - \theta^2 r^2}. \tag{3.88}$$

In this case the constant E can be explicitly computed in terms of r, τ and θ. Making $k = 0$ in relation (3.81) yields

$$ds = \frac{dr}{\sqrt{2E - 4\theta^2 r^2}}.$$

Integrating between 0 and τ yields

$$\int_0^\tau \frac{du}{\sqrt{1 - \frac{4\theta^2}{2E}u^2}} = \sqrt{2E}\,\tau.$$

Substituting $v = \theta u\sqrt{\frac{2}{E}}$ we obtain

$$\begin{aligned} & \int_0^{\sqrt{\frac{2}{E}}\theta r} \frac{dv}{\sqrt{1-v^2}} = 2\theta\tau \\ \Longleftrightarrow\quad & \sin^{-1}\left(\sqrt{\frac{2}{E}}\theta r\right) = 2\theta\tau \\ \Longleftrightarrow\quad & \sqrt{\frac{2}{E}}\theta r = \sin(2\theta\tau) \\ \Longleftrightarrow\quad & E = \frac{2\theta^2 r^2}{\sin^2(2\theta\tau)}. \end{aligned}$$

PROPOSITION 3.43. Given $E > 0$, there are finitely many solutions of the equation

$$E = \frac{2\theta^2 r^2}{\sin^2(2\theta\tau)} \tag{3.89}$$

if and only if $r \neq 0$.

PROOF. Equation (3.89) can be written as

$$\frac{r^2}{2\tau^2 E} = \left(\frac{\sin(2\theta\tau)}{2\theta\tau}\right)^2. \tag{3.90}$$

Let $\lambda(x) = (\frac{\sin x}{x})^2$. The graph of function $\lambda(x)$ is given in the Figure 3.7.

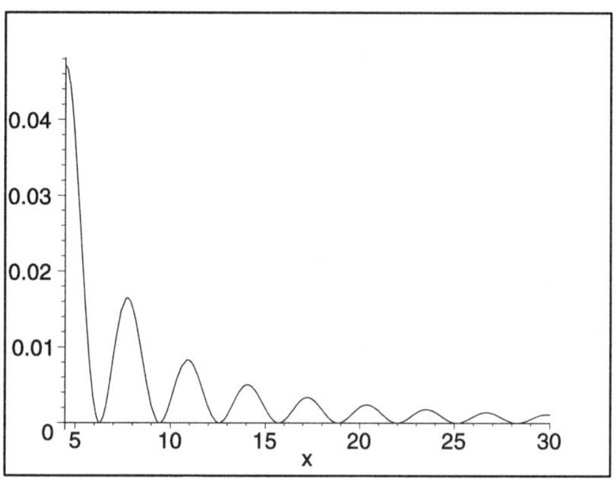

Figure 3.7: The graph of $\lambda(x)$.

If $r > 0$, the equation (3.90) has finitely many solutions, because $\lim_{x\to\infty} \lambda(x) = 0$. □

In the case $r = 0$, the equation (3.90) yields

$$\frac{\sin(2\theta\tau)}{2\theta\tau} = 0 \Longrightarrow 2\theta\tau = k\pi,\ k \in \mathbb{Z}$$

and hence there are infinitely many solutions

$$\theta = \theta_k = \frac{k\pi}{2\tau}.$$

The final computation of the action is given below. Using the equations (3.89) and (3.88)

$$\begin{aligned} S &= \theta t + r\sqrt{\frac{E}{2} - \theta^2 r^2} \\ &= \theta t + r\sqrt{\frac{\theta^2 r^2}{\sin^2(2\theta\tau)} - \theta^2 r^2} \\ &= \theta t + \theta r^2\sqrt{\frac{1}{\sin^2(2\theta\tau)} - 1} \\ &= \theta t + \theta r^2 \cot(2\theta\tau). \end{aligned}$$

Hence, the classical action in the Heisenberg group case is

$$S = \theta[t + r^2 \cot(2\theta\tau)], \tag{3.91}$$

where θ is a solution of the equation studied in chapter 1

$$\frac{t}{r^2} = \mu(2\theta\tau), \tag{3.92}$$

where

$$\mu(x) = \frac{x}{\sin^2 x} - \cot x.$$

From equations (3.91) and (3.92) we have

$$\begin{aligned} S &= \theta\Big[t + r^2\Big(\frac{2\theta\tau}{\sin^2(2\theta\tau)} - \mu(2\theta\tau)\Big)\Big] \\ &= \theta\Big(t + \frac{2\theta\tau r^2}{\sin^2(2\theta\tau)} - r^2\mu(2\theta\tau)\Big) \\ &= \theta\Big(t + \frac{2\theta\tau r^2}{\sin^2(2\theta\tau)} - t\Big) \\ &= \frac{2\theta^2\tau r^2}{\sin^2(2\theta\tau)}. \end{aligned}$$

We obtained the following result.

PROPOSITION 3.44. The classical action on the Heisenberg group is given by

$$S = \frac{2\theta^2\tau r^2}{\sin^2(2\theta\tau)},$$

where θ is a solution of (3.92). For each solution θ of (3.92), the action S has infinitely many singularities for $\tau_k = \frac{k\pi}{2\theta}$, $k \in \mathbb{N}$.

3.15. Exercises

EXERCISE 3.1. Let

$$X_1 = \partial_{x_1} + 2x_2|x|^{2k}\partial_t, \qquad X_2 = \partial_{x_2} + 2x_1|x|^{2k}\partial_t$$

be two vector fields with $|x|^2 = x_1^2 + x_2^2$. Show that

$$Z = \frac{1}{2}(X_1 - iX_2), \quad \text{and} \quad \bar{Z} = \frac{1}{2}(X_1 + iX_2)$$

form a basis of $T^{(1,0)} \oplus T^{(0,1)}$ of the surface

$$\partial\Omega_{k+1} = \Big\{(z_1, z_2) \in \mathbb{C}^2 : Im(z_2) = |z_1|^{2(k+1)}\Big\}.$$

Here we use the coordinates $z_1 = x_1 + ix_2$ and $Re(z_2) = t$.

EXERCISE 3.2. Show that the hypersurface $\partial\Omega_{k+1}$ is equivalent to the "ellipsoid"

$$E_{k+1} = \Big\{(w_1, w_2) \in \mathbb{C}^2 : |w_1|^{2(k+1)} + |w_2|^2 = 1\Big\}$$

via the "generalized Caley transform":

$$z_1 = \frac{w_1}{(1+w_2)^{2(k+1)}}, \qquad z_2 = \frac{i(1-w_2)}{1+w_2}.$$

EXERCISE 3.3. Show that

$$X_1^2 + X_2^2 = \frac{\partial^2}{\partial x_1^2} + \frac{\partial^2}{\partial x_2^2} + 4(k+1)^2|x|^{2k+2}\frac{\partial^2}{\partial t^2} + 4(k+1)|x|^{2k}\Big(x_2\frac{\partial}{\partial x_1} - x_1\frac{\partial}{\partial x_2}\Big)\frac{\partial}{\partial t},$$

and

$$[X_2, X_1] = 4k(k+1)|x|^{2k-2}(x_1^2 + x_2^2)\frac{\partial}{\partial t} + 4(k+1)|x|^{2k}\frac{\partial}{\partial t}.$$

EXERCISE 3.4. We exploit the partial Fourier transform in the variable t

$$\mathcal{F}(f)(x,\tau) = \widetilde{f}(x,\tau) = \int_{-\infty}^{\infty} e^{-i\tau t} f(t)dt, \qquad f(x,t) = \int_{-\infty}^{\infty} e^{it\tau}\widetilde{f}(x,\tau)\frac{d\tau}{2\pi}.$$

Show that

$$\mathcal{F}\Big(\frac{\partial f}{\partial t}\Big)(\tau) = i\tau\widetilde{f}(\tau), \qquad \mathcal{F}\Big(\frac{\partial^2 f}{\partial t^2}\Big)(\tau) = -\tau^2\widetilde{f}(\tau).$$

Consequently,

$$\begin{aligned}\widetilde{(\Delta f)}(\tau) &= \frac{1}{2}\Big[\frac{\partial^2}{\partial x_1^2} + \frac{\partial^2}{\partial x_2^2} + 4i(k+1)\tau|x|^{2k}\big(x_2\frac{\partial}{\partial x_1} - x_1\frac{\partial}{\partial x_2}\big) \\ &\quad -4(k+1)^2\tau^2|x|^{4k+2}\Big]\widetilde{f}.\end{aligned}$$

Here $\Delta = X_1^2 + X_2^2$ is the sub-Laplacian.

CHAPTER 4

Geometry on Higher Dimensional Heisenberg Groups

4.1. The Heisenberg group $\mathbf{H}_n$

In this chapter we shall study the behavior of subRiemannian geodesics on higher dimensional Heisenberg groups, which are step 2 subRiemannian manifolds. Let $\mathbb{R}^{2n+1} = \mathbb{R}^{2n}_{(x,y)} \times \mathbb{R}_t$ be endowed with the following non-commutative group law

$$(x, y, t) \circ (x', y', t') = \Big(x + x', y + y', t + t' + 2\sum_{j=1}^{n} a_j(x'_j y_j - x_j y'_j)\Big), \tag{4.1}$$

where $x = (x_1, \dots, x_n)$, $x' = (x'_1, \dots, x'_n)$, $y = (y_1, \dots, y_n)$, $y' = (y'_1, \dots, y'_n)$ and $a_1, \dots, a_n$ are positive constants.

The Lie group $\mathbf{H}_n = (\mathbb{R}^{2n+1}, \circ)$ is called the $2n + 1$ dimensional Heisenberg group. In this chapter we shall deal with the geometry induced by the following $2n$ vector fields

$$X_j = \frac{\partial}{\partial x_j} + 2a_j y_j \frac{\partial}{\partial t}, \qquad Y_j = \frac{\partial}{\partial y_j} - 2a_j x_j \frac{\partial}{\partial t}, \qquad j = 1, \dots, n. \tag{4.2}$$

The vector fields (4.2) are linearly independent, because the matrix coefficients

$$\begin{pmatrix} I_n & O_n & 2a\mathbf{y} \\ O_n & I_n & -2a\mathbf{x} \end{pmatrix} \tag{4.3}$$

has rank $2n$, where

$$\mathbf{x} = \begin{pmatrix} x_1 \\ \dots \\ x_n \end{pmatrix}, \qquad \mathbf{y} = \begin{pmatrix} y_1 \\ \dots \\ y_n \end{pmatrix}.$$

Since the vector fields $X_1, \dots, X_n, Y_1, \dots, Y_n$ and their first brackets

$$[X_j, Y_j] = -4a_j \frac{\partial}{\partial t}$$

span $T(\mathbb{R}^{2n+1})$, then the induced geometry is step 2 at every point .

Let $(a, b, s) \in \mathbf{H}_n$ be arbitrary fixed. The left translation $L_{(a,b,s)} : \mathbf{H}_n \to \mathbf{H}_n$ is defined as

$$L_{(a,b,s)}(x, y, t) = (a, b, s) \circ (x, y, t).$$

The left translation of function f on $\mathbf{H}_n$ is defined by

$$\Big(L_{(a,b,s)} f\Big)(x, y, t) = f\Big((a, b, s) \circ (x, y, t)\Big), \quad \forall (x, y, t) \in \mathbf{H}_n. \tag{4.4}$$

LEMMA 4.1. *The vector fields X_i, Y_i, $i = 1, \dots, n$ are left invariant, i.e.*

$$\begin{aligned} X_j(L_{(a,b,s)}f) &= L_{(a,b,s)}(X_jf), \\ Y_j(L_{(a,b,s)}f) &= L_{(a,b,s)}(Y_jf), \qquad \forall f \text{ function.} \end{aligned}$$

PROOF. An explicit application of the chain rule yields

$$\begin{aligned} X_j(L_{(a,b,s)}f)(x,y,t) &= \Big(\frac{\partial}{\partial x_j} + 2a_jy_j\frac{\partial}{\partial t}\Big) f\big((a,b,s)\circ(x,y,t)\big) \\ &= \Big(\frac{\partial}{\partial x_j} + 2a_jy_j\frac{\partial}{\partial t}\Big) f\Big(a+x, b+y, s+t+2\sum_i a_i(b_ix_i - y_ia_i)\Big) \\ &= f_{x_j}\big((a,b,s)\circ(x,y,t)\big) + 2a_jb_jf_t\big((a,b,s)\circ(x,y,t)\big) \\ &\quad +2a_jy_jf_t\big((a,b,s)\circ(x,y,t)\big) \\ &= f_{x_j}\big((a,b,s)\circ(x,y,t)\big) + 2a_j(b_j+y_j)f_t\big((a,b,s)\circ(x,y,t)\big) \\ &= L_{(a,b,s)}\Big(f_{x_j}(x,y,t) + 2y_jf_t(x,y,t)\Big) \\ &= L_{(a,b,s)}(X_jf)(x,y,t). \end{aligned}$$

The proof of the second relation is similar. □

It follows that we need to study the subRiemannian geodesics starting from the origin only. A left translation will provide geodesics starting at any other points.

4.2. Hamiltonian mechanics on $\mathbf{H}_n$

We shall relabel the vector fields (4.2) as follows

$$X_{2j-1} = \frac{\partial}{\partial x_{2j-1}} + 2a_jx_{2j}\frac{\partial}{\partial t}, \quad X_{2j} = \frac{\partial}{\partial x_{2j}} - 2a_jx_{2j-1}\frac{\partial}{\partial t}. \tag{4.5}$$

Setting

$$\Lambda = diag[\Lambda_1, \dots, \Lambda_n], \qquad \Lambda_j = \begin{pmatrix} 0 & 2a_j \\ -2a_j & 0 \end{pmatrix},$$

we have

$$\begin{aligned} \Lambda + \Lambda^T &= 0, \\ \Lambda^2 &= diag(-4a_1^2, -4a_1^2, -4a_2^2, \dots, -4a_n^2). \end{aligned}$$

The vector fields (4.5) may be denoted collectively by

$$X = \frac{\partial}{\partial x} + \Lambda x\frac{\partial}{\partial t},$$

and the group law (4.1) may be written in the form

$$(x,t)\circ(x',t') = (x+x', t+t'+\langle\Lambda x, x\rangle),$$

where

$$\langle x, y\rangle = \sum_{j=1}^{2n} x_jy_j.$$

We consider the Hamiltonian mechanics associated to the symbol of $-\Delta_H$:

$$\begin{aligned} H(x,\xi,\theta) &= \frac{1}{2}\sum_{j=1}^{n}[(\xi_{2j-1}+2a_jx_{2j}\theta)^2+(\xi_{2j}-2a_jx_{2j-1}\theta)^2] \\ &= \frac{1}{2}\langle \xi+\theta\Lambda x, \xi+\theta\Lambda x\rangle \\ &= \frac{1}{2}\langle \zeta,\zeta\rangle, \end{aligned}$$

where $\zeta = \xi + \theta\Lambda x$. In this notation the Hamiltonian equations for a curve $(x(s), t(s), \xi(s), \theta(s))$ take the form:

$$\begin{aligned} \dot{x}(s) &= \frac{\partial H}{\partial \xi} = \zeta(s), \\ \dot{\xi}(s) &= -\frac{\partial H}{\partial x} = \theta\Lambda\zeta(s), \\ \dot{t}(s) &= \frac{\partial H}{\partial t} = \langle \zeta(s), \Lambda x(s)\rangle, \\ \dot{\theta}(s) &= \frac{\partial H}{\partial t} = 0, \ \ i.e., \ \theta(s)=\theta(0), \end{aligned} \tag{4.6}$$

where the dot denotes d/ds. We let the parameter s run from 0 to $\tau > 0$. Because of the group invariance we shall study the paths relative to the origin and a point (x, t). The boundary conditions are

$$x(0)=0, \qquad x(\tau)=x, \qquad t(\tau)=t, \qquad \theta(s)=\theta. \tag{4.7}$$

The solutions $(x(s), t(s))$ of the Hamiltonian's system satisfying the boundary conditions (4.7) will be called subRiemannian geodesics.

As usual, the Hamiltonian itself is constant along a curve satisfying (4.6):

$$H(x(s),\xi(s),\theta) \equiv H_0 \equiv \frac{1}{2}\langle \zeta(0),\zeta(0)\rangle. \tag{4.8}$$

The following theorem provides the subRiemannian geodesics on $\mathbf{H}_n$, see Beals Gaveau and Greiner [**3**].

THEOREM 4.2. *The solution of* (4.6) *with the boundary conditions* (4.7) *is*

$$x(s) = e^{(s-\tau)\theta\Lambda}\frac{\sinh(s\theta\Lambda)}{\sinh(\tau\theta\Lambda)}x; \tag{4.9}$$

$$\begin{aligned} t-t(s) &= \frac{\tau-s}{2\theta}\langle\frac{(\theta\Lambda)^2}{\sinh^2(\tau\theta\Lambda)}x,x\rangle - \frac{1}{4\theta}\langle\frac{[\sinh(2\tau\theta\Lambda)-\sinh(2s\theta\Lambda)]\theta\Lambda}{\sinh^2(\tau\theta\Lambda)}x,x\rangle \\ &= (\tau-s)\sum_{j=1}^{n}\frac{2a_j^2\theta}{\sin^2(2a_j\tau\theta)}r_j^2 \\ &\quad -\sum_{j=1}^{n}\frac{a_j}{2}\frac{\sin(4a_j\tau\theta)-\sin(4a_js\theta)}{\sin^2(2a_j\tau\theta)}r_j^2, \quad r_j^2 = x_{2j-1}^2+x_{2j}^2. \end{aligned} \tag{4.10}$$

The value of the Hamiltonian H on this path is:

$$H_0 = \frac{1}{2}\langle\frac{(\theta\Lambda)^2}{\sinh^2(\tau\theta\Lambda)}x,x\rangle = \sum_{j=1}^{n}\frac{2a_j^2\theta^2}{\sin^2(2a_j\tau\theta)}r_j^2. \tag{4.11}$$

PROOF. The equations (4.6) imply that

$$\dot{\zeta}(s) = \dot{\xi}(s) + \theta\Lambda\dot{x}(s) = 2\theta\Lambda\zeta(s),$$

so one can solve for $x(s)$ as a function of x, τ and θ, and then solve for $t(s)$ as a function of x, t, τ and θ. We have

$$\zeta(s) = e^{2s\theta\Lambda}\zeta(0), \tag{4.12}$$

$$x(s) = \int_0^s \zeta(r)\,dr = (2\theta\Lambda)^{-1}\{e^{2s\theta\Lambda} - I\}\zeta(0)$$

$$= (2\theta\Lambda)^{-1}\{\zeta(s) - \zeta(0)\}, \tag{4.13}$$

$$\begin{aligned} t(s) &= t(0) + \int_0^s \langle\zeta(r), \Lambda x(r)\rangle\,dr = t(0) + \frac{1}{2\theta}\int_0^s \langle\zeta(r), \zeta(r) - \zeta(0)\rangle\,dr \\ &= t(0) + \frac{s}{\theta}H_0 - \frac{1}{2\theta}\int_0^s \langle\dot{x}(r), \zeta(0)\rangle\,dr \\ &= t(0) + \frac{s}{\theta}H_0 - \frac{1}{2\theta}\langle x(s), \zeta(0)\rangle. \end{aligned} \tag{4.14}$$

Then

$$\zeta(0) = 2\theta\Lambda(e^{2\tau\theta\Lambda} - I)^{-1}x = \frac{\theta\Lambda e^{-\tau\theta\Lambda}}{\sinh(\tau\theta\Lambda)}x, \quad x(s) = e^{(s-\tau)\theta\Lambda}\frac{\sinh(s\theta\Lambda)}{\sinh(\tau\theta\Lambda)}x. \tag{4.15}$$

We note that in the case $\theta = 0$ we have

$$\zeta(s) = \zeta(0), \qquad x(s) = \zeta(0)s \quad \text{and} \quad t(s) = t(0).$$

Using that in any expression

$$\langle F(\Lambda)x, x\rangle$$

we may replace F by its even part

$$\frac{1}{2}[F(\Lambda) + F(\Lambda^t)] = \frac{1}{2}[F(\Lambda) + F(-\Lambda)].$$

yields

$$\langle x(s), \zeta(0)\rangle = \frac{1}{2}\langle\frac{\sinh(2s\theta\Lambda)}{\sinh^2(\tau\theta\Lambda)}\theta\Lambda x, x\rangle. \tag{4.16}$$

Using

$$\exp\begin{pmatrix} 0 & b \\ -b & 0 \end{pmatrix} = \begin{pmatrix} \cos b & \sin b \\ -\sin b & \cos b \end{pmatrix}, \tag{4.17}$$

relations (4.8), (4.12) - (4.16) yield (4.9) - (4.11). □

REMARK 4.3. It is interesting to note that $\tau \neq 0$ implies that $t(0) \neq 0$ as long as $(x, t) \neq (0, 0)$.

COROLLARY 4.4. We have

$$t - t(0) = \sum_{j=1}^n a_j\mu(2a_j\tau\theta)r_j^2,$$

where we have set $r_j^2 = x_{2j-1}^2 + x_{2j}^2$ and $\mu(\varphi) = \frac{\varphi}{\sin^2\varphi} - \cot\varphi$.

4.3. The classical action

The *action integral* associated to the Hamiltonian curve is

$$S(x,t,\tau;\theta) = \int_0^\tau \{\langle \xi(s), \dot{x}(s)\rangle + \theta \dot{t}(s) - H(x(s), \xi(s), \theta)\}\, ds.$$

Because H is homogeneous of degree 2 with respect to (ξ, θ), we observe that

$$S = \int_0^\tau \{\langle \xi, \frac{\partial H}{\partial \xi}\rangle + \theta\frac{\partial H}{\partial \theta} - H\}\, ds = \int_0^\tau (2H - H)\, ds = \tau H_0,$$

where H_0 is the constant value of the Hamiltonian along the solution. Using Theorem 4.2 we obtain various forms for the action.

THEOREM 4.5. *The action integral* $S(x,t,\tau;\theta)$ *is given by:*

$$\begin{aligned} S(x,t,\tau;\theta) &= \frac{1}{2}\langle \tau(\theta\Lambda)^2[\sinh(\tau\theta\Lambda)]^{-2}x, x\rangle \\ &= [t - t(0)]\theta + \frac{1}{2}\langle \theta\Lambda\coth(\tau\theta\Lambda)x, x\rangle \\ &= \sum_{j=1}^n \frac{\tau(2a_j\theta)^2}{2\sin^2(2a_j\tau\theta)} r_j^2, \qquad \theta \in [0, \pi/2a_n), \\ &= [t - t(0)]\theta + \sum_{j=1}^n a_j\theta\cot(2a_j\tau\theta) r_j^2. \end{aligned}$$

4.4. The horizontal distribution

We shall work with the skew-symmetric model

$$X = \partial_x + \Lambda x \partial_t,$$

which means

$$X_j = \partial_{x_j} + 2\sum_{k=1}^{2n} a_{jk}x_k\partial_t, \qquad j = 1, \dots, 2n.$$

The horizontal distribution is defined by $\mathcal{H} = span\{X_1, \dots, X_{2n}\}$. A curve is called *horizontal* if its velocity vector belongs to the horizontal distribution. Let $c(s) = (x_1(s), x_2(s), \dots, x_{2n}(s), t(s))$ be a curve in $\mathbb{R}^{2n+1}$. The velocity vector field is

$$\begin{aligned} \dot{c} &= (\dot{x}_1, \dots, \dot{x}_{2n}, \dot{t}) = \sum_{j=1}^{2n} \dot{x}_j\partial_{x_j} + \dot{t}\partial_t \\ &= \langle \dot{x}, \partial_x\rangle + \dot{t}\partial_t = \langle \dot{x}, \partial_x + \Lambda x\partial_t - \Lambda x \partial_t\rangle + \dot{t}\partial_t \\ &= \langle \dot{x}, X\rangle - \langle \dot{x}, \Lambda x\rangle\partial_t + \dot{t}\partial_t. \end{aligned} \tag{4.18}$$

We arrive at the following result:

LEMMA 4.6. *A curve* $c = (x,t)$ *in* $\mathbb{R}^{2n+1}$ *is horizontal if and only if*

$$\dot{t} = \langle \dot{x}, \Lambda x\rangle. \tag{4.19}$$

PROOF. Since $\langle \dot{x}, X\rangle = \sum \dot{x}_j X_j \in \mathcal{H}$, the equation (4.18) yields

$$\dot{c} = \langle \dot{x}, X\rangle + (\dot{t} - \langle \dot{x}, \Lambda x\rangle)\partial_t.$$

Since $\partial_t \notin \mathcal{H}$, it follows that $\dot{c}$ is horizontal if and only if the coefficient of ∂_t vanishes, *i.e.* equation (4.19) holds. □

COROLLARY 4.7. Let

$$\omega = dt - \langle dx, \Lambda x\rangle = dt - \sum_{j,k} 2a_{jk} x_k \, dx_j.$$

Then $\omega(\partial_t) = 1$, $\omega(X_j) = 0$, $\forall j = 1, \dots, 2n$, i.e. $\mathcal{H} = \ker \omega$.

The 1-form ω is called the one-connection form. The following result provides a relation between the subRiemannian geodesics and the horizontal curves.

PROPOSITION 4.8. i) Any subRiemannian geodesic is a horizontal curve.
ii) There are horizontal curves, which are not subRiemannian geodesics.

PROOF. *i)* Let $c(s) = (x(s), t(s))$ be a subRiemannian geodesic. From the first equation of the Hamiltonian system (4.6) we have $\dot{x}(s) = \zeta(s)$. Substituting in the third equation of (4.6) yields

$$\dot{t}(s) = \langle \zeta(s), \Lambda x(s)\rangle = \langle \dot{x}(s), \Lambda x(s)\rangle,$$

which is (4.19). Hence $c(s) = (x(s), t(s))$ is horizontal.

ii) Set $x_{2j-1}(s) = s^2/2$, $x_{2j}(s) = s$, $j = 1, \dots, n$ and $t(s) = \frac{s^3}{3}\sum_{j=1}^n a_j$. We shall show that $c(s) = (x_1(s), \dots, x_{2n}(s), t(s))$ is a horizontal curve, which is not a geodesic. We have

$$\begin{aligned} \langle \dot{x}, \Lambda x\rangle &= 2a_1(x_2\dot{x}_1 - x_1\dot{x}_2) + 2a_2(x_4\dot{x}_3 - x_3\dot{x}_4) + \cdots + 2a_n(x_{2n}\dot{x}_{2n-1} - x_{2n-1}\dot{x}_{2n}) \\ &= 2a_1(s^2 - \frac{s^2}{2}) + 2a_2(s^2 - \frac{s^2}{2}) + \cdots + 2a_n(s^2 - \frac{s^2}{2}) \\ &= s^2\sum_{j=1}^n a_j = \dot{t}(s), \end{aligned}$$

and Lemma 4.6 yields $c = (x, t)$ horizontal curve.

Since the Hamiltonian

$$H = \frac{1}{2}|\dot{c}(s)|^2 = \frac{1}{2}\sum \dot{x}_j^2(s) = \frac{n}{2}(s^2 + 1)$$

is not constant along the curve, the curve cannot be a subRiemannian geodesic. □

4.5. The Carnot-Carathéodory distance

We obtain classical Hamiltonian mechanics by replacing the boundary condition $\theta(s) = \theta$ with the condition $t(0) = 0$, and setting $\tau = 1$. Then necessarily

$$t = \sum_{j=1}^n a_j \mu(2a_j\theta) r_j^2.$$

The behavior of the term on the right, as function of θ depends on both r_j and the distribution of the a_j. In the *isotropic* case $a_1 = a_2 = \cdots = a_n$, the results of Chapter 1 carry over with no change, except that each family of geodesics from $(0,0)$ to $(0,t)$ is parametrized by the $(2n-1)$-sphere.

We shall adopt the following convention regarding the function μ:

$$\mu(m\pi) = +\infty, \qquad m = 1,2,3,\dots, \text{and } 0\cdot\infty = 0. \tag{4.20}$$

In particular this means that

$$\sum_{j=1}^{n} a_j\mu(2a_j\pi/2a_n)r_j^2 < +\infty \text{ if and only if } \; r_j = 0 \; \text{ when } \; a_j = a_n.$$

In order to deal with the non-isotropic case, we assume that there is an index p with the property

$$0 < a_1 \le a_2 \le \cdots \le a_p < a_{p+1} = \cdots = a_n,$$

and set

$$x = (x', x''), \quad x'' = (x_{2p+1}, x_{2p+2}, \dots, x_{2n}).$$

The following treatment of the non-isotropic case can be found in Beals, Gaveau and Greiner [**3**].

THEOREM 4.9. *Suppose $x'' \ne 0$. Then there are finitely many geodesics from $(0,0)$ to (x,t). The geodesics are indexed by the solutions of*

$$|t| = \sum_{j=1}^{n} a_j\mu(2a_j\theta)r_j^2. \tag{4.21}$$

In particular, the Carnot-Carathédory distance from (x,t) to the origin is

$$d(x,t)^2 = 2S(x,|t|,1;\theta_c), \tag{4.22}$$

where θ_c is the unique solution on (4.21) *in the interval* $[0,\pi/2a_n)$.

PROOF. The second part comes from the fact that the subRiemannian geodesics are traces of Hamiltonian paths, see Bismut [**10**] and Strichartz [**50**]. The associated action is one-half the square of the length.
If $x'' \ne 0$, the right side of (4.21) is strictly increasing on $[0,\pi/2a_n)$, so there is an unique solution θ_c of (4.22) in the interval $[0,\pi/2a_n)$. In every other pole free interval the function on the right side of (4.21) decreases from $+\infty$ to a positive minimum at θ_m, $m = 1,2,\dots$, then it increases to $+\infty$. Furthermore the right side of (4.21) is bounded from below by a straight line through the origin with a positive slope, since this is true for the individual summands. Hence there are finitely many solutions of (4.21). □

REMARK 4.10. The minima of the right side of (4.21) do not necessarily increase with m. For instance in the case of $\mathbf{H}_2$, choosing $0 < a_1 < a_2$ with $a_2 - a_1$ small enough, we can arrange that

$$\sum_{j=1}^{2} a_j\mu(2a_j\theta_1)r_j^2 > \sum_{j=1}^{2} a_j\mu(2a_j\theta_2)r_j^2.$$

REMARK 4.11. The lengths of the geodesics associated to a solution increases with θ. We have done this in the case of $\mathbf{H}_1$ in Chapter 1 and it can be done directly for the isotropic $\mathbf{H}_n$, since the dependence in (4.21) is much simpler when the a_j are identical. In the non-isotropic case the behavior depends crucially on the relative sizes of the a_j and on the relative sizes of the r_j. This case will be approached in Chapter 5 by means of the modified complex action f.

In the following we shall study the geodesics when $x'' = 0$. We shall do a careful enumeration of all geodesics connecting $(0,0)$ and (x,t). For this, we need a few preliminary results.

LEMMA 4.12. *Suppose $x' \neq 0$, $x'' = 0$ and $t > 0$. Then there exist a finite number of geodesics connecting $(0,0)$ and (x,t) which are indexed by the solutions of*

$$t = \sum_{j=1}^{n} a_j \mu(2a_j\theta) r_j^2, \tag{4.23}$$

and their lengths increase with θ. The square of the length of the geodesic associated with a solution of (4.23) is $2S(x,t,1;\theta)$.

PROOF. If we replace a_n by the largest a_j such that $r_j \neq 0$, then the lemma is an immediate consequence of Theorem 4.9 and Remark 4.11. If $\theta_1 < \theta_2 \leq \ldots$ denote the solutions of (4.23), then $0 < \theta_1 < \pi/2a_1$. □

Next we shall find the geodesics which are not associated with a solution θ of (4.23). We assumed $x' \neq 0$. Then $t > 0$ yields $\theta > 0$, and then

$$\zeta_j(0) = \zeta_j(1), \quad j = 2p+1, \ldots, 2n. \tag{4.24}$$

Therefore (4.12) and (4.17) imply

$$\begin{bmatrix} \cos(4a_n\theta) & \sin(4a_n\theta) \\ -\sin(4a_n\theta) & \cos(4a_n\theta) \end{bmatrix} = \begin{bmatrix} 1 & 0 \\ 0 & 1 \end{bmatrix}, \tag{4.25}$$

so we have

$$2a_n\theta = m\pi, \quad m = 1, 2, \ldots \tag{4.26}$$

Assuming $t(0) = 0$, (4.14) yields

$$\theta t = \frac{1}{2}\|\zeta(0)'\|^2 + \frac{1}{2}\|\zeta(0)''\|^2 - \frac{1}{2}\langle x', \zeta(0)'\rangle. \tag{4.27}$$

Let

$$\zeta^{(j)} = (\zeta_{2j-1}, \zeta_{2j}), \quad x^{(j)} = (x_{2j-1}, x_{2j}), \quad j = 1, 2, \ldots, n. \tag{4.28}$$

Then

$$\frac{1}{2}[\|\zeta^j(0)\|^2 - \langle x^{(j)}, \xi^{(j)}(0)\rangle] = a_j\theta\mu(2a_j\theta)r_j^2. \tag{4.29}$$

Therefore θ has to be a solution of

$$\theta t = \frac{1}{2}\|\zeta(0)''\|^2 + \sum_{j=1}^{n} a_j\theta\mu(2a_j\theta)r_j^2. \tag{4.30}$$

This can be turned around by saying that we may set $t(0) = 0$ for a given θ if we choose $\zeta(0)''$ so that

$$\frac{1}{2}\|\zeta(0)''\|^2 = \theta\Big[t - \sum_{j=1}^{n} a_j\theta\mu(2a_j\theta)r_j^2\Big]. \tag{4.31}$$

Relation (4.14) yields the length $d_m(x,t)$ of such a geodesic:

$$d_m^2(x,t) = 2H_0 = 2t\theta + \langle x, \zeta(0)\rangle = 2\theta\Big[t + \sum_{j=1}^{n} a_j \cot(2a_j\theta)r_j^2\Big]. \tag{4.32}$$

We have proved the following result:

LEMMA 4.13. *Given $(x,t) \in \mathbf{H}_n$, $x' \neq 0$, $x'' = 0$ and $t \neq 0$, let m denote a positive integer, such that*

$$|t| \geq \sum_{j=1}^{n} a_j\mu(2a_j m\pi/2a_n)r_j^2. \tag{4.33}$$

Then there are infinitely many geodesics of the same length $d_m(x,t)$,

$$d_m^2(x,t) = 2S(x,|t|,1;m\pi/2a_n) = 2\frac{m\pi}{2a_n}\Big[|t| + \sum_{j=1}^{n} a_j \cot(2a_j m\pi/2a_n)r_j^2\Big], \tag{4.34}$$

between $(0,0)$ and (x,t). They are parametrized by the sphere

$$\frac{1}{2}\|\zeta(0)''\|^2 = \frac{m\pi}{2a_n}\Big[|t| - \sum_{j=1}^{n} a_j\mu(2a_j m\pi/2a_n)r_j^2\Big], \quad \zeta(0)'' \in \mathbb{R}^{2n-2p}. \tag{4.35}$$

REMARK 4.14. We note that (4.33) implies that $x'' = 0$ and that $a_n \neq a_j m$, when $r_j \neq 0$.

The Carnot-Carathéodory distance $d(x,t)$, *i.e.* the length of the shortest geodesic between $(0,0)$ and (x,t), will turn out to be associated to a θ in the interval $(0,\pi/2a_n]$. A geodesic associated to such a θ always exists. There are two cases:

(i) If we have

$$t < \sum_{j=1}^{n} a_j\mu(2a_j\pi/2a_n)r_j^2, \quad t > 0, \tag{4.36}$$

then the proof of Theorem 4.9 yields the existence of a geodesic with length $d(x,t)$, where

$$d(x,t)^2 = 2\theta\Big[t + \sum_{j=1}^{n} a_j \cot(2a_j\theta)r_j^2\Big], \tag{4.37}$$

is associated with θ, the unique solution of

$$t = \sum_{j=1}^{n} a_j\mu(2a_j\theta)r_j^2 \tag{4.38}$$

in the interval $(0,\pi/2a_n)$.

(ii) If

$$t \geq \sum_{j=1}^{n} a_j \mu(2a_j\pi/2a_n) r_j^2, \tag{4.39}$$

then Lemma 4.13 yields a geodesic with length $d(x,t)$, where

$$\begin{aligned} d(x,t)^2 &= 2\frac{\pi}{2a_n}\Big[t + \sum_{j=1}^{n} a_j \cot(2a_j\pi/2a_n) r_j^2\Big] \\ &= 2\frac{\pi}{2a_n}\Big[t - \sum_{j=1}^{n} a_j \mu(2a_j\pi/2a_n) r_j^2 + \sum_{j=1}^{n} a_j \frac{2a_j\pi/2a_n}{\sin^2(2a_j\pi/2a_n)} r_j^2\Big] \\ &= \|\zeta(0)''\|^2 + \sum_{j=1}^{n} \frac{(2a_j\pi/2a_n)^2}{\sin^2(2a_j\pi/2a_n)} r_j^2. \end{aligned} \tag{4.40}$$

In particular, $d(x,t)$ always has the form (4.37) with the understanding that $\theta = \pi/2a_n$ when (4.39) holds.

LEMMA 4.15. *Let $x' \neq 0$, $x'' = 0$, $t > 0$ and suppose that* (4.33) *holds for some positive integer m. Then*

$$d(x,t) \leq d_m(x,t), \tag{4.41}$$

where $d(x,t) = d_1(x,t)$.

PROOF. Let $\theta'' \in ((k-1)\pi/2a_n, k\pi/2a_n)$ be the solution of (4.38) which increases as $x'' \to 0$. We shall prove that for x'' small enough we have

$$\lim_{x''\to 0} 2S(x', x'', t, 1; \theta'') = d_k(x', 0, t)^2, \tag{4.42}$$

where $k = 1$ or $k = m$.

(i) If $k = 1$ and (4.36) holds, then (4.42) follows from the behavior of the function μ in the interval $[0, \pi/2a_n)$.

(ii) Otherwise we assume that (4.33) holds, and $d_k(x,t)$, $k = 1$ or m, is given by (4.34). With $x'' \neq 0$ we assume that θ'' is a solution of

$$t - \sum_{j=1}^{p} a_j \mu(2a_j\theta'') r_j^2 = a_n \mu(2a_n\theta'') \|x''\|^2 \tag{4.43}$$

in the interval $((k-1)\pi/2a_n, k\pi/2a_n)$, which increases to $k\pi/2a_n$ as $x'' \to 0$. Then

$$\begin{aligned} \lim_{x''\to 0} a_n \mu(2a_n\theta'') \|x''\|^2 &= \lim_{x''\to 0} a_n \frac{2a_n\theta''}{\sin^2(2a_n\theta'')} \|x''\|^2 \\ &= t - \sum_{j=1}^{p} a_j \mu(2a_j k\pi/2a_n) r_j^2. \end{aligned} \tag{4.44}$$

The square of the distance associated with θ'' is given by

$$2S(x', x'', t, 1; \theta'') = 2\sum_{j=1}^{p} \frac{(2a_j\theta'')^2}{2\sin^2(2a_j\theta'')} r_j^2 + 2\frac{(2a_n\theta'')^2}{2\sin^2(2a_n\theta'')} \|x''\|^2. \tag{4.45}$$

Then (4.44) yields

$$\begin{aligned}\lim_{x''\to 0} 2S(x',x'',t,1;\theta'') &= 2\sum_{j=1}^{p}\frac{(2a_jk\pi/2a_n)^2}{2\sin^2(2a_jk\pi/2a_n)}r_j^2\\ &\quad +2\frac{k\pi}{2a_n}\Big[t-\sum_{j=1}^{p}a_j\mu(2a_jk\pi/2a_n)r_j^2\Big]\\ &= 2\frac{k\pi}{2a_n}\Big[t+\sum_{j=1}^{p}a_j\cot(2a_jk\pi/2a_n)r_j^2\Big]\\ &= d_k(x',0,t)^2=d_k(x,t)^2.\end{aligned}$$

Therefore the distances $d_1(x,t)$ and $d_m(x,t)$ with $x''=0$, are limits of distances $d(y,t)$, $y''\neq 0$, as $y''\to 0$. When $y''\neq 0$, the distances are increasing with increasing θ, see Remark 4.11. Hence this is also true for limits which implies (4.41). □

In the following we shall introduce more terminology and notation. Let $-4\alpha_1^2$, $-4\alpha_2^2,\dots,-4\alpha_m^2$ denote the distinct eigenvalues of Λ^2, where $\alpha_1<\alpha_2<\dots a<\alpha_m$. Let V_k denote the eigenspace of Λ^2, which is associated with the eigenvalue $-4\alpha_k^2$. We set

$$W_k=\bigcup_{j=1}^{k}V_j,\qquad R_k^2=\sum_{a_j=a_k}r_j^2. \tag{4.46}$$

The following results deals with all the geodesics between $(0,0)$ and a given point (x,t), $x\neq 0$, $t>0$.

LEMMA 4.16. *Let $x\in W_k$, $x\notin W_{k-1}$. Suppose there is an integer q, $k<q\leq n$, and a positive integer m, such that:*

$$t\geq\sum_{j=1}^{n}a_j\mu(2a_jm\pi/2a_q)r_j^2. \tag{4.47}$$

Then there exist infinitely many geodesics of the same length, connecting $(0,0)$ and (x,t). There is only one geodesic if (4.47) is an equality. This set of geodesics is parametrized by the sphere:

$$\frac{1}{2}\sum_{a_j>\alpha_k,a_jm=\ell a_q}\big(|\zeta_{2j-1}(0)|^2+|\zeta_{2j}(0)|^2\big)=\frac{m\pi}{2a_q}\Big[|t|-\sum_{j=1}^{n}a_j\mu(2a_jm\pi/2a_q)r_j^2\Big], \tag{4.48}$$

where ℓ is an arbitrary positive integer, and the square of their length is given by

$$2S(x,t,1;m\pi/2a_q)=2\frac{m\pi}{2a_q}\Big[|t|+\sum_{j=1}^{n}a_j\cot(2a_jm\pi/2a_q)r_j^2\Big]. \tag{4.49}$$

Moreover, if $m=m_1$ and $m=m_2$ both satisfy (4.47), then $m_1<m_2$ implies

$$2S(x,t,1;m_1\pi/2a_q)\leq 2S(x,t,1;m_2\pi/2a_q). \tag{4.50}$$

PROOF. It suffices to prove only (4.50). This follows from taking the limit of $S(y,t,1;\theta)$ as $y \to x$, where y has a nonzero projection into V_q and θ is a solution of

$$t = \sum_{j=1}^{n} a_j \mu(2a_j\theta) r_j^2, \quad \theta \in ((m-1)\pi/2a_q, m\pi/2a_q), \tag{4.51}$$

which increase to $m\pi/2a_q$ as $y \to x$. □

4.6. The shortest geodesic

Next we shall deal with the shortest geodesic between $(0,0)$ and a given point (x,t), $x \neq 0$, $t > 0$. Suppose $x \in W_k$, $x \notin W_{k-1}$. Among the geodesic provided by Lemma 4.12 the shortest is associated with $\theta \in (0, \pi/2a_k)$. Among the geodesics given by Lemmas 4.13 and 4.16 the shortest occurs when $m = 1$, and it is associated with θ, where $\theta = \pi/2a_q$, $q > k$. Therefore the shortest geodesic is associated with a $\theta \in (0, \pi/2a_k)$.

LEMMA 4.17. *Suppose $x \neq 0$, $x \in W_k$, $x \notin W_{k-1}$. Then the shortest geodesic connecting $(0,0)$ and (x,t) is associated with a $\theta \in (0, \pi/2a_n]$, and its length, the Carnot-Carathéodory distance of (x,t) from the origin, is given by $d(x,t)$, where:*

$$d(x,t)^2 = 2\theta\Big[|t| + \sum_{j=1}^{n} a_j \cot(2a_j\theta) r_j^2\Big]. \tag{4.52}$$

PROOF. (i) Suppose $|t| > 0$ is large, so that (4.33) holds with $m = 1$, and also (4.50) holds with $m = 1$ and with $q > k$. We need to show that

$$2S(x,t,1;\pi/2a_n) < 2S(x,t,1;\pi/2a_q),$$

that is

$$\begin{aligned} &2\frac{\pi}{2a_n}|t| + \sum_{j=1}^{n}\Big[\frac{(2a_j\pi/2a_n)^2}{\sin^2(2a_j\pi/2a_n)} - \frac{2a_j\pi}{2a_n}\mu(2a_j\pi/2a_n)\Big] r_j^2 \\ &< 2\frac{\pi}{2a_q}|t| + \sum_{j=1}^{n}\Big[\frac{(2a_j\pi/2a_q)^2}{\sin^2(2a_j\pi/2a_q)} - \frac{2a_j\pi}{2a_q}\mu(2a_j\pi/2a_q)\Big] r_j^2, \end{aligned} \tag{4.53}$$

see (4.40). Moving $(2\pi/2a_n)|t|$ to the right side and replacing $|t|$ by the smaller quantity

$$\sum_{j=1}^{n} a_j \mu(2a_j\pi/2a_q) r_j^2.$$

Then (4.53) follows from

$$\frac{(2a_j\pi/2a_n)^2}{\sin^2(2a_j\pi/2a_n)} - \frac{2a_j\pi}{2a_n}\mu(2a_j\pi/2a_n) < \frac{(2a_j\pi/2a_q)^2}{\sin^2(2a_j\pi/2a_q)} - \frac{2a_j\pi}{2a_n}\mu(2a_j\pi/2a_q). \tag{4.54}$$

Let $\alpha = 2a_j\pi/2a_n$ and $x = 2a_j\pi/2a_q$. Then (4.54) takes the form:

$$\frac{\alpha^2 - \alpha(\alpha - \sin\alpha\cos\alpha)}{\sin^2\alpha} < \frac{x^2 - \alpha(x - \sin x \cos x)}{\sin^2\alpha}, \quad \alpha < x. \tag{4.55}$$

We set

$$\psi(x) = \frac{2x^2 - \alpha(2x - \sin 2x)}{2\sin^2 x} \tag{4.56}$$

and note that

$$\psi'(x) = (x - \alpha)\mu'(x) > 0 \quad \text{if} \quad \alpha < x < \pi.$$

This proves (4.55).

(ii) The only other possibility which needs to be considered is that (4.33) holds with $m = 1$, and there is a $\theta \in (\pi/2a_n, \pi/2a_k)$ such that

$$|t| = \sum_{j=1}^{n} a_j \mu(2a_j\theta) r_j^2. \tag{4.57}$$

In this case we need to show that

$$2\frac{\pi}{2a_n}|t| + \sum_{j=1}^{n}\Big[\frac{(2a_j\pi/2a_n)^2}{\sin^2(2a_j\pi/2a_n)} - \frac{2a_j\pi}{2a_n}\mu(2a_j\pi/2a_n)\Big]r_j^2 < \sum_{j=1}^{n}\frac{(2a_j\theta)^2}{\sin^2(2a_j\theta)}r_j^2,$$

which may be written as

$$\begin{aligned}&\sum_{j=1}^{n}\Big[\frac{(2a_j\pi/2a_n)^2}{\sin^2(2a_j\pi/2n)} - \frac{2a_j\pi}{2a_n}\mu(2a_j\pi/2a_n)\Big]r_j^2\\ &< \sum_{j=1}^{n}\Big[\frac{(2a_j\theta)^2}{\sin^2(2a_j\theta)} - \frac{2a_j\pi}{2a_n}\mu(2a_j\theta)\Big]r_j^2.\end{aligned} \tag{4.58}$$

This follows from (4.55), and hence the lemma is proved. □

COROLLARY 4.18. Let $x = 0$, $t \neq 0$, and fix a pair of integers (j, m), $j = 1, 2, \ldots, n$, $m = 1, 2, 3, \ldots$. There are infinitely many geodesics of the same length $d_{jm}(0, t)$ between $(0, 0)$ and $(0, t)$, where

$$d_{jm}(0,t)^2 = \frac{2m\pi|t|}{2a_j}.$$

The Carnot-Caratheodory distance of $(0, t)$ from the origin is

$$d(0,t)^2 = d_{n1}(0,t)^2 = \frac{\pi|t|}{a_n}.$$

REMARK 4.19. We note that different pairs (j, m) may yield the same geodesic.

The previous lemmas lead to the following result:

THEOREM 4.20. *Suppose $x' \neq 0$, $x'' = 0$.*
(i) If

$$|t| < \sum_{j+1}^{n} a_j \mu(2a_j\pi/2a_n) r_j^2, \tag{4.59}$$

then the Carnot-Caratheodory distance $d(x, t)$ is given by

$$d(x,t)^2 = 2S(x,t,1;\theta) = 2\theta\Big[|t| + \sum_{j=1}^{n} a_j \cot(2a_j\theta) r_j^2\Big], \tag{4.60}$$

where θ is the unique solution of

$$|t| = \sum_{j+1}^{n} a_j \mu(2a_j\theta) r_j^2$$

in the interval $(0, \pi/2a_n)$.
(ii) If (4.59) fails, then there are infinitely many shortest geodesics of the same length from $(0,0)$ to (x,t), parametrized by the sphere

$$\frac{1}{2}\|\zeta''(0)\| = \frac{\pi}{2a_n}\Big[|t| - \sum_{j=1}^{n} a_j \mu(2a_j\pi/2a_n) r_j^2\Big], \tag{4.61}$$

where we set $\zeta = (\zeta', \zeta'')$, $\zeta'' = (\zeta_{2p+1}, \dots, \zeta_{2p})$. The length of these geodesics yield the Carnot-Caratheodory distance of (x,t) from the origin:

$$\begin{aligned} d(x,t)^2 &= 2S(s,t,1;\pi/2a_n) \\ &= 2\pi/2a_n\Big[|t| + \sum_{j=1}^{n} a_j \cot(2a_j\pi/2a_n) r_j^2\Big]. \end{aligned} \tag{4.62}$$

The distance $d(x,t)$ is homogeneous of degree 1 with respect to the dilations

$$(x,t) \to (\lambda x, \lambda^2 t), \quad \lambda > 0, \tag{4.63}$$

it is bounded above and below by multiples of $|x| + |t|^{1/2}$ and it is a continuous function of (x,t), although it is not differentiable.

PROOF. We note that it suffices to prove the result for $t > 0$, since for $t < 0$ we simply switch to $-\pi/2a_n$ in our arguments. We also note that formulas (4.61) and (4.62) are justified by the convention (4.20), since $r_{p+1} = \dots = r_n = 0$. Lemma 4.12 proves the statement (4.62), and Lemmas 4.47, 4.15, 4.16 and 4.17 imply (4.61) and (4.62). To prove that $d(x,t)$ is continuous on $\mathbf{H}_n$ we only need to show that

$$\lim_{x''\to 0} d(x', x'', t) = d(x', 0, t). \tag{4.64}$$

We obtain the above limit from the part (ii) of the proof of Lemma 4.15, if we set $k = 1$.
Homogeneity with respect to the dilations (4.63) is a consequence of the fact that $\theta_c(x,t)$ is homogeneous of degree 0. The quotient $d(x,t)/(|x| + |t|^{1/2})$ is homogeneous of degree 0 and positive on $H_n \backslash \{0\}$, so it is bounded above and below. Thus to finish the proof we are left with showing that $d(x,t)$ is not differentiable. Let

$$E_p(x,t) = |t| - \sum_{j=1}^{n} a_j \mu(2a_j\pi/2a_n) r_j^2 \tag{4.65}$$

denote the *excess* function defined on

$$H_p = \{(x,t) \in \mathbf{H}_n, x = (x', 0)\}. \tag{4.66}$$

Lemmas 4.13 and 4.16 show that for points in H_p, the initial momentum $\xi(0)$ is not uniquely defined, and we obtain a multitude of bicharacteristics. We may refer to such points as *caustic* points.

We need the following result:

LEMMA 4.21. *All those first derivatives of $S(x,t,1;\theta_c)$ which are normal to H_p are discontinuous on $\{(x,t) \in \mathbf{H}_n; E_p(x,t) > 0\}$.*

PROOF. It suffices to prove the lemma for $t > 0$, which we shall assume in the rest of the proof. Let $(x, t) \in H_p$ and let h denote one of the variables $x_{2p+1}, \dots, x_{2n}$. Then S is given by

$$\begin{aligned} S(h) &= S(x_1, \dots, x_{2p}, 0, \dots, 0, h, 0, \dots, 0, t, 1, \theta_h) \\ &= \frac{1}{2} \sum_{j=1}^{p} \frac{(2a_j\theta_h)^2}{\sin^2(2a_j\theta_h)} r_j^2 + \frac{1}{2} \frac{(2a_j\theta_h)^2}{\sin^2(2a_j\theta_h)} h^2, \end{aligned} \tag{4.67}$$

where

$$t = \sum_{j=1}^{p} a_j \mu(2a_j\theta_h) r_j^2 + a_n \mu(2a_h\theta_h) h^2. \tag{4.68}$$

We shall calculate $\lim_{h\to 0} S'(h)$. Using

$$\frac{d}{d\varphi}\left(\frac{\varphi^2}{\sin^2\varphi}\right) = \varphi\mu'(\varphi)$$

we obtain

$$\begin{aligned} S'(h) &= 2\theta_h \Big[\sum_{j=1}^{p} a_j^2 \mu'(2a_j\theta_h) \frac{d\theta_h}{dh} r_j^2 + a_n \mu(2a_n\theta_h) h \\ &\quad + a_n \cot(2a_n\theta_h) h + a_n^2 \mu(2a_n\theta_h) \frac{d\theta_h}{dh} h^2 \Big]. \end{aligned} \tag{4.69}$$

Differentiating (4.68) with respect to h, yields

$$0 = \sum_{j=1}^{p} a_j^2 \mu'(2a_j\theta_h) \frac{d\theta_h}{dh} r_j^2 + a_n \mu(2a_n\theta_h) h + a_n^2 \mu'(2a_n\theta_h) \frac{d\theta_h}{dh} h^2. \tag{4.70}$$

Substituting (4.70) into (4.69), we find

$$S'(h) = 2a_n\theta_h \cot(2a_n\theta_h) h. \tag{4.71}$$

Since $h \to 0$, $\theta_h \to \pi/2a_n$, then (4.67) we have:

$$\begin{aligned} \lim_{h\to 0} \left[\frac{2a_n\theta_h}{\sin(2a_n\theta_h)} h\right] &= \lim_{h\to 0} \left(2S(h) - \sum_{j=1}^{p} \frac{(2a_j\theta_h)^2}{\sin^2(2a_j\theta_h)} r_j^2 \right) \\ &= 2\frac{\pi}{2a_n} \Big[t + \sum_{j=1}^{p} a_j \cot(2a_j\pi/2a_n) r_j^2 \Big] - \sum_{j=1}^{p} \frac{(2a_j\pi/2a_n)^2}{\sin^2(2a_j\pi/2a_n)} r_j^2 \\ &= \frac{\pi}{a_n} \Big[t - \sum_{j=1}^{p} a_j \mu(2a_j\pi/2a_n) r_j^2 \Big] = \frac{\pi}{a_n} E_p. \end{aligned} \tag{4.72}$$

Since θ_h is symmetric in h, yields

$$\lim_{h\to 0\pm} \frac{dS(h)}{dh} = \pm \left(\frac{\pi E_p}{a_n}\right)^{1/2} \neq 0, \tag{4.73}$$

when $E_p > 0$. Hence $S'(h)$ is discontinuous at $h = 0$. This proves Lemma 4.21. □

With Lemma 4.21 we have completed the proof of Theorem 4.20. □

4.7. Caustics

Consider the set of *caustics*:

$$C_p = \{(x,t) \in H_p; E_p(x,t) > 0\}. \tag{4.74}$$

Let $S_p(x,t)$ be the classical action on H_p:

$$S_p(x,t) = \theta_p|t| + \sum_{j=1}^{p} a_j\theta_p \cot(2a_j\theta_p)r_j^2, \qquad (x,t) \in H_p, \tag{4.75}$$

whenever $\theta_p \in (0, \pi/2a_p)$ is a solution of

$$|t| = \sum_{j=1}^{p} a_j\mu(2a_j\theta_p)r_j^2; \tag{4.76}$$

otherwise we extend (4.75) to all of H_p by continuity. Then (ii) in the proof of Lemma 4.17 implies that

$$S_p(x,t) > S(x,t), \qquad \text{if } (x,t) \in H_p \text{ and } E_p(x,t) > 0. \tag{4.77}$$

On the other hand we always have

$$S_p(x,t) = S(x,t), \qquad \text{if } (x,t) \in H_p \text{ and } E_p(x,t) < 0. \tag{4.78}$$

Consequently,

$$S_p(x,t) \geq S(x,t), \qquad (x,t) \in H_p. \tag{4.79}$$

Suppose that

$$0 < a_1 \leq \cdots \leq a_{p_2} < a_{p_2+1} = \cdots = a_{p_1} < a_{p_1+1} = \cdots = a_p < \ldots.$$

We set

$$H_{p_1} = \{(x,t) \in \mathbf{H}_n; \sum_{j=p_1+1}^{n} r_j^2 = 0\}, \tag{4.80}$$

$$E_{p_1}(x,t) = |t| - \sum_{j=1}^{p_1} a_j\mu(2a_j\pi/2a_p)r_j^2, \quad (x,t) \in H_{p_1}, \tag{4.81}$$

and

$$C_{p_1} = \{(x,t) \in H_{p_1}; E_{p_1}(x,t) > 0\}. \tag{4.82}$$

Since $E_p(x,t) > E_{p_1}(x,t)$ on H_{p_1} we obtain

$$C_{p_1} \subset C_p. \tag{4.83}$$

The points of C_{p_1} are more caustic than the points of $C_p \backslash C_{p_1}$, in the sense that they have more indeterminacy in their initial momenta, which leads to a larger set of bicharacteristics.

Let $S_{p_1}(x,t)$ be the classical (continuous) action on H_{p_1}:

$$S_{p_1}(x,t) = \theta_{p_1}|t| + \sum_{j=1}^{p_1} a_j\theta_{p_1} \cot(2a_j\theta_{p_1})r_j^2, \quad (x,t) \in H_{p_1}, \tag{4.84}$$

whenever $\theta_{p_1} \in (0, \pi/2a_{p_1})$ is a solution of

$$|t| = \sum_{j=1}^{p_1} a_j \mu(2a_j\theta_{p_1}) r_j^2; \tag{4.85}$$

otherwise S_{p_1} is extended by continuity to all H_{p_1} Again the proof of Lemma 4.17 applies and we have

$$S(x,t) \leq S_p(x,t) \leq S_{p_1}(x,t) \leq \dots, \tag{4.86}$$

where (x,t) belongs to the appropriate domain of definition of the corresponding actions. The sequence of inequalities (4.86) may be continued to S_0, the action on $\{(0,t); y \in \mathbb{R}\}$. We refer to $S(x,t)$ as the *minimal action.* Extending (4.83) we obtain a decreasing sequence of sets:

$$C_p \supset C_{p_1} \supset C_{p_2} \supset \dots,$$

whose intersection is the t-axis.

4.8. Exercises

EXERCISE 4.1. Let $n \geq 1$ and let $\mathbf{B}^{n+1} \subset \mathbb{C}^{n+1}$ be the unit ball. Let

$$\mathcal{U}_{n+1} = \Big\{ z \in \mathbb{C}^{n+1} : \quad Im(z_{n+1}) > \sum_{j=1}^{n} |z_j|^2 \Big\}.$$

Define $F(w) = (z_1, \dots, z_{n+1})$ with

$$z_j = \frac{w_j}{1 + w_{n+1}}, \qquad j = 1, \dots, n, \qquad z_{n+1} = i\frac{(1 - w_{n+1})}{(1 + w_{n+1})}.$$

Verify that F maps $\mathbf{B}^{n+1}$ biholomorphically onto $\mathcal{U}_{n+1}$. The domain Ω_{n+1} is called a *Siegel upper half-space.*

EXERCISE 4.2. The set $\mathbb{C}^n \times \mathbb{R}$ can be equipped with the group law

$$(z,t) \cdot (w,s) = \big(z + w, t + s + 2Im(z \cdot \bar{w}) \big),$$

where $z \cdot \bar{w} = \sum_{j=1}^{n} z_j \bar{w}_j$. Show that this binary operator makes $\mathbb{C}^n \times \mathbb{R}$ into a noncommutative group which is the Heisenberg group $\mathbf{H}_n$ as we discussed in this chapter.

EXERCISE 4.3. If $(z_1, \dots, z_n, t + i|z'|^2) \in \partial\mathcal{U}_{n+1}$, where $z' = (z_1, \dots, z_n)$, identify the boundary $\partial\mathcal{U}_{n+1}$ with the Heisenberg group $\mathbf{H}_n$. Show that in this way $\partial\mathcal{U}_{n+1}$ becomes a group in a natural way.

EXERCISE 4.4. If $z \in \mathcal{U}_{n+1}$, denote $\rho(z) = Im(z_{n+1}) - |z'|^2$ be the "height" function on $\mathcal{U}_{n+1}$ and

$$\begin{aligned} z &= \big(z_1, \dots, z_n, (\mathrm{Re}(z_{n+1}) + i|z'|^2) + i(Im(z_{n+1}) - |z'|^2\big) \\ &= (\xi_1, \dots, x_{n+1}) + (0, \dots, 0, i\rho(z)). \end{aligned}$$

If $(\zeta, t) = g \in \mathbf{H}_n$, then define

$$g \circ (z_1, \dots, z_{n+1}) = g \circ (\xi_1, \dots, x_{n+1}) + (0, \dots, 0, i\rho(z)),$$

where $g \circ (z_1, \dots, z_{n+1})$ is the group action of $\mathbf{H}_n$ on $\partial\mathcal{U}_{n+1}$. In other words, each $g \in \mathbf{H}_n$ induces an automorphism on $\mathcal{U}_{n+1}$, which preserves the level set

$\{z \in \mathcal{U}_{n+1} : \rho(z) = \text{constant}\}$. Denote this subgroup of the automorphism group $\text{Aut}\,\mathcal{U}_{n+1}$ by N.

i) To what subgroup of the automorphism group on $\mathbf{B}^{n+1}$ does $N \subset \text{Aut}\,\mathcal{U}_{n+1}$ correspond by way of the map F?

ii) Compute the subgroup K of $\text{Aut}\,\mathcal{U}_{n+1}$ consisting of elements that fix $(0, \dots, 0, i)$. To what subgroup of $\text{Aut}\,\mathbf{B}^{n+1}$ does K correspond?

EXERCISE 4.5. Let $z = (z_1, \dots, z_{n+1})$ be a point in $\mathcal{U}_{n+1}$ and let $\varepsilon > 0$. Define the non-isotropic dilation as follows

$$\delta_\varepsilon(z) = (\varepsilon z_1, \dots, \varepsilon z_n, \varepsilon^2 z_{n+1}).$$

Show that $\delta_\varepsilon \in \text{Aut}\,\mathcal{U}_{n+1}$ and the set of all dilations $\{\delta_\varepsilon\}$ form a subgroup of $\text{Aut}\,\mathcal{U}_{n+1}$. Denote this subgroup by A. To what subgroup of $\text{Aut}\,\mathbf{B}^{n+1}$ does it correspond?

EXERCISE 4.6. Show that $\text{Aut}\,\mathcal{U}_{n+1} = K \cdot A \cdot N$. See [**36**].

EXERCISE 4.7. Let $\Omega \subset \mathbb{C}^n$. The Bergman kernel is the function $K(z, w)$, $z, w \in \Omega$, which has the reproducing property

$$f(z) = \int_\Omega K(z, w) f(w) dV(w),$$

for all $f \in A^2(\Omega)$, where

$$A^2(\Omega) = \left\{ f \text{ holomorphic on } \Omega : \int_\Omega |f(z)|^2 dV(z) = \|f\|^2_{A^2(\Omega)} < \infty \right\}$$

is the Bergman space. (For more detailed discussion of the Bergman kernel, readers may consult [**41**].)

(*i*) Show that the Bergman kernel $K(z, w)$ is conjugate symmetric, *i.e.*, $K(z, w) = \overline{K(w, z)}$.

(*ii*) Prove that the Bergman kernel is uniquely determined by the properties that it is an element of $A^2(\Omega)$ in z, is conjugate symmetric, and reproduces $A^2(\Omega)$.

(*iii*) Let Ω_1 and Ω_2 be two domains in $\mathbb{C}^n$. Let $f : \Omega_1 \to \Omega_2$ be biholomorphic. Show that

$$\det J_{\mathbb{C}} f(z) K_{\Omega_2}(f(z), f(w)) \det \overline{J_{\mathbb{C}} f(w)} = K_{\Omega_1}(z, w).$$

Here $\det J_{\mathbb{C}} f(z)$ is the complex Jacobian of the mapping f.

EXERCISE 4.8. Since $A^2(\Omega)$ is a subspace of the Hilbert space $L^2(\Omega)$, thus there exists a complete orthonormal basis $\{\phi_k\}_{k \in \mathbf{N}}$ for $A^2(\Omega)$. Let V be a compact subset of Ω.

(*i*) Show that the series

$$\sum_{k=1}^{\infty} \phi_k(z) \overline{\phi_k(w)}$$

converges uniformly on $V \times V$ to the Bergman kernel $K(z, w)$.

(*ii*) Let $\Omega = \mathbf{B}^n$ be the unit ball in $\mathbb{C}^n$. By the uniqueness of the power series expansion for an element of $A^2(\mathbf{B}^n)$, the element $z^{\mathbf{k}} = z_1^{k_1} z_2^{k_2} \dots z_n^{k_n}$ form a complete orthonormal system on $\mathbf{B}^n$. Setting

$$\alpha_{\mathbf{k}} = \int_{\mathbf{B}^n} |z^{\mathbf{k}}|^2 dV(z),$$

show that $\{z^{\mathbf{k}}/\sqrt{\alpha_{\mathbf{k}}}\}$ is an orthonormal basis in $A^2(\mathbf{B}^n)$. Thus by (i) of Exercise 4.8,

$$K(z,w)=\sum_{|\mathbf{k}|=0}^{\infty}\frac{z^{\mathbf{k}}\bar{w}^{\mathbf{k}}}{\alpha_{\mathbf{k}}}.$$

(iii) Show that

$$\int_{\mathbb{R}^{2n}}e^{-\pi|x|^2}dx=1.$$

(iv) Show that

$$\begin{aligned}\int_{\partial\mathbf{B}^n}|z_1|^{2k}d\sigma &= \frac{2\pi^n(k!)^2}{(n+k-1)!},\\ \int_{\mathbf{B}^n}|z_1|^{2k}dV(z) &= \frac{\pi^n k!}{(n+k)!}.\end{aligned}$$

(v) Let $z\in\mathbf{B}^n$ and $0<r<1$. Denote $\mathbf{e}$ the point $(1,0,\dots,0)$. Then show that

$$K_{\mathbf{B}^n}(z,r\mathbf{e})=\frac{n!}{\pi^n}\frac{1}{(1-rz_1)^{n+1}}.$$

(vi) Now applying part (iii) of Exercise 4.8, one may conclude that

$$K_{\mathbf{B}^n}(z,w)=\frac{n!}{\pi^n}\frac{1}{(1-z\cdot\bar{w})^{n+1}}$$

where $z\cdot\bar{w}=\sum_{k=1}^n z_k\bar{w}_k$.

EXERCISE 4.9. Let $\{a_1,\dots,a_n\}$ be a set of n positive constants. Let

$$\mathcal{U}_{n+1}^{(a)}=\Big\{z\in\mathbb{C}^{n+1}:\ 2Im(z_{n+1})>\sum_{k=1}^{n}a_j|z_j|^2\Big\}.$$

Prove that the Bergman kernel for $\mathcal{U}_{n+1}^{(a)}$ is given by

$$K(z,w)=\frac{(n+1)!\prod_{k=1}^n a_k}{\pi^{n+1}(2Im(z_{n+1})-\sum_{k=1}^n a_k z_k\bar{w}_k)^{n+2}}.$$

For more details the reader may consult [**9**].

EXERCISE 4.10. Let $A=\{a_k^j\}$ be an $m\times n$ real matrix. Let

$$A_k=(a_k^1,a_k^2,\cdots,a_k^m)\neq 0$$

for $1\le k\le n$, so that $\{A_1,A_2,\cdots,A_n\}\subset\mathbb{R}^m$ is a set of n non-zero vectors. Write

$$A(z)=\left(\sum_{k=1}^n a_k^1|z_k|^2,\cdots,\sum_{k=1}^n a_k^m|z_k|^2\right)\in\mathbb{R}^m.$$

The quadratic manifold associated to A is the set

$$\Sigma_A=\{(z,w)\in\mathbb{C}^n\times\mathbb{C}^m:\quad Im(w)=A(z)\}.$$

Show that for $m=1$ with $A_k=\{a_k\}$, $a_k>0$, $\Sigma_A=\partial\mathcal{U}_{n+1}$. Furthermore, show that when $\{A_1,\dots,A_n\}$ is the standard basis of $\mathbb{R}^n$, then Σ_A is the Cartesian product of n copies of $\partial\mathcal{U}_2$. For more details wee [**47**].

EXERCISE 4.11. We introduce coordinates on Σ_A by identifying the point $(z, t+iA(z)) \in \Sigma_A$ with the point $(z,t) \in \mathbb{C}^n \times \mathbb{R}^m$, so that $t = \mathrm{Re}(w)$. When using these coordinates, we have the following identifications:

$$\frac{\partial}{\partial w_j} \leftrightarrow \frac{1}{2}\frac{\partial}{\partial t_j} \quad \text{and} \quad \frac{\partial}{\partial \bar{w}_j} \leftrightarrow \frac{1}{2}\frac{\partial}{\partial t_j}.$$

Show that Σ_A is a generic CR submanifold of $\mathbb{C}^n \times \mathbb{C}^m$ with real dimension $2n+m$, CR dimension n, and m "missing directions".

EXERCISE 4.12. Define real vector fields by

$$X_k = \frac{\partial}{\partial x_k} + 2x_{k+n}A_k \cdot \nabla_t, \quad X_{k+n} = \frac{\partial}{\partial x_{k+n}} - 2x_k A_k \cdot \nabla_t, \quad T_j = \frac{\partial}{\partial t_j}$$

where $\nabla_t = \left(\frac{\partial}{\partial t_1}, \dots, \frac{\partial}{\partial t_m}\right)$. Calculate the commutator structure of the vector fields X_k, $k = 1, \dots, 2n$ and T_j, $j = 1, \dots, m$ and to conclude that Σ_A is a CR submanifold of step 2.

EXERCISE 4.13. Let (w,s) and (z,t) be two points in $C^n \times \mathbb{R}^m$. Define a product on $\mathbb{C}^n \times \mathbb{R}^m$ by setting

$$(w,s) \cdot (z,t) = \left(w+z, s+t+2Im\left[\sum_{k=1}^{n} A_k w_k \bar{z}_k\right]\right). \tag{4.87}$$

Prove that the product given in (4.87) defines a Lie group structure G_A on $\mathbb{C}^n \times \mathbb{R}^m$. Show also that the vector fields $\{X_1, \dots, X_{2n}, T_1, \dots, T_m\}$ are left invariant and hence can be identified with the Lie algebra of the Lie group G_A.

CHAPTER 5

Complex Hamiltonian Mechanics

We recall that second order elliptic operators, essentially Laplace-Beltrami operators, induce a Riemannian geometry, and the attached inverses, *i.e.*, fundamental solutions, heat kernels and wave kernels have analytic expressions in terms of the underlying geometric concepts, mainly in terms of the lengths of the appropriate geodesics.

We strongly believe that the structure of the inverses attached to sub-elliptic operators should be understood just as precisely as the well known structure of the kernels attached to Laplace-Beltrami operators is understood. One may say that elliptic operators arise from classical mechanics and second order sub-elliptic operators have connections to quantum mechanics.

In this chapter we shall describe the complex geodesics from the origin and we shall deal with the lengths of the real geodesics starting at the origin by means of complex Hamiltonian mechanics. An understanding of the dynamics in the complexified phase space is essential also for quantum mechanics, see [**54**].

5.1. The harmonic oscillator and the Heisenberg group

We shall show that the harmonic oscillator leads to the Heisenberg operator by certain complex quantization. Consider a unit mass particle under the influence of force $F(x) = x$. Newton's equation is $\ddot{x} = x$. This is the equation which describes the dynamics of an inverse pendulum in an unstable equilibrium, for small angle x, see Figure 5.1.
The potential energy is

$$U(x) = -\int_0^x F(u)\, du = -\frac{x^2}{2}.$$

The Lagrangian $L : T\mathbb{R} \to \mathbb{R}$ is the difference between the kinetic and the potential energy

$$L(x, \dot{x}) = K - U = \frac{1}{2}\dot{x}^2 + \frac{1}{2}x^2.$$

The momentum is $p = \frac{\partial L}{\partial \dot{x}} = \dot{x}$. The Hamiltonian associated with the above Lagrangian is obtained using the Legendre transform: $H : T^*\mathbb{R} \to \mathbb{R}$

$$H(x,p) = p\dot{x} - L(x,\dot{x}) = p^2 - \frac{1}{2}p^2 - \frac{1}{2}x^2 = \frac{1}{2}p^2 - \frac{1}{2}x^2.$$

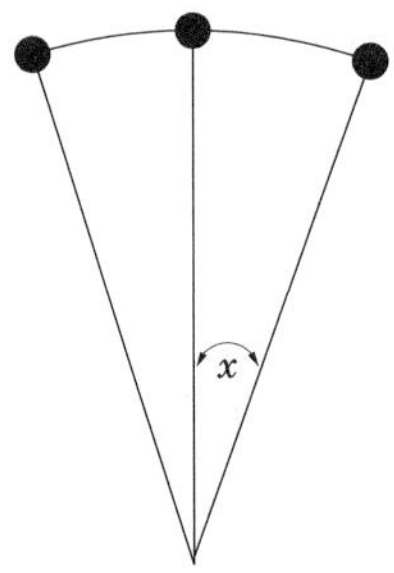

Figure 5.1: *The inverse pendulum problem.*

Using ideas from Xavier [**54**], we consider the following complexification

$$x = x_1 + ip_2, \quad p = p_1 + ix_2. \tag{5.1}$$

Hence $H : T^*\mathbb{C} \to \mathbb{C}$ and

$$\begin{aligned} H(x,p) &= \frac{1}{2}p^2 - \frac{1}{2}x^2 \\ &= \frac{1}{2}(p_1 + ix_2)^2 - \frac{1}{2}(x_1 + ip_2)^2 \\ &= \frac{1}{2}(p_1 + ix_2)^2 + \frac{1}{2}i^2(x_1 + ip_2)^2 \\ &= \frac{1}{2}(p_1 + ix_2)^2 + \frac{1}{2}(ix_1 - p_2)^2 \\ &= \frac{1}{2}(p_1 + ix_2)^2 + \frac{1}{2}(p_2 - ix_1)^2. \end{aligned}$$

Replacing $\theta = -i$,

$$H(x,p;\theta) = \frac{1}{2}(p_1 - \theta x_2)^2 + \frac{1}{2}(p_2 + \theta x_1)^2. \tag{5.2}$$

Quantizing, *i.e.*, making the changes $p_1 \to \partial_{x_1}$, $p_2 \to \partial_{x_2}$, $\theta \to \partial_t$, we get $H \to \Delta_X$, where

$$\Delta_X = \frac{1}{2}\big(\partial_{x_1} - x_2\partial_t\big)^2 + \frac{1}{2}\big(\partial_{x_2} + x_1\partial_t\big)^2$$

is the Heisenberg operator.

REMARK 5.1. a) If we choose the complexification

$$x = 2x_1 + ip_2, \qquad p = p_1 + 2ix_2$$

then we arrive at the operator

$$\Delta_X = \frac{1}{2}\big(\partial_{x_1} - 2x_2\partial_t\big)^2 + \frac{1}{2}\big(\partial_{x_2} + 2x_1\partial_t\big)^2.$$

b) The expression $\theta = -i$ was used in the literature for a long time without an explanation. We can see that this value for θ is the natural one and it is related to quantum behaviors.

5.2. The quantum Hamiltonian

The previous assumption $\theta = -i$ can be considered in the general case of the study of complex dynamics on the model defined by the vector fields

$$X_1 = \partial_{x_1} + 2x_2|x|^{2k}\partial_t, \quad X_2 = \partial_{x_2} - 2x_1|x|^{2k}\partial_t. \tag{5.3}$$

In this case, the Hamiltonian becomes

$$H(x,\xi) = \frac{1}{2}\Big(\xi_1 - 2ix_2|x|^{2k}\Big)^2 + \frac{1}{2}\Big(\xi_2 + 2ix_1|x|^{2k}\Big)^2, \tag{5.4}$$

and it will be called *quantum Hamiltonian.*

DEFINITION 5.2. The $\mathcal{PT}$-symmetry operations are defined as follows. Combined parity:

$$\mathcal{P}: \quad (x_1, x_2, \xi_1, \xi_2) \to (-x_1, -x_2, -\xi_1, -\xi_2).$$

Time reversal:

$$\mathcal{T}: \quad (x_1, x_2, \xi_1, \xi_2) \to (x_1, x_2, -\xi_1, -\xi_2),\ i \to -i.$$

A rigorous study of the one-dimensional complex Hamiltonian systems with $\mathcal{PT}$-symmetry can be found in the recent work of Bender and coworkers [**8**]. We shall deal with a two-dimensional Hamiltonian in the variable x. The following result is obvious.

PROPOSITION 5.3. The quantum Hamiltonian (5.4) is $\mathcal{PT}$-symmetric.

This symmetry will influence the symmetry of the solutions of the Hamiltonian equations. Xavier and Aguiar [**54**] used complex solutions of the Hamiltonian system to study the semiclassical approximation of the coherent-state propagator in the case of a quartic potential.

Consider complex solutions $x(s) = (x_1(s), x_2(s)) \in \mathbb{C}^2$ for the Hamiltonian system with Hamiltonian (5.4)

$$\dot{x} = H_\xi, \qquad \dot{\xi} = -H_x, \tag{5.5}$$

with real boundary condition

$$x(0) = x_0 \in \mathbb{R}^2, \qquad x(1) = x \in \mathbb{R}^2.$$

PROPOSITION 5.4. The solutions of system (5.5) satisfy the system

$$\begin{cases} \ddot{x}_1 = -4i(k+1)|x|^{2k}\dot{x}_2 \\ \ddot{x}_2 = 4i(k+1)|x|^{2k}\dot{x}_1. \end{cases} \tag{5.6}$$

PROOF. It comes from the system (3.6) of chapter 3, with the substitution $\theta = -i$. □

5.3. Complex subRiemannian geodesics

In the case of a complex subRiemannian geodesic we shall ask that the momentum θ is complex and constant along the geodesic, with $\theta = -i$. To eliminate the over-determination of the Hamiltonian system, we shall assume non-standard boundary conditions, *i.e.,* $t(0)$ is not specified, since we fixed θ.

DEFINITION 5.5. A complex subRiemannian geodesic is the projection on the (x,t)-space of a solution of the Hamiltonian system

$$\dot{x} = H_\xi, \qquad \dot{t} = H_\theta,$$
$$\dot{\xi} = -H_x, \qquad \dot{\theta} = -H_t$$

with non-standard boundary conditions

$$x(0) = x_0, \qquad \theta = \theta_0 = -i, \qquad x(s_f) = x_f, \qquad t(s_f) = t_f,$$

and Hamiltonian

$$H(x,t,\xi,\theta) = \frac{1}{2}\Big(\xi_1 + 2\theta x_2|x|^{2k}\Big)^2 + \frac{1}{2}\Big(\xi_2 - 2\theta x_1|x|^{2k}\Big)^2.$$

Note the boundary conditions are real, *i.e.*, $x_0, x_f \in \mathbb{R}^2$, $t_f \in \mathbb{R}$, while the geodesic components between the end-points might be complex.
The following theorem is useful in future computations.

THEOREM 5.6. *A curve $c(s) = (x(s), t(s))$ is a complex subRiemannian geodesic (in the sense of Definition* 5.5*) if and only if the x-component satisfies the system* (5.6) *with boundary conditions*

$$x(0) = x_0, \qquad x(s_f) = x_f,$$

and the t-component is the unique solution of the equation

$$\dot{t} = 2|x|^{2k}(\dot{x}_1 x_2 - x_1 \dot{x}_2), \tag{5.7}$$

with the final condition $t(0) = t_f$.

PROOF. " $\Longrightarrow$ " Let $c(s) = (x(s), t(s))$ be a complex subRiemannian geodesic. Applying Proposition 5.4, the x-component satisfies the system (5.6). The Hamiltonian equation $\dot{t} = H_\theta$ yields the second part.
" $\Longleftarrow$ " Is the same proof as in Theorem 2.4 of chapter 3, with $\theta = -i$. □

Complex symmetries
From the $\mathcal{PT}$-symmetry of the quantum Hamiltonian (see Proposition 5.3), if $c(s) = (x(s), t(s))$ is a complex geodesic, then $\mathcal{P}(c(s)) = (-x(s), t(s))$ and $\mathcal{T}(c(s)) = \overline{c}(s)$ are also complex geodesics, where $\overline{c}(s) = (\overline{x_1}, \overline{x_2}, \overline{t})$ denotes the complex conjugation. Hence, we deal with quadruples of geodesics $(\pm x, t)$, $(\pm \overline{x}, \overline{t})$.

5.4. Complex subRiemannian geodesics on the Heisenberg group

The x-component
We are interested in finding complex subRiemannian geodesics with real boundary conditions $x_0 = 0$, $x_f = x$ and $t_f = t$ on the Heisenberg group $\mathbf{H}_1$. To find them, we use Theorem 5.6. In this case, $k = 0$ and the x-component satisfies the complex system

$$\begin{cases} \ddot{x}_1 = -4i\dot{x}_2 \\ \ddot{x}_2 = 4i\dot{x}_1, \end{cases} \tag{5.8}$$

where $x_1 = a_1 + ib_1$, $x_2 = a_2 + ib_2$, with $a_i = a_i(s)$ and $b_i = b_i(s)$ real functions. The system (5.8) can be decomposed in the following two real systems:

$$(I)\begin{cases} \ddot{a}_1 = 4\dot{b}_2 \\ \ddot{b}_2 = 4\dot{a}_1, \end{cases} \qquad (II)\begin{cases} \ddot{a}_2 = -4\dot{b}_1 \\ \ddot{b}_1 = -4\dot{a}_2, \end{cases} \tag{5.9}$$

with boundary conditions

$$a_i(0)=0,\quad a_i(1)=x_i,\quad b_i(0)=0,\qquad b_i(1)=0,\qquad i=\overline{1,2}. \tag{5.10}$$

• *Solving the system (I).* Multiplying the first equation by $\dot a_1$, the second by $\dot b_2$ and subtracting, yields

$$\ddot a_1\dot a_1-\ddot b_2\dot b_2=0 \implies \dot a_1^2-\dot b_2^2=\pm R^2 \text{ (constant)}.$$

∘ *Case* $\dot a_1^2-\dot b_2^2=R^2$. There is a real function $\alpha=\alpha(s)$ such that

$$\dot a_1(s)=R\cosh\alpha(s),\qquad \dot b_2(s)=R\sinh\alpha(s). \tag{5.11}$$

Differentiating and substituting in system (I), yields

$$\dot\alpha\sinh\alpha=4\sinh\alpha,\qquad \dot\alpha\cosh\alpha=4\cosh\alpha. \tag{5.12}$$

Then

$$\dot\alpha(s)=4 \implies \alpha(s)=4s+\alpha_0, \tag{5.13}$$

see Exercise 5.1. Using (5.13), and integrating the relations (5.11), yields

$$\begin{aligned} a_1(s) &= a_1(0)+R\int_0^s \cosh(4u+\alpha_0)\,du\\ &= a_1(0)+\frac{R}{4}\Big(\sinh(4s+\alpha_0)-\sinh\alpha_0\Big). \end{aligned}$$

$$\begin{aligned} b_2(s) &= b_2(0)+R\int_0^s \sinh(4u+\alpha_0)\,du\\ &= b_2(0)+\frac{R}{4}\Big(\cosh(4s+\alpha_0)-\cosh\alpha_0\Big). \end{aligned}$$

From the boundary conditions $b_2(0)=b_2(1)=0$, it follows from the above relation that α_0 verifies the equation

$$\cosh(4+\alpha_0)-\cosh\alpha_0=0, \tag{5.14}$$

which has the unique solution $\alpha_0=-2$, see Exercise 5.2.
Using the boundary conditions $a_1(0)=0$, $a_1(1)=x_1$ and $\alpha_0=-2$, the expression for $a_1(s)$ yields

$$x_1=\frac{R}{4}\Big(\sinh(4-2)-\sinh(-2)\Big) \implies R=\frac{2x_1}{\sinh 2}.$$

Hence, the solutions for the system (I) are

$$\begin{aligned} a_1(s) &= \frac{x_1}{2\sinh 2}\Big(\sinh(4s-2)+\sinh 2\Big),\\ b_2(s) &= \frac{x_1}{2\sinh 2}\Big(\cosh(4s-2)-\cosh 2\Big), \qquad s\in[0,1]. \end{aligned}$$

Other solutions can be obtained if in (5.11) we choose combinations of

$$\dot a_1(s)=\pm R\cosh\alpha(s),\qquad \dot b_2=\pm R\sinh\alpha(s). \tag{5.15}$$

The corresponding solutions are as above but with sign combinations.

∘ *Case* $\dot a_1^2-\dot b_2^2=-R^2$. We shall show that this case is impossible. In this case

$$\dot b_2(s)=R\cosh\alpha(s),\qquad \dot a_1(s)=R\sinh\alpha(s),$$

with $\alpha(s) = 4s + \alpha_0$. Integrating,

$$\begin{aligned} b_2(s) &= b_2(0) + R\int_0^s \cosh(4u+\alpha)\,du \\ &= b_2(0) + \frac{R}{4}\Big(\sinh(4s+\alpha_0) - \sinh\alpha_0\Big). \end{aligned}$$

From the boundary conditions $b_2(0) = b_2(1) = 0$ we get

$$\sinh(4+\alpha_0) = \sinh(\alpha_0),$$

which is a contradiction, because sinh is one-to-one.

• *Solving the system (II).* Multiplying the first equation by $\dot{a}_2$, the second by $\dot{b}_1$ and subtracting, yields

$$\ddot{a}_2\dot{a}_2 - \ddot{b}_1\dot{b}_1 = 0 \implies \dot{a}_2^2 - \dot{b}_1^2 = \pm\rho^2 \text{ (constant)}.$$

∘ *Case* $\dot{a}_2^2 - \dot{b}_1^2 = \rho^2$. There is a real function $\beta = \beta(s)$ such that

$$\dot{a}_2(s) = \rho\cosh\beta, \qquad \dot{b}_1(s) = \rho\sinh\beta. \tag{5.16}$$

Differentiating and substituting in system (II), yields

$$\dot{\beta}\cosh\beta = -4\cosh\beta, \tag{5.17}$$

and then $\beta = -4s + \beta_0$. Integrating in (5.16) and using $a_2(0) = b_1(0) = 0$ and $\beta = -4s + \beta_0$, yields

$$\begin{aligned} a_2(s) &= -\frac{\rho}{4}\Big(\sinh(-4s+\beta_0) - \sinh\beta_0\Big) \\ &= \frac{\rho}{4}\Big(\sinh(4s-\beta_0) + \sinh\beta_0\Big) \qquad (5.18) \\ b_1(s) &= -\frac{\rho}{4}\Big(\cosh(-4s+\beta_0) - \cosh\beta_0\Big). \qquad (5.19) \end{aligned}$$

The boundary condition $b_1(1) = 0$ yields

$$\cosh(-4+\beta_0) - \cosh\beta_0 = 0 \Longrightarrow \beta_0 = 2.$$

Substituting $a_2(1) = x_2$ in formula (5.18)

$$x_2 = \frac{\rho}{4}\Big(\sinh(4-\beta_0) + \sinh\beta_0\Big) = \frac{\rho}{2}\sinh 2 \Longrightarrow \rho = \frac{2x_2}{\sinh 2}.$$

Hence, the solutions of the system (II) are

$$\begin{aligned} a_2(s) &= \frac{x_2}{2\sinh 2}\Big(\sinh(4s-2) + \sinh 2\Big), \\ b_1(s) &= -\frac{x_2}{2\sinh 2}\Big(\cosh(4s-2) - \cosh 2\Big), \qquad s \in [0,1]. \end{aligned}$$

Changing signs in (5.16)

$$\dot{a}_2(s) = \pm\rho\cosh\beta, \qquad \dot{b}_1(s) = \pm\rho\sinh\beta, \tag{5.20}$$

leads to solutions $\pm a_2(s)$ and $\pm b_1(s)$, with $a_2(s)$ and $b_1(s)$ provided above.

∘ *Case* $\dot{a}_2^2 - \dot{b}_1^2 = -\rho^2$. This case is impossible, see the proof in the previous case.

The above discussion can be summarized into the following result:

PROPOSITION 5.7. The solutions of system (5.8) with the boundary conditions

$$x(0) = 0, \qquad x(1) = x$$

are $\pm x(s)$, with

$$x(s) = A(s)\widehat{x} \pm iB(s)J\widehat{x}, \quad s \in [0,1], \tag{5.21}$$

where $A(s) = \sinh(4s-2) + \sinh 2$, $B(s) = \cosh(4s-2) - \cosh 2$ and

$$J = \begin{bmatrix} 0 & 1 \\ -1 & 0 \end{bmatrix}, \qquad \widehat{x} = \frac{x}{2\sinh 2}. \tag{5.22}$$

PROOF. If we choose all the possible sign combinations in

$$\begin{aligned} a_1(s) &= \pm\widehat{x}_1 A(s), \qquad a_2(s) = \pm\widehat{x}_2 A(s), \\ b_2(s) &= \pm\widehat{x}_1 B(s), \qquad b_1(s) = -\Big(\pm\widehat{x}_2 B(s)\Big), \end{aligned}$$

we get

$$x(s) = \pm(a(s) \pm ib(s)) = \pm(\widehat{x}A(s) \pm iJ\widehat{x}B(s)).$$

□

This result is consistent with the $\mathcal{PT}$-symmetries of the quantum Hamiltonian

$$H(x,\xi) = \frac{1}{2}\Big(\xi_1 - 2ix_2\Big)^2 + \frac{1}{2}\Big(\xi_2 + 2ix_1\Big)^2.$$

The t-component
From the boundary condition $t(1) = t \in \mathbb{R}$, while $t(0)$ is not fixed. In this section we shall study the behavior of the t-component, and show that $t(0)$ cannot be a real number unless $x = 0$.
Integrating in $\dot{t} = 2(\dot{x}_1 x_2 - x_1\dot{x}_2)$, yields

$$t - t(s) = 2\int_s^1 (\dot{x}_1 x_2 - x_1\dot{x}_2). \tag{5.23}$$

Using the geodesics

$$\begin{aligned} x_1(s) &= a_1(s) + ib_1(s) = \widehat{x}_1 A(s) - i\widehat{x}_2 B(s), \\ x_2(s) &= a_2(s) + ib_2(s) = \widehat{x}_2 A(s) + i\widehat{x}_1 B(s), \end{aligned}$$

a computation provides

$$2(\dot{x}_1 x_2 - x_1\dot{x}_2) = 8i|x|^2(1 - \cosh 4s).$$

Substituting in (5.23), yields

$$\begin{aligned} t - t(s) &= 8i|\widehat{x}|^2 \int_s^1 (1 - \cosh(4u))\, du \\ &= 8i|\widehat{x}|^2\Big((1-s) - \frac{\sinh 4 - \sinh 4s}{4}\Big) \\ &= 2i|\widehat{x}|^2\Big((\sinh 4s - 4s) - (\sinh 4 - 4)\Big). \end{aligned}$$

We arrive at the following result.

PROPOSITION 5.8. The t-component is given by

$$t(s) = t - 2i|\widehat{x}|^2\Big((\sinh 4s - 4s) - (\sinh 4 - 4)\Big),$$

where

$$\widehat{x} = \frac{x}{2\sinh 2}.$$

Hence, for $x \neq 0$, $t(s)$ is never real for any $s \in [0, 1)$.
In particular,

$$t(0) = t - i|x|^2\Big(\frac{2}{\sinh^2 2} - \coth 2\Big).$$

PROOF. The function $\psi(s) = \sinh 4s - 4s$ has the derivative $\psi'(s) = 4(\cosh s - 1) \geq 0$, and hence ψ is increasing and one-to-one. $Im\, t(s) = -2|\widehat{x}|^2(\psi(s) - \psi(1)) > 0$, for any $0 \leq s < 1$.

$$\begin{aligned} t(0) &= t - 2i|\widehat{x}|^2(4 - \sinh 4) = t - \frac{i|x|^2}{\sinh^2 2}(2 - \sinh 2\cosh 2) \\ &= t - i|x|^2\Big(\frac{2}{\sinh^2 2} - \coth 2\Big). \end{aligned}$$

□

LEMMA 5.9. *Let*

$$\begin{aligned} \epsilon(x) &= \frac{x}{\sinh^2 x} - \coth x, \\ \mu(x) &= \frac{x}{\sin^2 x} - \cot x, \end{aligned}$$

be two odd functions. Then

$$\epsilon(ix) = -i\mu(x), \qquad \mu(ix) = -i\epsilon(x).$$

PROOF.

$$\begin{aligned} \epsilon(ix) &= \frac{ix}{\sinh^2(ix)} - \coth(ix) \\ &= \frac{ix}{(i\sin x)^2} - (-i\cot x) = \frac{ix}{-\sin^2 x} + i\cot x \\ &= -i\Big(\frac{x}{\sin^2 x} - \cot x\Big) = -i\mu(x). \end{aligned}$$

$$\begin{aligned} \mu(ix) &= \frac{ix}{(\sin ix)^2} - \cot(ix) = \frac{ix}{(i\sinh x)^2} - (-i\coth x) \\ &= \frac{ix}{-\sinh^2 x} + i\coth x = -i\Big(\frac{x}{\sinh^2} - \coth x\Big) = -i\epsilon(x). \end{aligned}$$

□

Lemma 5.9 and Proposition 5.8 yield

$$t(0) = t - i|x|^2\epsilon(2) = t + |x|^2\mu(2i),$$

which can be written as

$$\frac{t - t(0)}{|x|^2} = \mu(-2i).$$

If we consider $\theta = -i$, we arrive at the same relation as in the case of *real geodesics*, see chapter 1:

$$\frac{t}{|x|^2} = \mu(2\theta s_f),$$

where $t(0) = 0$ and $s_f = 1$. From this point of view, real and complex geodesics behave similarly. The symmetries for real geodesics induced by changing θ into $-\theta$, became time reversal or conjugation $\mathcal{T}$-symmetry for complex geodesics.

5.5. Complex connectivity

Due to the non-standard boundary conditions, the usual definition of connectivity is not working in the case of complex subRiemannian geodesics. The difficulty arises from the fact that the equation $\dot{t} = 2|x|^{2k}(\dot{x}_1 x_2 - x_1 \dot{x}_2)$ may become supra-determinate if the following two boundary conditions are required

$$t(0) = t_0, \qquad t(1) = t_f.$$

On the other hand, we have seen in the previous section that in the case $x \neq 0$, $t(s)$ is real only for $s = 1$ and complex for $0 \le s < 1$. We have the following non-connectivity result on the Heisenberg group.

PROPOSITION 5.10. Given two distinct points $P(x,t)$ and $Q(0,t_0)$ in $\mathbb{R}^3$, with $t \neq t_0$, there are no complex subRiemannian geodesics between Q and P.

PROOF. If $x = 0$, from Propositions 5.8 and 5.7, $x(s) = 0$, and $t(s) = t$, constant. P and Q cannot be connected because $t \neq t_0$.
If $x \neq 0$, from Proposition 5.8, $t(0)$ cannot be real, and hence $t(0) = t_0 \in \mathbb{R}$ doesn't hold.

□

Hence, even in the step 2 case, the connectivity by complex geodesics fails. We need to give an appropriate definition for the connectivity in the complex case. For this we shall give up the t-component and we shall consider only the connectivity in the x-plane.

DEFINITION 5.11. Let H be the quantum Hamiltonian (5.4). The complex connectivity property holds if given any two points $P(x_0)$ and $Q(x_f)$ in the x-plane, there is a solution of the system

$$\dot{x} = H_\xi, \qquad \dot{\xi} = -H_x, \tag{5.24}$$

satisfying the real boundary conditions $x(0) = x_0$, $x(s_f) = x_f$.

In the following we shall prove the geodesic connectivity from the origin. In polar coordinates (r,ϕ), the system

$$\begin{cases} \ddot{x}_1 = -4i(k+1)|x|^{2k}\dot{x}_2 \\ \ddot{x}_2 = 4i(k+1)|x|^{2k}\dot{x}_1 \end{cases} \tag{5.25}$$

becomes

$$\begin{cases} \ddot{r} = r\dot{\phi}(\dot{\phi} - 4(k+1)ir^{2k}) \\ r^2(\dot{\phi} - 2ir^{2k}) = C(\text{constant}). \end{cases} \tag{5.26}$$

As the solution starts from the origin, we have $r(0) = 0$. Hence $C = 0$ and $\dot{\phi} = 2ir^{2k}$. Substituting in the first equation, yields

$$\begin{aligned} \ddot{r} &= r \cdot 2ir^{2k}(2ir^{2k} - 4(k+1)ir^{2k}) \\ &= -2ir^{2k+1}(4k+2)ir^{2k}) \\ &= 4(2k+1)r^{4k+1} = -V'(r), \end{aligned}$$

where $V(r) = -2r^{4k+2}$. The following conservation of energy holds

$$\frac{1}{2}\dot{r}^2 + V(r) = E, \tag{5.27}$$

where the constant of total energy may be complex. Sometimes, for some particular boundary conditions, the energy is real. Indeed, choose the boundary conditions

$$x_1(0) = x_2(0) = 0, \qquad x_1(1) > 0, \qquad x_2(1) = 0. \tag{5.28}$$

Assume that $c(s) = \big(x_1(s), x_2(s)\big)$ is a solution of the Euler-Lagrange system (5.25) with the boundary conditions (5.28). Taking conjugation in the system (5.25), yields

$$\begin{cases} (\overline{x}_1)^{\cdot\cdot} = -4i(k+1)(\overline{x}_1^2 + (-\overline{x}_2)^2)^k(-\overline{x}_2)^{\cdot} \\ (-\overline{x}_2)^{\cdot\cdot} = 4i(k+1)(\overline{x}_1^2 + (-\overline{x}_2)^2)^k(\overline{x}_1)^{\cdot}, \end{cases} \tag{5.29}$$

which is the Euler-Lagrange system for $\gamma(s) = (\overline{x}_1(s), -\overline{x}_2(s))$. Thus $c(s)$ and $\gamma(s)$ satisfy the same system of differential equations and the same boundary conditions (5.28). Consequently, if they exist, for $x_1(1)$ small enough the solution is unique, *i.e.*

$$x_1(s) = \overline{x}_1(s),\ x_2(s) = -\overline{x}_2(s) \Longrightarrow r^2(s) \in \mathbb{R}.$$

From (5.27) we get $E \in \mathbb{R}$. Using the rotational symmetry of the Euler-Lagrange system we can always choose, using a rotation, $x_2(1) = 0$. Hence from now on we shall assume the energy E real. We have the following result:

LEMMA 5.12. *Given a real number $R > 0$, there is a real function $r : [0,1] \to \mathbb{R}$, which is a solution for the boundary value problem*

$$\begin{cases} \ddot{r} = 4(2k+1)r^{4k+1} \\ r(0) = 0,\ r(1) = R. \end{cases} \tag{5.30}$$

The energy function $E = E(R)$ is increasing as a function of R.

PROOF. Assume $r = r(s)$ is a real valued function. Then E will be real. The equation (5.27) becomes

$$\dot{r}^2 - 4r^{4k+2} = 2E \iff \dot{r} = \sqrt{2E + 4r^{4k+2}}$$

$$\frac{dr}{\sqrt{2E + 4r^{4k+2}}} = ds \iff \int_0^{r(s)} \frac{du}{\sqrt{2E + 4u^{4k+2}}} = s.$$

For $s = 1$,

$$\int_0^R \frac{du}{\sqrt{2E + 4u^{4k+2}}} = 1. \tag{5.31}$$

We shall show that there is an unique real number $E > 0$ which satisfies (5.31). Consider the function

$$\eta(E) = \int_0^R \frac{du}{\sqrt{2E + 4u^{4k+2}}},$$

which is decreasing because

$$\frac{d\eta}{dE} = -\int_0^R \frac{du}{\sqrt{(2E + 4u^{4k+2})^3}} < 0.$$

Using

$$\begin{aligned}\lim_{E\nearrow\infty} \eta(E) &= 0,\\ \lim_{E\searrow 0} \eta(E) &= \int_0^R \frac{du}{2u^{2k+1}} = -\frac{1}{4ku^{2k}}\Big|_{u=0}^{u=R} = +\infty,\end{aligned}$$

and the continuity of η, the equation $\eta(E) = 1$ will have an unique solution $E = E(R)$, see Figure 5.2.

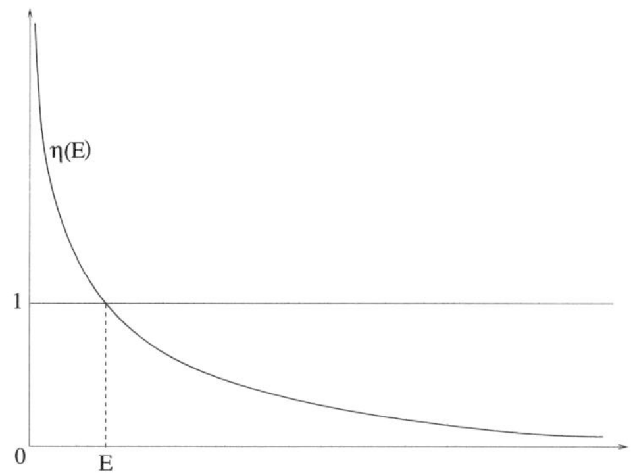

Figure 5.2: The uniqueness of E.

When R increases, in order to get the same value for the integral, E has to increase, too. With the previous E, let

$$\Psi(r) = \int_0^r \frac{du}{\sqrt{2E + 4u^{4k+2}}}.$$

As $\Psi'(r) > 0$, Ψ is increasing and hence it is one-to-one. As $\Psi(0) = 0$, $\Psi(R) = 1$, Ψ is onto on $[0, 1]$. Then the solution is defined by $r(s) = \Psi^{-1}(s)$, $s \in [0, 1]$. □

The connectivity is given in the following result.

THEOREM 5.13. *Given a point $P(x)$ in the plane, there is at least one complex geodesic in $\mathbb{R}^3$ such that its projection onto the x-plane joins the origin and the point P.*

PROOF. Let $R = |x|$ and ϕ_f be the radius and argument of P, respectively, *i.e.* $x = Re^{i\phi_f}$. The numbers R and ϕ_f are real. Let $r(s)$ be given by the Lemma 5.12. Then the function $\phi(s)$ is a solution of the Cauchy problem

$$\dot\phi = 2ir^{2k}, \qquad \phi(1) = \phi_f,$$

which is

$$\phi(s) = 2i\int_1^s r^{2k}(u)\,du + \phi_f.$$

The radius function $r(s)$ is real but the argument $\phi(s)$ is complex. Hence the x-projection of the complex geodesic will be complex with real end points, which are the origin and the point $P(R, \phi_f)$. □

An explicit computation of the radius $r(s)$ and argument $\phi(s)$ in general step is almost impossible. It can be done explicitly in the particular cases $k = 0$ and $k = 1$, which correspond to step 2 and step 4 cases. We shall do this computation in the following sections.

The case $k = 0$ (Heisenberg group case). We shall follow the next few steps:

• *Finding the argument function.* In this case the argument is given by $\phi(s) = 2i(s-1) + \phi_f$.
• *Finding the energy.* The energy E is defined by

$$\int_0^R \frac{du}{\sqrt{2E + 4u^2}} = 1.$$

A computation of the integral yields

$$\frac{1}{2}\sinh^{-1}\left(\sqrt{\frac{2}{E}}\,u\right)\Bigg|_{u=0}^{u=R} = 1 \Longleftrightarrow \sinh^{-1}\left(\sqrt{\frac{2}{E}}\,R\right) = 2,$$

and hence the energy depends only on the boundary condition

$$E = \frac{2R^2}{\sinh^2 2}. \tag{5.32}$$

We note the energy E is an increasing function of R.

• *Finding the radius $r(s)$.* The radius is defined by the integral equation

$$\int_0^{r(s)} \frac{du}{\sqrt{2E + 4u^2}} = s \Longleftrightarrow \sinh^{-1}\left(\sqrt{\frac{2}{E}}\,r(s)\right) = 2s.$$

Hence, using (5.32)

$$r(s) = \sqrt{\frac{E}{2}}\sinh(2s) = \frac{\sinh(2s)}{\sinh 2}R.$$

To conclude the case $k = 0$, we have the following result:

THEOREM 5.14. *Given a point $P(x)$ in the plane, there is exactly one complex geodesic in $\mathbb{R}^3$, such that the projection onto the x-plane joins the origin and the point P.*

The case $k = 1$. The radius $r(s)$ satisfies the integral equation

$$\int_0^{r(s)} \frac{du}{\sqrt{2E + 4u^6}} = s \Longleftrightarrow \int_0^{r(s)} \frac{du}{\sqrt{1 + \frac{2}{E}u^6}} = \sqrt{2E}\,s.$$

The substitution $v = (2/E)^{1/6}\,u$ yields

$$\int_0^{\sigma(s)} \frac{dv}{\sqrt{1+v^6}} = 2^{2/3}\,E^{1/3}s,$$

where $\sigma(s) = (2/E)^{1/6}\,r(s)$. Making $v^2 = t$, yields

$$\int_0^{\sigma^2(s)} \frac{dt}{\sqrt{t(1+t^3)}} = 2^{5/3}\,E^{1/3}s. \tag{5.33}$$

We need the following result:

LEMMA 5.15. *Let*

$$g(\sigma) = \frac{\sqrt{3}}{\sigma^2 + \frac{\sqrt{3}-1}{2}} - 1, \qquad b = 2 + \sqrt{3}, \qquad k = \frac{\sqrt{b}}{2}.$$

Then

$$\int_0^{\sigma^2(s)} \frac{dt}{\sqrt{t(1+t^3)}} = 3^{-1/4}\, sn^{-1}\left(\sqrt{1 - \left(\frac{g(\sigma)}{b}\right)^2}, k\right).$$

PROOF. Let $f(t) = t(1+t^3) = t(t+1)(t^2-t+1) = (t^2+t)(t^2-t+1)$. Making the substitution

$$t = \frac{pu+q}{u+1}, \tag{5.34}$$

$$\begin{aligned} f(t(u)) &= \frac{p(1+p)u^2 + [q(1+p) + p(q+1)]u + q(q+1)}{(u+1)^4} \\ &\quad + \frac{(p^2-p+1)u^2 + (2pq - p + 2 - q)u + (q^2-q+1)}{(u+1)^4}. \end{aligned}$$

Choose p and q such that the coefficients of u vanish

$$\begin{aligned} q(1+p) + p(q+1) &= 0 \\ 2pq - p + 2 - q &= 0. \end{aligned}$$

The above system has the symmetric solution

$$p = \frac{1-\sqrt{3}}{2}, \qquad q = \frac{1+\sqrt{3}}{2}.$$

Then

$$\begin{aligned} f(t(u)) &= \frac{\left[\left(\frac{3}{2} - \sqrt{3}\right)u^2 + \frac{3}{2} + \sqrt{3}\right]\left[\frac{3}{2}u^2 + \frac{3}{2}\right]}{(u+1)^4} \\ &= \frac{3}{2}\frac{1}{(u+1)^4}\left(\frac{3}{2} - \sqrt{3}\right)\left[u^2 + \frac{3+2\sqrt{3}}{3-2\sqrt{3}}\right](u^2+1) \\ &= \frac{3}{2}\frac{1}{(u+1)^4}\left(\frac{3}{2} - \sqrt{3}\right)\left[u^2 - (2+\sqrt{3})^2\right](u^2+1) \\ &= \frac{3}{2}\frac{1}{(u+1)^4}\left(\sqrt{3} - \frac{3}{2}\right)(b^2 - u^2)(u^2+1), \end{aligned}$$

and hence

$$\sqrt{f} = \sqrt{\frac{3}{2}\left(\sqrt{3} - \frac{3}{2}\right)}\, \frac{\sqrt{(b^2-u^2)(u^2+1)}}{(u+1)^2}.$$

Using the substitution (5.34) and

$$dt = \frac{-\sqrt{3}}{(u+1)^2}\, du, \qquad \frac{\sqrt{3}}{\sqrt{\frac{3}{2}(\sqrt{3} - \frac{3}{2})}} = \frac{2}{\sqrt{2\sqrt{3} - 3}},$$

yields

$$\begin{aligned}\int_0^{\sigma^2}\frac{dt}{\sqrt{f(t)}} &= -\frac{2}{\sqrt{2\sqrt{3}-3}}\int_b^{g(\sigma)}\frac{du}{\sqrt{(b^2-u^2)(u^2+1)}}\\ &= \frac{2}{\sqrt{2\sqrt{3}-3}}\int_{g(\sigma)}^{b}\frac{du}{\sqrt{(b^2-u^2)(u^2+1)}}.\end{aligned}$$

The limits of integration of the above integral are computed as follows.

$$t=\frac{pu+q}{u+1}\Longleftrightarrow u=\frac{q-t}{t-p}.$$

For $t=0$,

$$u = -\frac{q}{p}=-\frac{1+\sqrt{3}}{1-\sqrt{3}}=2+\sqrt{3}=b.$$

For $t=\sigma^2$,

$$\begin{aligned}u &= \frac{q-\sigma^2+\sigma^2-p}{\sigma^2-p}-1=\frac{q-p}{\sigma^2-p}-1\\ &= \frac{\sqrt{3}}{\sigma^2-p}-1=g(\sigma).\end{aligned}$$

Following [**42**], p. 52,

$$\int_x^b\frac{du}{\sqrt{(b^2-u^2)(a^2+u^2)}}=\frac{1}{\sqrt{a^2+b^2}}\operatorname{sn}^{-1}\left(\frac{\sqrt{b^2-x^2}}{b},\frac{b}{\sqrt{a^2+b^2}}\right).$$

In our case $a=1$, $x=g(\sigma)$,

$$\frac{\sqrt{b^2-x^2}}{b}=\sqrt{1-\left(\frac{x}{b}\right)^2},$$

$$\begin{aligned}\sqrt{a^2+b^2} &= \sqrt{1+b^2}=\sqrt{1+(2+\sqrt{3})^2}=\sqrt{8+4\sqrt{3}}\\ &= 2\sqrt{2+\sqrt{3}}=2\sqrt{b}.\\ k &= \frac{b}{\sqrt{a^2+b^2}}=\frac{\sqrt{b}}{2}.\\ \frac{2}{\sqrt{2\sqrt{3}-3}}\cdot\frac{1}{\sqrt{a^2+b^2}} &= \frac{2}{\sqrt{2\sqrt{3}-3}}\cdot\frac{1}{2\sqrt{b}}=\frac{1}{\sqrt{b(2\sqrt{3}-3)}}\\ &= \frac{1}{\sqrt{(2+\sqrt{3})(2\sqrt{3}-3)}}=3^{-1/4}.\end{aligned}$$

Hence

$$\int_0^{\sigma^2}\frac{dt}{\sqrt{f(t)}}=3^{-1/4}\operatorname{sn}^{-1}\left(\sqrt{1-\left(\frac{g(\sigma)}{b}\right)^2},k\right).$$

□

• *Quantification of the energy E*
Using Lemma 5.15, equation (5.33) yields

$$\begin{aligned}\operatorname{sn}^{-1}\left(\sqrt{1-\left(\frac{g(\sigma(s))}{b}\right)^2},k\right) &= 3^{1/4}2^{5/3}E^{1/3}\,s\\ \Longleftrightarrow 1-\left(\frac{g(\sigma(s))}{b}\right)^2 &= \operatorname{sn}^2\left(3^{1/4}2^{5/3}E^{1/3}\,s,k\right)\\ \Longleftrightarrow \left(\frac{g(\sigma(s))}{b}\right)^2 &= \operatorname{cn}^2\left(3^{1/4}2^{5/3}E^{1/3}\,s,k\right).\end{aligned}$$

If $s=0$, then $\sigma(0)=\left(\frac{2}{E}\right)^{1/6}r(0)=0$, and $g(0)=\frac{2\sqrt{3}}{\sqrt{3}-1}-1=2+\sqrt{3}=b$. Hence both left and right terms in the above equation are equal to 1.
If $s=1$, then $\sigma(1)=\left(\frac{2}{E}\right)^{1/6}r(1)=\left(\frac{2}{E}\right)^{1/6}R$, and

$$\begin{aligned}g(\sigma(1)) &= \frac{\sqrt{3}}{\sigma^2(1)+\frac{\sqrt{3}-1}{2}}-1=\frac{\sqrt{3}}{\left(\frac{2}{E}\right)^{1/3}R^2+\frac{\sqrt{3}-1}{2}}-1,\\ \left(\frac{g(\sigma(1))}{b}\right)^2 &= \frac{1}{7+4\sqrt{3}}\left[\frac{\sqrt{3}}{\left(\frac{2}{E}\right)^{1/3}R^2+\frac{\sqrt{3}-1}{2}}-1\right]^2.\end{aligned}$$

Let $\tau=E^{1/3}$, and

$$F(\tau)=\frac{1}{7+4\sqrt{3}}\left[\frac{\sqrt{3}}{\frac{2^{1/3}R^2}{\tau}+\frac{\sqrt{3}-1}{2}}-1\right]^2.$$

PROPOSITION 5.16. The function $F(\tau)$ has the following properties:
(i) $\lim_{\tau\to\infty}F(\tau)=1$, i.e. the line $y=1$ is a horizontal asymptote for the graph of F;
(ii) There is an unique $\tau_0>0$ such that $F(\tau_0)=0$;
(iii) F is decreasing on $(0,\tau_0)$ and increasing on $(\tau_0,+\infty)$.

PROOF. (*i*) We have

$$\begin{aligned}\lim_{\tau\to\infty}F(\tau) &= \frac{1}{7+4\sqrt{3}}\left[\frac{2\sqrt{3}}{\sqrt{3}-1}-1\right]^2=\frac{1}{7+4\sqrt{3}}\left[\frac{\sqrt{3}+1}{\sqrt{3}-1}\right]^2\\ &= \frac{1}{7+4\sqrt{3}}(2+\sqrt{3})^2=1.\end{aligned}$$

(*ii*) $F(\tau_0)=0$ if and only if

$$\frac{2^{1/3}R^2}{\tau_0}+\frac{\sqrt{3}-1}{2}=\sqrt{3}\Longleftrightarrow\tau_0=\frac{2^{4/3}R^2}{\sqrt{3}+1}.$$

(*iii*) See Figure 5.3 below. □

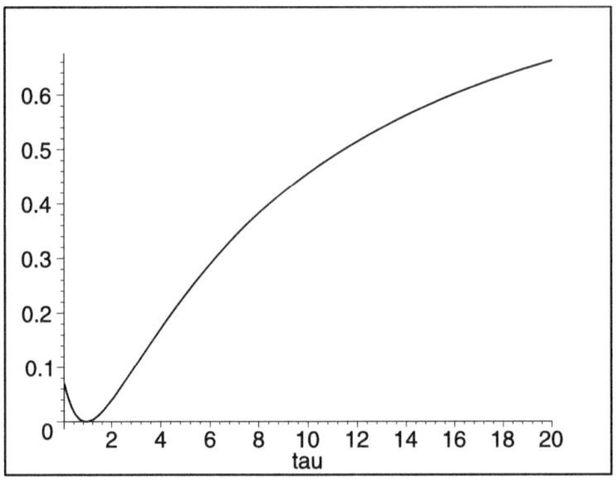

Figure 5.3: The graph of $F(\tau)$.

THEOREM 5.17. *Given $R > 0$, there is an infinite sequence of energies*

$$0 < E_1 < E_2 < E_3 < \cdots < E_n < \cdots < +\infty,$$

parametrized by the solutions $\tau = E^{1/3}$ of the equation

$$F(\tau) = cn^2(3^{1/4}2^{5/3}\tau, k). \tag{5.35}$$

PROOF. As $F(\tau) \nearrow 1$ and cn^2 oscillates between 0 and 1, the equation (5.35) has an infinite number of solutions. See Figure 5.4. $\square$

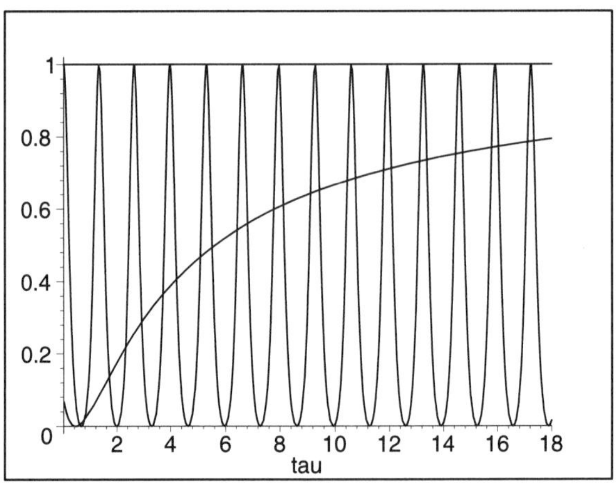

Figure 5.4: The graphs of $F(\tau)$ and $cn^2(3^{1/4}2^{5/3}\tau, k)$.

REMARK 5.18. The energy E provided by Lemma 5.12 is E_1 (the smallest one).

• *Finding $r_n(s)$*
There is a radius function $r_n(s)$ associated with each energy level E_n

$$\int_0^{r_n(s)} \frac{du}{\sqrt{2E_n + 4u^6}} = s.$$

Let $\sigma_n(s) = \left(\frac{2}{E_n}\right)^{1/6} r_n(s)$. As before, an algebraic calculation provides

$$\left(\frac{g(\sigma_n(s))}{b}\right)^2 = \mathrm{cn}^2\Big(3^{1/4}2^{5/3}E_n^{1/3}\, s, k\Big)$$

and then

$$g(\sigma_n(s)) = \pm b\,\mathrm{cn}\Big(3^{1/4}2^{5/3}E_n^{1/3}\, s, k\Big). \tag{5.36}$$

On the other hand

$$\sigma_n(s)^2 = \frac{\sqrt{3}}{1+g(\sigma_n(s))} - \frac{\sqrt{3}-1}{2}, \tag{5.37}$$

and

$$r_n^2(s) = \left(\frac{E_n}{2}\right)^{1/3}\sigma_n^2(s). \tag{5.38}$$

Substituting (5.36) and (5.37) in (5.38), yields

$$r_n^2(s) = \left(\frac{E_n}{2}\right)^{1/3}\left[\frac{\sqrt{3}}{1\pm b\,\mathrm{cn}\Big(3^{1/4}2^{5/3}E_n^{1/3}\, s, k\Big)} - \frac{\sqrt{3}-1}{2}\right], \qquad s\in[0,1]. \tag{5.39}$$

- *Finding the argument $\phi_n(s)$*

Let $\gamma_n = 3^{1/4}2^{5/3}E_n^{1/3}$. Then we have

$$\begin{aligned}\phi_n(s) &= \phi_f + 2i\int_1^s r_n^2(u)\,du\\ &= \phi_f + 2i\left(\frac{E_n}{2}\right)^{1/3}\left[\sqrt{3}\int_1^s \frac{du}{1\pm b\,\mathrm{cn}(\gamma_n u)} - \frac{\sqrt{3}-1}{2}(s-1)\right]\\ &= \phi_f + \frac{3^{1/4}i}{2}\int_{\gamma_n}^{\gamma_n s}\frac{dv}{1\pm b\,\mathrm{cn}v} + i\left(\frac{E_n}{2}\right)^{1/3}(\sqrt{3}-1)(1-s).\end{aligned}$$

We shall show that the integral in the above expression can be computed explicitly using elliptic functions theory.

Following [**42**], p. 93, for $\alpha^2>1$

$$\begin{aligned}\int\frac{du}{1+\alpha\,\mathrm{cn}u} &= -\frac{1}{\alpha^2-1}\Lambda\Big(u, \frac{\alpha}{\sqrt{(\alpha^2-1)}}, k\Big) + \frac{\alpha}{\sqrt{(\alpha^2-1)(k^2+k'^2\alpha^2)}}\\ &\qquad\times\tanh^{-1}\left(\sqrt{\frac{k^2+k'^2\alpha^2}{\alpha^2-1}}\,\mathrm{sd}u\right),\end{aligned}$$

where

$$\Lambda\Big(u,\beta,k\Big) = \int_0^u \frac{dv}{1-\beta^2\mathrm{sn}^2 v}.$$

In our case $\alpha=\pm b$, and $\alpha^2 = (2+\sqrt{3})^2 > 1$.

$$\begin{aligned}\beta &= \frac{\alpha}{\sqrt{\alpha^2-1}} = \frac{\pm b}{\sqrt{b^2-1}} = \pm\frac{2+\sqrt{3}}{\sqrt{6+4\sqrt{3}}} \approx \pm 0.8578,\\ k &= \frac{\sqrt{b}}{2} = \frac{\sqrt{2+\sqrt{3}}}{2} \approx 0.9659.\end{aligned}$$

Following [**42**] p. 69, in the case $0 < \beta < k < 1$ we have

$$\Lambda\Big(u,\beta,k\Big) = u + \frac{\beta}{\sqrt{(k^2-\beta^2)(1-\beta^2)}}\Big[\frac{1}{2}\ln\frac{\Theta(u-F(\theta,k))}{\Theta(u+F(\theta,k))} + uD(\theta,k) - \frac{E}{K}uF(\theta,k)\Big],$$

where $\Theta(u) = \theta_4(x)$, with $x = 2Kx/\pi$ and

$$\theta_4(z,q) = 1 + 2\sum_{n=1}^{\infty}(-1)^n q^{n^2}\cos 2nz = \sum_{n=-\infty}^{\infty}(-1)^n q^{n^2} e^{2inz},$$

and

$$F(\theta,k) = \mathrm{sn}^{-1}(\beta/k), \qquad \sin\theta = \beta/k, \qquad D(\theta,k) = E(F(\theta,k)),$$

$$E(u,k) = \int_0^u \mathrm{dn}^2(v,k)\,dv, \qquad E = E(K,k),$$

$$K = \int_0^{\pi/2}(1-k^2\sin^2\theta)^{-1/2}\,d\theta = \int_0^1 \frac{dt}{\sqrt{(1-t^2)(1-k^2t^2)}} = \frac{3^{1/4}}{2\sqrt{\pi}}\Gamma\Big(\frac{1}{6}\Big)\Gamma\Big(\frac{1}{3}\Big),$$

see [**42**] p. 86.
To conclude the case $k = 1$, we have the following result:

THEOREM 5.19. *Given a point $P(x)$ in the plane, there are infinitely many complex geodesics in $\mathbb{R}^3$ such that their projections onto the x-plane join the origin and the point P.*

5.6. The complex action

Let $(x_1(s), x_2(s), t(s), \xi_1(s), \xi_2(s), \theta(s))$, $s \in [0,\tau]$ be a solution for the Hamiltonian system given by the quantum Hamiltonian (5.4) with the non-standard boundary conditions

(5.40)
$$x_1(0) = 0, \qquad x_2(0) = 0, \qquad x_1(\tau) = x_1, \qquad x_2(\tau) = x_2, \qquad t(\tau) = t, \qquad \theta = -i.$$

DEFINITION 5.20. The complex action associated with the above solution is

$$g = g(x,t,\tau) = -it + \int_0^\tau \Big[\langle\xi,\dot{x}\rangle - H(s)\Big]\,ds. \tag{5.41}$$

One may write a formal relation between the complex action and the classical action S

$$g = S - it(0),$$

with

$$S = S(x,t,\tau,\theta) = \int_0^\tau \Big[\langle\xi,\dot{x}\rangle + \theta\dot{t} - H(s)\Big]\,ds,$$

where $\theta = -i$. The Lagrangian is given by

$$L = \langle\xi,\dot{x}\rangle + \theta\dot{t} - H.$$

A computation (see chapter 3) yields

$$L = \frac{1}{2}(\dot{x}_1^2 + \dot{x}_2^2) + \theta\Big(\dot{t} - 2|x|^{2k}(x_2\dot{x}_1 - x_1\dot{x}_2)\Big). \tag{5.42}$$

The Euler-Lagrange system is

$$\text{(5.43)} \qquad \begin{cases} \ddot{x}_1 = 4\theta(k+1)|x|^{2k}\dot{x}_2 \\ \ddot{x}_2 = -4\theta(k+1)|x|^{2k}\dot{x}_1, \end{cases}$$

with boundary conditions (5.40). Multiplying the first equation by $\dot{x}_1$ and the second by $\dot{x}_2$ and adding, yields $\ddot{x}_1\dot{x}_1 + \ddot{x}_2\dot{x}_2 = 0$, *i.e.*

$$\frac{1}{2}(\dot{x}_1^2 + \dot{x}_2^2) = E,$$

where the energy constant E is real. Multiplying the first equation of the system (5.43) by x_2 and the second equation by x_1, yields

$$\begin{aligned} \ddot{x}_1 x_2 &= 4\theta(k+1)|x|^{2k}\dot{x}_2 x_2 \\ \ddot{x}_2 x_1 &= -4\theta(k+1)|x|^{2k}\dot{x}_1 x_1. \end{aligned}$$

Let $r = |x|$. Subtracting, we have

$$\begin{aligned} \ddot{x}_1 x_2 - \ddot{x}_2 x_1 &= 4\theta(k+1)|x|^{2k}(\dot{x}_2 x_2 + \dot{x}_1 x_1) \\ \Longleftrightarrow (\dot{x}_1 x_2 - \dot{x}_2 x_1)^{\cdot} &= 2\theta(k+1)r^{2k}(2x_2\dot{x}_2 + 2x_1\dot{x}_1) \\ &= 2\theta(k+1)r^{2k}(x_1^2 + x_2^2)^{\cdot} \\ &= 2\theta(k+1)r^{2k}(r^2)^{\cdot} \\ &= 2\theta\Big((r^2)^{k+1}\Big)^{\cdot} \end{aligned}$$

and there is a constant Ω, called momentum, such that

$$\dot{x}_1 x_2 - \dot{x}_2 x_1 = 2\theta r^{2(k+1)} + \Omega.$$

Since $x(0) = 0$, then $\Omega = 0$. Hence, the Lagrangian (5.42) becomes

$$\begin{aligned} L &= E + \theta(\dot{t} - 2r^{2k}\, 2\theta r^{2k+2}) \\ &= E - i\dot{t} + 4r^{4k+2}, \end{aligned}$$

where we considered $\theta = -i$. The complex action becomes

$$\begin{aligned} g &= S - it(0) = \int_0^\tau L\, ds - it(0) \\ &= \int_0^\tau \Big(E - i\dot{t} + 4r^{4k+2}(s)\Big)\, ds - it(0) \\ &= E\tau - it(\tau) + 4\int_0^\tau r^{4k+2}(s)\, ds. \end{aligned}$$

We need to work out the last term of the above relation. This will be done in the following result.

LEMMA 5.21. *Let* $r(s) = \sqrt{x_1^2(s) + x_2^2(s)}$ *and* $r(\tau) = r$. *Then*

$$\text{(5.44)} \qquad 4\int_0^\tau r^{4k+2}(s)\, ds = \frac{r}{k+1}\sqrt{\frac{E}{2} + r^{4k+2}} - \frac{E\tau}{k+1}.$$

PROOF. Let

$$J = \int_0^\tau r^{4k+2}(s)\, ds.$$

From the proof of Lemma 5.12

$$ds = \frac{dr}{\sqrt{2E + 4r^{4k+2}}},$$

and substituting in J we have

$$J=\int_0^r \frac{u^{4k+2}}{\sqrt{2E+4u^{4k+2}}}\,du.$$

With $v=\left(2/E\right)^{\frac{1}{4k+2}}u$ and $\sigma=\left(2/E\right)^{\frac{1}{4k+2}}r$,

$$\begin{aligned}
4J &= \frac{4}{\sqrt{2E}}\int_0^r \frac{u^{4k+2}}{\sqrt{1+\frac{2}{E}u^{4k+2}}}\,du = \frac{4}{\sqrt{2E}}\int_0^\sigma \frac{\frac{E}{2}v^{4k+2}\left(\frac{E}{2}\right)^{\frac{1}{4k+2}}}{\sqrt{1+v^{4k+2}}}\,dv\\
&= \frac{4E^{\frac{k+1}{2k+1}}}{2^{\frac{3k+2}{2k+1}}}\int_0^\sigma \frac{v^{4k+2}}{\sqrt{1+v^{4k+2}}}\,dv = 2^{\frac{k}{2k+1}}\,E^{\frac{k+1}{2k+1}}\int_0^\sigma \frac{v^{4k+2}}{\sqrt{1+v^{4k+2}}}\,dv.
\end{aligned}$$

Let

$$I=\int_0^\sigma \frac{v^{4k+2}}{\sqrt{1+v^{4k+2}}}\,dv.$$

We shall compute I by cyclic integration. Before that we need to compute the integral $\int_0^\sigma \frac{dv}{\sqrt{1+v^{4k+2}}}$. It is known that

$$\int_0^r \frac{du}{\sqrt{2E+4u^{4k+2}}}=\tau.$$

With the substitutions $v=\left(2/E\right)^{\frac{1}{4k+2}}u$ and $\sigma=\left(2/E\right)^{\frac{1}{4k+2}}r$ the above integral becomes

$$\int_0^\sigma \frac{dv}{\sqrt{1+v^{4k+2}}}=2^{\frac{k+1}{2k+1}}\,E^{\frac{k}{2k+1}}\,\tau. \tag{5.45}$$

Cyclic integration yields

$$\begin{aligned}
I &= \int_0^\sigma \frac{v^{4k+2}}{\sqrt{1+v^{4k+2}}}\,dv = \frac{1}{2k+1}\int_0^\sigma v\cdot\frac{(2k+1)v^{4k+1}}{\sqrt{1+v^{4k+2}}}\,dv\\
&= \frac{1}{2k+1}\int_0^\sigma v\left(\sqrt{1+v^{4k+2}}\right)'dv\\
&= \frac{1}{2k+1}\left(\sigma\sqrt{1+\sigma^{4k+2}}-\int_0^\sigma \sqrt{1+v^{4k+2}}\,dv\right)\\
&= \frac{1}{2k+1}\left(\sigma\sqrt{1+\sigma^{4k+2}}-\int_0^\sigma \frac{1+v^{4k+2}}{\sqrt{1+v^{4k+2}}}\,dv\right).
\end{aligned}$$

Hence

$$(2k+1)I=\sigma\sqrt{1+\sigma^{4k+2}}-\int_0^\sigma \frac{dv}{\sqrt{1+v^{4k+2}}}-I.$$

Solving for I and using equation (5.45), yields

$$\begin{aligned}
I &= \frac{1}{2(k+1)}\left(\sigma\sqrt{1+\sigma^{4k+2}}-2^{\frac{k+1}{2k+1}}\,E^{\frac{k}{2k+1}}\,\tau\right)\\
&= \frac{1}{2(k+1)}\sigma\sqrt{1+\sigma^{4k+2}}-\frac{E^{\frac{k}{2k+1}}}{2^{\frac{k}{2k+1}}}\frac{\tau}{k+1}.
\end{aligned}$$

Then

$$\begin{aligned} 4J &= 2^{\frac{k}{2k+1}} E^{\frac{k+1}{2k+1}} I \\ &= 2^{\frac{k}{2k+1}} E^{\frac{k+1}{2k+1}} \frac{1}{2(k+1)} \sigma\sqrt{1+\sigma^{4k+2}} - \frac{E\tau}{k+1}. \end{aligned}$$

Next we shall compute $\sigma\sqrt{1+\sigma^{4k+2}}$.

$$\begin{aligned} \sigma\sqrt{1+\sigma^{4k+2}} &= \Big(\frac{2}{E}\Big)^{\frac{1}{4k+2}} r\sqrt{1+\frac{2}{E}r^{4k+2}} = \Big(\frac{2}{E}\Big)^{\frac{1}{4k+2}} \frac{1}{\sqrt{E}} r\sqrt{E+2r^{4k+2}} \\ &= \Big(\frac{2}{E}\Big)^{\frac{1}{4k+2}} \frac{\sqrt{2}}{\sqrt{E}} r\sqrt{\frac{E}{2}+r^{4k+2}} = \Big(\frac{2}{E}\Big)^{\frac{k+1}{2k+1}} r\sqrt{\frac{E}{2}+r^{4k+2}}. \end{aligned}$$

Then

$$\begin{aligned} 4J &= 2^{\frac{k}{2k+1}} E^{\frac{k+1}{2k+1}} \frac{1}{2(k+1)} \Big(\frac{2}{E}\Big)^{\frac{k+1}{2k+1}} r\sqrt{\frac{E}{2}+r^{4k+2}} - \frac{E\tau}{k+1} \\ &= \frac{r}{k+1}\sqrt{\frac{E}{2}+r^{4k+2}} - \frac{E\tau}{k+1}. \end{aligned}$$

□

Lemma 5.21 is useful for computing the complex action.

THEOREM 5.22. *The complex action is given by*

$$g = -it + \frac{1}{k+1}\left(kE\tau + r\sqrt{\frac{E}{2}+r^{4k+2}}\right).$$

PROOF. A straightforward computation shows

$$\begin{aligned} g &= E\tau - it(\tau) + 4\int_0^\tau r^{4k+2}(s) \\ &= E\tau - it(\tau) + \frac{r}{k+1}\sqrt{\frac{E}{2}+r^{4k+2}} - \frac{E\tau}{k+1} \\ &= -it + \frac{1}{k+1}\left(kE\tau + r\sqrt{\frac{E}{2}+r^{4k+2}}\right), \end{aligned}$$

where we used the boundary condition $t(\tau) = t$.

□

The complex action from the origin is defined for any $\tau > 0$ for which $r(\tau)$ makes sense.

The complex action depends on the energy parameter E. Only in the step 2 case one may compute the energy E explicitly as a unique function of the boundary condition $r = r(\tau)$. In the case $k = 1$ this is not possible because there are many energies E, which form an infinite increasing sequence. In the case when the complex action is not starting from the origin, there is the second parameter, the momentum Ω, see [**4**]. When the action starts from the origin, Ω vanishes. We shall deal more with this issue in section 5.10.

In the step 2 case the energy E depends on $r(\tau) = r$ as follows:

$$\tau = \int_0^r \frac{du}{\sqrt{2E+4u^2}} = \frac{1}{2}\int_0^{\sqrt{\frac{2}{E}}r} \frac{dv}{\sqrt{1+v^2}} = \frac{1}{2}\sinh^{-1}\Big(\sqrt{\frac{2}{E}}r\Big).$$

Solving for E yields

$$E = \frac{2r^2}{\sinh^2(2\tau)}. \tag{5.46}$$

Using the above expression of the energy we can compute the complex action in the step 2 case without using the parameter E.

COROLLARY 5.23. When $k = 0$, the complex action is

$$g = -it + r^2 \coth(2\tau).$$

PROOF. Making $k = 0$ in Theorem 5.22 yields

$$g = -it + r\sqrt{\frac{E}{2} + r^2}.$$

Using (5.46) we can work out the term under the square root.

$$\begin{aligned}\frac{E}{2} + r^2 &= \frac{r^2}{\sinh^2(2\tau)} + r^2 = \frac{1 + \sinh^2(2\tau)}{\sinh^2(2\tau)} r^2 \\ &= r^2 \frac{\cosh^2(2\tau)}{\sinh^2(2\tau)} = r^2 \coth^2(2\tau).\end{aligned}$$

Hence

$$g = -it + r\sqrt{\frac{E}{2} + r^2} = -it + r^2 \coth(2\tau).$$

□

5.7. Hamilton-Jacobi equation

In this section we shall deal with the Hamilton-Jacobi equation for the complex action. We shall describe the complex geodesic flow as a horizontal gradient of the complex action. The following computations follow from classical ideas of Lagrangian and Hamiltonian formalism. The momentum θ will be considered arbitrary, if otherwise stated. Recall the Hamiltonian

$$H(x,t,\xi,\theta) = \frac{1}{2}\Big(\xi_1 + 2\theta x_2|x|^{2k}\Big)^2 + \frac{1}{2}\Big(\xi_2 - 2\theta x_1|x|^{2k}\Big)^2,$$

which is associated with the Lagrangian

$$L(x,t,\dot{x},\dot{t}) = \frac{1}{2}(\dot{x}_1^2 + \dot{x}_2^2) + \theta\Big(\dot{t} - 2|x|^{2k}(\dot{x}_1 x_2 - x_1\dot{x}_2)\Big). \tag{5.47}$$

The variables x_1, x_2, t are the coordinates and ξ_1, ξ_2, θ are the momenta. The following result provides a relation between the Lagrangian and the momenta.

LEMMA 5.24. *Along the solution (x,t,ξ,θ) of the Hamiltonian system, we have*

$$\theta = \frac{\partial L}{\partial \dot{t}}, \qquad \xi_i = \frac{\partial L}{\partial \dot{x}_i}, \quad i = 1, 2.$$

PROOF. The first relation is obvious. Differentiating in the Lagrangian (5.47)

$$\frac{\partial L}{\partial \dot{x}_1} = \dot{x}_1 - 2\theta x_2|x|^{2k} = \xi_1,$$

where we used Hamiltonian equation

$$\dot{x}_1 = H_{\xi_1} = \xi_1 + 2\theta x_2|x|^{2k}.$$

A similar computation provides $\xi_2 = \frac{\partial L}{\partial \dot{x}_2}$.

□

The following lemma gives a relation between the momenta and the action.

LEMMA 5.25. *Along the solution (x, t, ξ, θ) of the Hamiltonian system we have*

$$\begin{aligned} (1) \qquad \frac{\partial g}{\partial x_j} &= \xi_j, \qquad j = 1, 2 \\ (2) \qquad \frac{\partial g}{\partial t} &= \theta \; (= -i). \end{aligned}$$

PROOF. (1) Using Euler-Lagrange equations and Lemma 5.24

$$\begin{aligned} \frac{\partial g}{\partial x_j} &= \frac{\partial}{\partial x_j} \int L(s)\, ds = \int \frac{\partial L}{\partial x_j}\, ds \\ &= \int \frac{d}{ds}\Big(\frac{\partial L}{\partial \dot{x}_j}\Big)\, ds = \frac{\partial L}{\partial \dot{x}_j} = \xi_j. \end{aligned}$$

(2) Replace x_j by t in the calculation of (1) or just differentiate in the formula of g given by the Theorem 5.22.

□

THEOREM 5.26. *The complex action given in Theorem 5.22 satisfies the Hamilton-Jacobi equation*

$$\frac{\partial g}{\partial \tau} + \frac{1}{2}(X_1 g)^2 + \frac{1}{2}(X_1 g)^2 = 0. \tag{5.48}$$

PROOF. Consider the total derivative with respect to τ of the complex action $g = g(x, t, \tau)$

$$\frac{dg}{d\tau} = \frac{dg}{dx_j}\dot{x}_j + \frac{dg}{dt}\dot{t} + \frac{\partial g}{\partial \tau}, \tag{5.49}$$

where $\dot{x} = dx(\tau)/d\tau$ and $\dot{t} = dt(\tau)/d\tau$. On the other hand,

$$\frac{dg}{d\tau} = \frac{d}{d\tau}\int_0^\tau L(x, t, \xi, \theta)\, ds = L(x, t, \xi, \theta). \tag{5.50}$$

Substituting in (5.49) and using Lemma 5.25 yields

$$\begin{aligned} \frac{\partial g}{\partial \tau} &= \frac{dg}{d\tau} - \Big(\frac{\partial g}{\partial x_j}\dot{x}_j + \frac{\partial g}{\partial t}\dot{t}\Big) \\ &= L(x, t, \dot{x}, \dot{t}) - (\xi_j \dot{x}_j + \theta \dot{t}) \\ &= -\Big((\xi_j \dot{x}_j + \theta \dot{t}) - L(x, t, \dot{x}, \dot{t})\Big) = -H(x, t, \xi, \theta) \\ &= -H\Big(x, t, \frac{\partial g}{\partial x}, \frac{\partial g}{\partial t}\Big). \end{aligned}$$

Hence

$$\frac{\partial g}{\partial \tau} + H\Big(x, t, \frac{\partial g}{\partial x}, \frac{\partial g}{\partial t}\Big) = 0. \tag{5.51}$$

The Hamiltonian can be also written as

$$\begin{aligned} H\Big(x,t,\frac{\partial g}{\partial x},\frac{\partial g}{\partial t}\Big) &= \frac{1}{2}\Big(\xi_1+2\theta x_2|x|^{2k}\Big)^2+\frac{1}{2}\Big(\xi_2-2\theta x_1|x|^{2k}\Big)^2 \\ &= \frac{1}{2}\Big(\frac{\partial g}{\partial x_1}+2x_2|x|^{2k}\frac{\partial g}{\partial t}\Big)^2+\frac{1}{2}\Big(\frac{\partial g}{\partial x_2}-2x_1|x|^{2k}\frac{\partial g}{\partial t}\Big)^2 \\ &= \frac{1}{2}(X_1g)^2+\frac{1}{2}(X_2g)^2, \end{aligned}$$

and the Hamilton-Jacobi equation (5.51) becomes (5.48).

□

• *Application to step 2 case*

We shall use the Hamilton-Jacobi equation to compute the energy in the step 2 case. Unfortunately, this is the only case when the computation can be performed explicitly. It is known that the action in this case is

$$g(\tau)=-it+r^2\coth(2\tau),$$

and it satisfies the Hamilton-Jacobi equation

$$\frac{\partial g}{\partial \tau}+H=0.$$

As $dH/d\tau=0$, the Hamiltonian is constant along the solutions and its value will be the energy $E=H$. Then

$$\begin{aligned} E &= -\frac{\partial g}{\partial \tau}=-\frac{\partial}{\partial \tau}\Big(-it+r^2\coth(2\tau)\Big) \\ &= -r^2\frac{\partial}{\partial \tau}\coth(2\tau)=-2r^2(1-\coth^2(2\tau)) \\ &= -2r^2\frac{\sinh^2(2\tau)-\cosh^2(2\tau)}{\sinh^2(2\tau)} \\ &= \frac{2r^2}{\sinh^2(2\tau)}. \end{aligned}$$

• *Application to complex geodesic flow*

By similarity with the definition of the gradient in the elliptic case, we have the following definition.

DEFINITION 5.27. Let $f:[0,\tau]\to\mathbb{C}$ be a function. The horizontal gradient of f is defined by

$$\nabla_X f=(X_1f)X_1+(X_2f)X_2.$$

The following result provides the complex geodesic flow when the complex action is known.

PROPOSITION 5.28. If $c=(x_1,x_2,t)$ is a complex geodesic, then the velocity is

$$\dot c=\nabla_X g.$$

PROOF. From the Hamiltonian equation $\dot{t} = H_\theta = 2|x|^{2k}(\dot{x}_1 x_2 - x_1\dot{x}_2)$. Following a computation made at the beginning of chapter 3

$$\begin{aligned}
\dot{c} &= \dot{x}_1\partial_{x_1} + \dot{x}_2\partial_{x_2} + \dot{t}\partial_t \\
&= \dot{x}_1 X_1 + \dot{x}_2 X_2 + \Big(\dot{t} - 2|x|^{2k}(\dot{x}_1 x_2 - x_1\dot{x}_2)\Big)\partial_t \\
&= \dot{x}_1 X_1 + \dot{x}_2 X_2.
\end{aligned}$$

On the other hand, using Hamiltonian equations and Lemma 5.25

$$\begin{aligned}
\dot{x}_1 &= H_{\xi_1} = \xi_1 + 2\theta x_2|x|^{2k} \\
&= \frac{\partial g}{\partial x_1} + 2x_2|x|^{2k}\frac{\partial g}{\partial t} \\
&= X_1 g. \\
\dot{x}_2 &= H_{\xi_2} = \xi_2 - 2\theta x_1|x|^{2k} \\
&= \frac{\partial g}{\partial x_2} - 2x_1|x|^{2k}\frac{\partial g}{\partial t} \\
&= X_2 g.
\end{aligned}$$

Hence

$$\dot{c} = \dot{x}_1 X_1 + \dot{x}_2 X_2 = (X_1 g)X_1 + (X_2 g)X_2 = \nabla_X g.$$

□

5.8. Geodesic completeness

DEFINITION 5.29. If for any point P in the x-plane, any complex geodesic $c(s) = (x(s), t(s)) \in \mathbb{C}^3$ emanating from P is defined for all $s \in \mathbb{R}$, then we say the geometry is geodesic complete.

It suffice to show the geodesic completeness just for the x-component, because the t component will follow. In the following we shall study the geodesic completeness for complex geodesics. We shall investigate whether the solutions of system (5.6) are defined for all $s \in \mathbb{R}$. We shall use the action g and the tools developed in the previous sections.

Considering the action g known, we shall investigate the system

$$\dot{x}_1 = X_1(g), \quad \dot{x}_2 = X_2(g).$$

In polar coordinates $x_1 = r\cos\phi$, $x_2 = r\sin\phi$, we have

$$\begin{aligned}
\partial_r &= \cos\phi\,\partial_{x_1} + \sin\phi\,\partial_{x_2}, \\
\partial_\phi &= -r\sin\phi\,\partial_{x_1} + r\cos\phi\,\partial_{x_2}.
\end{aligned}$$

Inverting, yields

$$\begin{aligned}
\partial_{x_1} &= \cos\phi\,\partial_r - \frac{\sin\phi}{r}\partial_\phi, \\
\partial_{x_2} &= \sin\phi\,\partial_r + \frac{\cos\phi}{r}\partial_\phi.
\end{aligned}$$

In polar coordinates, the vector fields become

$$\begin{aligned} X_1 &= \partial_{x_1} + 2x_2|x|^{2k}\partial_t \\ &= \cos\phi\,\partial_r - \frac{\sin\phi}{r}\partial_\phi + 2r^{2k+1}\sin\phi\,\partial_t. \\ X_2 &= \partial_{x_2} - 2x_1|x|^{2k}\partial_t \\ &= \sin\phi\,\partial_r + \frac{\cos\phi}{r}\partial_\phi - 2r^{2k+1}\cos\phi\,\partial_t. \end{aligned}$$

Relations $\partial_\phi g = 0$ and $\partial_t g = -i$ yield

$$\begin{aligned} \dot{x}_1 &= X_1 g = \cos\phi\,\partial_r g - 2ir^{2k+1}\sin\phi, \\ \dot{x}_2 &= X_2 g = \sin\phi\,\partial_r g + 2ir^{2k+1}\cos\phi. \end{aligned}$$

Differentiating in the polar coordinates formula,

$$\begin{aligned} \dot{x}_1 &= \dot{r}\cos\phi - r\sin\phi\,\dot{\phi}, \\ \dot{x}_2 &= \dot{r}\sin\phi + r\cos\phi\,\dot{\phi}. \end{aligned}$$

Equating the last two sets of identities, we obtain

$$\begin{aligned} \dot{r}\cos\phi - r\sin\phi\,\dot{\phi} &= \cos\phi\,\partial_r g - 2ir^{2k+1}\sin\phi, \\ \dot{r}\sin\phi + r\cos\phi\,\dot{\phi} &= \sin\phi\,\partial_r g + 2ir^{2k+1}\cos\phi \end{aligned}$$

$\Longleftrightarrow$

$$\begin{aligned} \cos\phi\Big(\partial_r g - \dot{r}\Big) + \sin\phi\Big(r\dot{\phi} - 2ir^{2k+1}\Big) &= 0, \\ \sin\phi\Big(\partial_r g - \dot{r}\Big) - \cos\phi\Big(r\dot{\phi} - 2ir^{2k+1}\Big) &= 0 \end{aligned}$$

$\Longleftrightarrow$

$$\underbrace{\begin{pmatrix} \cos\phi & \sin\phi \\ \sin\phi & -\cos\phi \end{pmatrix}}_{\det\neq 0}\begin{pmatrix} \partial_r g - \dot{r} \\ r\dot{\phi} - 2ir^{2k+1} \end{pmatrix} = \begin{pmatrix} 0 \\ 0 \end{pmatrix}. \tag{5.52}$$

Hence

$$\begin{aligned} \dot{r} &= \partial_r g, \\ \dot{\phi} &= 2ir^{2k}. \end{aligned}$$

Using the action g provided by Theorem 5.22,

$$\begin{aligned} \partial_r g &= \frac{1}{k+1}\left[\frac{1}{2}\sqrt{2E + 4r^{4k+2}} + \frac{(4k+2)r^{4k+2}}{\sqrt{2E+4r^{4k+2}}}\right] \\ &> \frac{1}{k+1}\cdot\frac{1}{2}\sqrt{2E+4r^{4k+2}} > \frac{1}{k+1}r^{2k+1}, \end{aligned}$$

and hence

$$\dot{r} > \frac{1}{k+1}r^{2k+1}. \tag{5.53}$$

For $k \geq 1$ the derivative of r is increasing faster than $\mathrm{O}(r^2)$, and we shall show that r cannot be defined globally for any $\tau \in \mathbb{R}$.

Separating and integrating, (5.53) yields

$$\begin{aligned}
\int \frac{dr}{r^{2k+1}} &> \int \frac{d\tau}{k+1}\\
\Longleftrightarrow \frac{r^{-2k}}{-2k} &> \frac{1}{k+1}(\tau - C)\\
\Longleftrightarrow \frac{1}{r^{2k}} &< \frac{2k}{k+1}(C-\tau)\\
\Longleftrightarrow r^{2k}(\tau) &> \frac{k+1}{2k(C-\tau)},
\end{aligned}$$

where C is the integration constant. Hence

$$\lim_{\tau \longrightarrow C} r^{2k}(\tau) = +\infty,$$

and $r(\tau)$ blows up for $\tau = C$.

In the case $k = 0$, using Corollary 5.23 yields

$$\begin{aligned}
\partial_r g &= 2r\coth(2\tau) \Longrightarrow\\
\dot r &= 2r\coth(2\tau),
\end{aligned}$$

which together with the initial solution $r(0) = 0$ has a global solution $r(\tau)$ with $\tau \in [0, \infty)$, see Exercise 5.3.

We arrived at the following result.

THEOREM 5.30. *Let $k \geq 1$ be an integer. Then the complex geodesics starting at the origin cannot be extended indefinitely on the whole real line $\mathbb{R}$. This can be done only for the case $k = 0$, i.e. in the step 2 case.*

With the general complex action formula developed in section 5.9 and similar computations, one may show that the complex geodesics starting outside of the origin cannot be extended indefinitely for the case $k \geq 1$.

5.9. Solving Hamilton-Jacobi equation

In this section we shall work out the complex action reducing the Hamilton-Jacobi equation to an eiconal equation. This works out well because the Hamiltonian is conservative (since does not contain τ explicitly).

• *The Hamiltonian in polar coordinates*

In this paragraph we shall write the Hamiltonian

$$H(x,t,\xi,\theta) = \frac{1}{2}\Big(\xi_1 + 2\theta x_2|x|^{2k}\Big)^2 + \frac{1}{2}\Big(\xi_2 - 2\theta x_1|x|^{2k}\Big)^2 \tag{5.54}$$

in polar coordinates. The associated Lagrangian is given by (5.47). In polar coordinates (ϕ, r) the Lagrangian is

$$L = \frac{1}{2}(\dot r^2 + r^2\dot\phi^2) + \theta(\dot t + 2r^{2k+2}\dot\phi).$$

Let p_ϕ, p_r, p_t be the momenta associated with the coordinates ϕ, r, t.

$$\begin{aligned} p_\phi &= \frac{\partial L}{\partial \dot\phi} = r^2\dot\phi + 2\theta r^{2k+2} \Longrightarrow \dot\phi = \frac{p_\phi}{r^2} - 2\theta r^{2k}, \\ p_r &= \frac{\partial L}{\partial \dot r} = \dot r \Longrightarrow \dot r = p_r, \\ p_t &= \frac{\partial L}{\partial \dot t} = \theta \Longrightarrow \theta = p_t. \end{aligned}$$

Legendre transform yields the Hamiltonian

$$\begin{aligned} H &= p_\phi\dot\phi + p_r\dot r + p_t\dot t - L \\ &= (r^2\dot\phi + 2\theta r^{2k+2})\dot\phi + \dot r^2 + \theta\dot t \\ &\quad -\frac{1}{2}\dot r^2 - \frac{1}{2}r^2\dot\phi^2 - \theta\dot t - 2\theta r^{2k+2}\dot\phi \\ &= \frac{1}{2}(\dot r^2 + r^2\dot\phi^2) \\ &= \frac{1}{2}p_r^2 + \frac{1}{2}\Big(\frac{p_\phi}{r} - 2p_t r^{2k+1}\Big)^2. \end{aligned}$$

PROPOSITION 5.31. The Hamiltonian (5.54) can be expressed in polar coordinates (ϕ, r) as

$$H(r,\phi,t,p_r,p_\phi,p_t) = \frac{1}{2}p_r^2 + \frac{1}{2}\Big(\frac{p_\phi}{r} - 2p_t r^{2k+1}\Big)^2. \tag{5.55}$$

The Hamiltonian does not depend on the variables t and ϕ. This yields the conservations

$$\begin{aligned} \dot p_t &= -H_t = 0 \Longrightarrow p_t = \theta \text{ (constant)}, \\ \dot p_\phi &= -H_\phi = 0 \Longrightarrow r^2(\dot\phi + 2\theta r^{2k}) = C \text{ (momentum constant)}. \end{aligned}$$

• *The eiconal equation*

We shall solve the following Hamilton-Jacobi equation in polar coordinates

$$\frac{\partial S}{\partial \tau} + H\Big(\phi, r, \frac{\partial S}{\partial \phi}, \frac{\partial S}{\partial r}, \frac{\partial S}{\partial t}\Big) = 0. \tag{5.56}$$

We shall look for a solution $S(\phi, r, t, \tau)$ of (5.56) with separable variables

$$S(\phi, r, t, \tau) = W(\phi, r) + U(\tau) + V(t). \tag{5.57}$$

Substituting in the Hamilton-Jacobi equation (5.56) yields

$$U'(\tau) + H = 0.$$

Using the constant value of the Hamiltonian along the solutions $H = E$ we have

$$U(\tau) = -E\tau,$$

where E is the energy constant. Consider E as a parameter. Substituting in (5.56) yields an eiconal equation for W

$$H\Big(\phi, r, \frac{\partial W}{\partial \phi}, \frac{\partial W}{\partial r}, \frac{\partial V}{\partial t}\Big) = E \Longleftrightarrow$$

$$\frac{1}{2}\Big(\frac{\partial W}{\partial r}\Big)^2 + \frac{1}{2}\Big(\frac{1}{r}\frac{\partial W}{\partial \phi} - 2\frac{\partial V}{\partial t}r^{2k+1}\Big)^2 = E. \tag{5.58}$$

We shall consider the ansatz: W does not depend on ϕ, *i.e.* $\partial W/\partial\phi = 0$, and $\partial V/\partial t = \theta$.
Then

$$V(t) = \theta t + \alpha,$$

where α is a constant.
The equation (5.58) becomes a first order ODE

$$\frac{\partial W}{\partial r} = \pm\sqrt{2E - 4\theta^2 r^{4k+2}}.$$

Taking $\theta = -i$ and the positive sign, we have

$$\frac{\partial W}{\partial r} = \sqrt{2E + 4r^{4k+2}}.$$

We assume the geodesic starts at $r(0) = r_0$. Then W will depend on parameters E and r_0

$$W(r, r_0, E) = \int_{r_0}^{r} \sqrt{2E + 4u^{4k+2}}\,du = \underbrace{\int_0^r \sqrt{2E + 4u^{4k+2}}\,du}_{=I_1} - \underbrace{\int_0^{r_0} \sqrt{2E + 4u^{4k+2}}\,du.}_{=I_0}$$

We shall compute each of the above integrals.

$$\begin{aligned} I_1 &= \int_0^r \frac{2E + 4u^{4k+2}}{\sqrt{2E + 4u^{4k+2}}}\,du \\ &= 2E\int_0^r \frac{1}{\sqrt{2E + 4u^{4k+2}}}\,du + 4\int_0^r \frac{u^{4k+2}}{\sqrt{2E + 4u^{4k+2}}}\,du \\ &= 2E\tau + 4J, \end{aligned}$$

with J defined in Lemma 5.21. Substituting the expression for J, yields

$$I_1 = 2E\tau + \frac{r}{k+1}\sqrt{\frac{E}{2} + r^{4k+2}} - \frac{E\tau}{k+1}.$$

Replacing τ by $\tau_0 = 0$ and r by r_0 yields an expression for I_0

$$I_0 = \frac{r_0}{k+1}\sqrt{\frac{E}{2} + r_0^{4k+2}}.$$

Summing up, yields

$$\begin{aligned} S(r, t, \tau) &= W(\phi, r) + U(\tau) + V(t) = I_1(r) - I_0(r_0) + U(\tau) + V(t) \\ &= 2E\tau + \frac{r}{k+1}\sqrt{\frac{E}{2} + r^{4k+2}} - \frac{E\tau}{k+1} - \frac{r_0}{k+1}\sqrt{\frac{E}{2} + r_0^{4k+2}} - E\tau + \theta t + \alpha \\ &= -it + \frac{1}{k+1}\left(kE\tau + r\sqrt{\frac{E}{2} + r^{4k+2}} - r_0\sqrt{\frac{E}{2} + r_0^{4k+2}} + \alpha\right). \end{aligned}$$

Let $S_0 = S(r_0, 0, 0)$. Then the constant $\alpha = S_0$. Taking $S_0 = 0$ we arrive at the following result, obtained also by Beals, Gaveau and Greiner [**4**] in a different way.

THEOREM 5.32. *The complex action, which does not start at the origin, is given by*

$$g(r, r_0, t, E, \tau) = -it + \frac{1}{k+1}\left(kE\tau + r\sqrt{\frac{E}{2} + r^{4k+2}} - r_0\sqrt{\frac{E}{2} + r_0^{4k+2}}\right). \tag{5.59}$$

When $r_0 = 0$, we obtain Theorem 5.22.

5.10. Theorem of Jacobi and applications

In this section we shall show the non-uniqueness of the complex geodesics between any two given points for the case $k \geq 1$. We provide the complete picture of the number of geodesics from the origin and from outside of the origin. In order to do this we need to take advantage of Jacobi's method of integration of the Hamiltonian system when a solution for the Hamilton-Jacobi equation is known. We start by recalling a few known definitions and results, see [**27**], vol. II, p.366. Consider a solution $S(\tau, x, a)$ of the Hamilton-Jacobi equation

$$\frac{\partial S}{\partial \tau} + H(x, S_x) = 0, \tag{5.60}$$

which depends on n parameters $a = (a^1, \ldots, a^n) \in \mathcal{P}$. In our case $n = 1$ and $a^1 = E$ the energy only, and $x = r$. Assume that S is defined for $(\tau, x) \in G$.

DEFINITION 5.33. A function $S(\tau, x, a)$ is called a complete solution of the Hamilton-Jacobi equation (5.60) if we have
(i) $S \in \mathcal{C}^2(G \times \mathcal{P})$ and $det(S_{x^i a^j}) \neq 0$ on $G \times \mathcal{P}$.
(ii) For any $a \in \mathcal{P}$ the function $S(\,\cdot\,,\,\cdot\,, a)$ is a solution of (5.60), i.e.

$$\frac{\partial S(\tau, x, a)}{\partial \tau} + H(x, S_x(\tau, x, a)) = 0.$$

We remark that a *complete solution* of (5.60) should not be mistaken for a *general solution*, see [**27**], vol. II, p.370.
Consider the Hamiltonian system

$$\dot{x} = H_p(x, p), \qquad \dot{p} = -H_x(x, p). \tag{5.61}$$

Jacobi's method of obtaining a *general solution* $x(\tau, a, b)$ and $p(\tau, a, b)$ of the system (5.61) depending on $2n$ parameters $a = (a^1, \ldots, a^n)$, $b = (b^1, \ldots, b^n)$ by means of a complete solution $S(\tau, x, a)$ of (5.60) consists of the following two steps:
• Solve the n implicit equations

$$S_{a^i}(\tau, x, a) = -b^i, \qquad 1 \leq i \leq n$$

with respect to x, obtaining a solution $x = x(\tau, a, b)$.
• Construct the function $p = p(\tau, a, b)$ defined by

$$p(\tau, a, b) = S_x(\tau, x(\tau, a, b), a).$$

The formal statement of Jacobi's result is given below.

THEOREM 5.34. *(Theorem of Jacobi)*
Let $S(\tau, x, a)$ be a complete solution of the Hamilton-Jacobi equation (5.60), and suppose that $x = x(\tau, a, b)$, $p = p(\tau, a, b)$ are functions of class $\mathcal{C}^1$ satisfying the equations

$$S_a(\tau, x(\tau, a, b), a) = -b, \qquad p(\tau, a, b) = S_x(\tau, x(\tau, a, b), a). \tag{5.62}$$

Then $x(\,\cdot, a, b)$, $p(\,\cdot, a, b)$ is a solution of the Hamiltonian system (5.61) depending on $2n$ parameters a and b.

PROOF. Differentiating with respect to a^i in the Hamilton-Jacobi equation

$$S_\tau(\tau, x, a) + H(x, S_x(\tau, x, a)) = 0, \tag{5.63}$$

and inserting $x = x(\tau, a, b)$, yields

$$S_{\tau a^i}(\tau, x(\tau, a, b), a) + H_{p^k}(x(\tau, a, b), p(\tau, a, b)) S_{x^k a^i}(\tau, x(\tau, a, b), a) = 0. \tag{5.64}$$

Differentiating the first equation of (5.62) with respect to τ, we obtain

$$S_{a^i\tau}(\tau, x(\tau, a, b), a) + S_{a^ix^k}(\tau, x(\tau, a, b), a))\dot{x}^k = 0. \tag{5.65}$$

Subtracting (5.64) and (5.65) yields

$$\Big[\dot{x}^k - H_{p^k}(x(\tau, a, b), a)\Big] S_{x^ka^i}(\tau, x(\tau, a, b), a) = 0.$$

The condition $\det S_{x^ka^i} \neq 0$ implies

$$\dot{x}^k = H_{p^k}(x(\tau, a, b), a),$$

which is the first equation of system (5.61). To derive the second equation of the Hamiltonian system (5.61), differentiate (5.63) with respect to x^i and obtain

$$S_{\tau x^i}(\tau, x, a) + H_{x^i}(\tau, x, S_x(\tau, x, a)) + H_{p^k}(\tau, x, S_x(\tau, x, a))S_{x^kx^i}(\tau, x, a) = 0,$$

which becomes after using the relations of $\dot{x}$ and p

$$S_{\tau x^i}(\tau, x, a) + S_{x^kx^i}(\tau, x, a)\dot{x}^k = -H_{x^i}(\tau, x, p(\tau, a, b)). \tag{5.66}$$

On the other hand, differentiating in the second relation of (5.62) with respect to τ and using (5.66), yields

$$\begin{aligned} \dot{p}^i(\tau, a, b) &= S_{x^i\tau}(\tau, x(\tau, a, b), a) + S_{x^ix^k}(\tau, x, a)\dot{x}^k \\ &= -H_{x^i}(\tau, x, p(\tau, a, b)). \end{aligned}$$

□

In our case a and b are 1-dimensional. The complex action is a complete solution because

$$\frac{\partial^2 g}{\partial E \partial r} = -\frac{-(2k+1)r^{4k+1}}{(E + 2r^{4k+2})^{3/2}} \neq 0, \text{ for } r > 0.$$

We should solve

$$\frac{\partial g}{\partial E} = -b$$

and get $r = r(\tau, E, b)$. This leads to

$$\frac{1}{k+1}\left[k\tau + \frac{r}{2}\frac{1}{\sqrt{2E + 4r^{4k+2}}} - \frac{r_0}{2}\frac{1}{\sqrt{2E + 4r_0^{4k+2}}}\right] = -b,$$

which can be rewritten as

$$\frac{r}{\sqrt{2E + 4r^{4k+2}}} = \frac{r_0}{\sqrt{2E + 4r_0^{4k+2}}} - 2k\tau - 2b(k+1), \tag{5.67}$$

with the solution $r = r(\tau, E, b)$. The following result deals with the number of parameters of a solution which starts from the origin.

PROPOSITION 5.35. Let $k \geq 1$. In the case when $r(0) = 0$ the solution depends only on the energy parameter E.

PROOF. Making $r_0 = 0$ in the equation (5.67), yields

$$\frac{r}{\sqrt{2E + 4r^{4k+2}}} = -2k\tau - 2b(k+1). \tag{5.68}$$

When $\tau = 0$, then $r = r(0) = r_0 = 0$, and hence $b = 0$ and b is not a parameter any more. Hence $r = r(\tau, E)$ depends only on the energy parameter E.

□

Let

$$\lambda(r) = \frac{r}{\sqrt{2E + 4r^{4k+2}}}.$$

The function $\lambda(r)$ has the following properties (see Figure 5.5).

(i) $\lambda(r) \geq 0$, with $\lambda(r) = 0 \Longleftrightarrow r = 0$.
(ii) There is an unique critical point $r_1 > 0$ for $\lambda(r)$.
(iii) $\lambda(r)$ is increasing on $(0, r_1)$ and decreasing on (r_1, ∞).
(iv) The line $y = 0$ is a horizontal asymptote to $+\infty$ for $\lambda(r)$.
(v) The maximum of λ depends on the energy E and it is given by

$$\max \lambda = \frac{\left(\frac{E}{4k}\right)^{\frac{1}{4k+2}}}{\sqrt{E(2 + \frac{1}{k})}}.$$

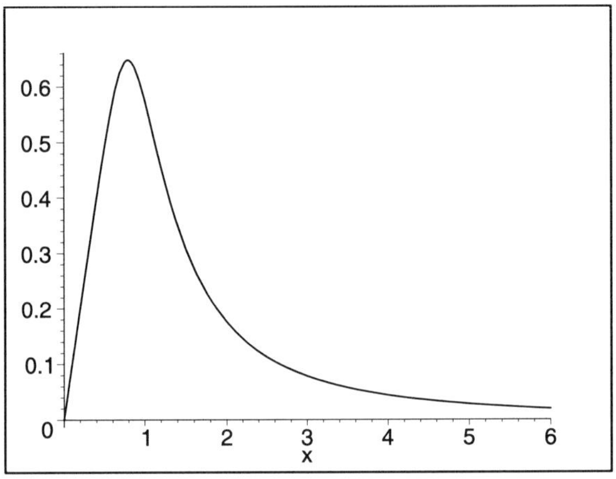

Figure 5.5: *The graph of* $\lambda(r)$.

In the following we shall keep the energy E fixed. Making $b = 0$ in equation (5.68) yields

$$\lambda(r) = -2k\tau,$$

which has zero, one or at most two real solutions for negative values of τ. (For positive values we need to take positive sign in W of the previous section). Hence we arrived at:

PROPOSITION 5.36. Given the energy E and a point x, there are at most two solutions joining the origin and the point x.

We shall consider now the case $r_0 \neq 0$. Given E, we shall find an interval where the parameter b belongs to. Equation (5.67) yields

$$0 < \frac{r_0}{\sqrt{2E + 4r_0^{4k+2}}} - 2k\tau - 2b(k+1) < \max \lambda,$$

which defines an interval for the parameter b by the double inequality

$$\lambda(r_0) - \max \lambda - 2k\tau < 2b(k+1) < \lambda(r_0) - 2k\tau. \tag{5.69}$$

PROPOSITION 5.37. Let $k \geq 1$. If $r_0 \neq 0$, the solution $r = r(\tau, E, b)$ depends on two parameters E and b. Given E, the parameter b belongs to the interval defined by (5.69). Hence, for each given E, there are infinitely many solutions $r(\tau, E, b)$ parametrized by b.

The parameter b is called momentum, see [**4**].

5.11. The lengths of the real geodesics

The classical action from the origin is given by

$$S = S(x, t, \tau, \theta) = \int_0^\tau \Big[\langle \xi, \dot{x}\rangle + \theta \dot{t} - H(s)\Big]\, ds. \tag{5.70}$$

S is the integral of the Lagrangian along the bicharacteristic with boundary conditions:

$$x(0) = 0, \qquad x(\tau) = x, \qquad t(0) = 0, \qquad t(\tau) = t. \tag{5.71}$$

Here we fix $t_0 = t(0)$, so θ cannot be chosen arbitrarily; it is a real constant along the path. In the case of the complex action, the boundary condition $t(0) = 0$ is replaced by $\theta = -i$. The complex action and the classical action are related by the formula

$$g = S - it(0). \tag{5.72}$$

The Hamiltonian H is homogeneous of degree 2 with respect to (ξ_1, ξ_2, θ), so

$$S(x, t, \tau) = \int_0^\tau \Big[\langle \xi, \dot{x}\rangle + \theta \dot{t} - H(s)\Big]\, ds = \int_0^\tau (2H - H)\, ds = \tau H_0,$$

where H_0 is the constant value of the Hamiltonian along the solution. The Hamilton-Jacobi equation may be rewritten as

$$\frac{\partial g}{\partial \tau} + \frac{S}{\tau} = 0, \tag{5.73}$$

so

$$\frac{\partial g}{\partial \tau} = -\frac{g + it(0)}{\tau}. \tag{5.74}$$

DEFINITION 5.38. The modified complex action function is defined as

$$f(x, t, \tau) = \tau g(x, t, \tau), \tag{5.75}$$

where $g(x, t, \tau)$ is the complex action from the origin.

We shall show that the critical points of f with respect to τ play an important role in finding the lengths of the (real) subRiemannian geodesics between the origin and the point (x, t). This was showed by Beals, Gaveau and Greiner (see [**3**]) for the step 2 case, where the modified complex action function is

$$f(x, t, \tau) = -i\tau t + \tau \coth(2\tau)(x_1^2 + x_2^2).$$

The general results of this section were obtained by Calin, Chang and Greiner (see [**17**]).

The geodesics starting from the origin are described by Theorem 3.17 of chapter 3:

THEOREM 5.39. *There are finitely many geodesics that join the origin to (x,t) if and only if $x \neq 0$. These geodesics are parametrized by the solutions ψ of:*

$$\frac{|t|}{|x|^{2(k+1)}} = \mu(\psi), \tag{5.76}$$

There is exactly one such geodesic if and only if:

$$|t| < \mu(c_1)|x|^{2(k+1)},$$

where c_1 is the first critical point for μ. The number of geodesics increases without bound as $\frac{|t|}{|x|^{2(k+1)}} \to \infty$. If $0 \le \psi_1 < \cdots < \psi_N$ are the solutions of (5.76), then there are exactly N geodesics, and their lengths are given by

$$\ell_m^{2(k+1)} = \nu(\psi_m)\Big(|t| + \|x\|^{2(k+1)}\Big), \qquad m = 1, \dots, N \tag{5.77}$$

where

$$\nu(\psi) = \frac{\Big[\int_0^{(2k+1)\psi} \sin^{-\frac{2k}{2k+1}}(v)\, dv\Big]^{2(k+1)}}{(k+1)^{2(k+1)}\big(1+\mu(\psi)\big)\sin^{\frac{2(k+1)}{2k+1}}\big((2k+1)\psi\big)}. \tag{5.78}$$

Differentiating (5.75), yields

$$\frac{\partial f}{\partial \tau} = \frac{\partial(\tau g)}{\partial \tau} = g + \tau\frac{\partial g}{\partial \tau} = g - (g + it(0)) = -it(0).$$

For $t(0) \neq 0$, relation (5.76) of Theorem 5.39 becomes

$$t - t(0) = -\mu(\widetilde{\phi})|x|^{2(k+1)}, \qquad \widetilde{\phi} = \phi - \phi_0, \tag{5.79}$$

and therefore

$$\frac{\partial f}{\partial \tau} = -i(t + \mu(\widetilde{\phi})|x|^{2(k+1)}). \tag{5.80}$$

Since $\frac{d\widetilde{\phi}}{ds} = -2\theta r^{2k}$ (see (3.18) of chapter 3), one has

$$\widetilde{\phi} = -2\theta \int_0^{\tau} r^{2k}(s)\, ds. \tag{5.81}$$

Now (5.80) implies

PROPOSITION 5.40. Let $x \neq 0$. τ_c is a critical point for the modified complex action $f(x,t,\tau)$ if and only if

$$\widetilde{\phi} = 2i \int_0^{\tau_c} r^{2k}(s)\, ds$$

is a solution for the equation

$$\frac{t}{|x|^{2(k+1)}} = -\mu(\widetilde{\phi}). \tag{5.82}$$

Theorem 5.39 implies

PROPOSITION 5.41. Let $\tau_1, \dots, \tau_m$ be the critical points of the modified complex action $f(x,t,\tau)$.
The numbers

$$\zeta_j = 2i \int_0^{\tau_j} r^{2k}(s)\, ds, \quad j = 1, \dots, m,$$

satisfy the equation

$$\frac{t}{|x|^{2(k+1)}} = \mu(-\zeta_j), \tag{5.83}$$

and for each ζ_j we have a geodesic connecting (x, t) to the origin. The length ℓ_j of the geodesic parametrizied by ζ_j is

$$\begin{aligned}
\ell_j^{2(k+1)} &= \nu(\zeta_j)\Big(|t| + \|x\|^{2(k+1)}\Big) \\
&= \nu\Big(2i\int_0^{\tau_j} r^{2k}(s)\,ds\Big)\Big(|t| + \|x\|^{2(k+1)}\Big).
\end{aligned}$$

Consequently, the knowledge of the critical points τ_c of the modified complex action function gives the lengths of the geodesics. The rest of this section contains the calculation of τ_j and the geodesic lengths for the step 2 and step 4 cases.

Step 2 case: The Heisenberg group. Here $k = 0$. In this case $\widetilde{\phi}_j = 2i\tau_j$ and τ_j is purely imaginary. The lengths of the geodesics between the origin and (x, t), $x \neq 0$, are given by (5.77):

$$\ell_j^2 = \nu(2i\tau_j)\Big(|t| + \|x\|^2\Big). \tag{5.84}$$

Step 4 case. This is a lengthy calculation. We need to compute the integral

$$2i\int_0^{\tau} r^2(s)\,ds, \tag{5.85}$$

and we shall do that in terms of elliptic functions. Recall the Hamiltonian function

$$H(\xi, \theta, x, t) = \frac{1}{2}\Big(\xi_1 + 2x_2|x|^2\theta\Big)^2 + \frac{1}{2}\Big(\xi_2 - 2x_1|x|^2\theta\Big)^2. \tag{5.86}$$

Using the Hamiltonian equations

$$\begin{aligned}
\dot{x}_1 &= \xi_1 + 2x_2|x|^2\theta, \\
\dot{x}_2 &= \xi_2 - 2x_1|x|^2\theta,
\end{aligned}$$

we have

$$H = \frac{1}{2}(\dot{x}_1^2 + \dot{x}_2^2) = \frac{1}{2}, \tag{5.87}$$

where we have used the arc length parametrization. In polar coordinates $x_1 = r\cos\widetilde{\phi}$, $x_2 = r\sin\widetilde{\phi}$ we have $\frac{d\widetilde{\phi}}{ds} = -2\theta r^{2k}$ and the conservation of energy (5.87) takes the form

$$\dot{r}^2 + 4\theta^2 r^6 = 1. \tag{5.88}$$

With $\theta = -i$,

$$\dot{r}^2 - 4r^6 = 1, \tag{5.89}$$

and

$$\frac{dr}{ds} = \pm\sqrt{1 + 4r^6}. \tag{5.90}$$

Since $r(0) = 0$, then $r(s)$ satisfies the integral equation

$$\int_0^r \frac{dx}{\sqrt{1 + 4x^6}} = \pm s. \tag{5.91}$$

The substitutions $x^2 = t$, $4t^3 = u^3$ and $u = -v$ yield

$$\begin{aligned}\int_0^r \frac{dx}{\sqrt{1+4x^6}} &= \frac{1}{2}\int_0^{r^2} \frac{dt}{\sqrt{t(1+4t^3)}}\\ &= \frac{1}{2^{4/3}}\int_0^{4^{1/3}r^2} \frac{du}{\sqrt{u(1+u^3)}}\\ &= \frac{i}{2^{4/3}}\int_0^{-4^{1/3}r^2} \frac{dv}{\sqrt{v(1-v^3)}}.\end{aligned}$$

Recall Lemma 2.16 of chapter 2:

LEMMA 5.42.

$$\int_0^a \frac{dx}{\sqrt{x(1-x^3)}} = \frac{1}{3^{1/4}} sn^{-1}\Big(\sqrt{1-\frac{h(a)^2}{b^2}}, k\Big),$$

where $b = 2-\sqrt{3}$, $k = \sqrt{b}/2$ *and*

$$h(a) = \frac{\sqrt{3}}{a+\frac{1+\sqrt{3}}{2}} - 1.$$

Consequently,

$$\begin{aligned}\int_0^r \frac{dx}{\sqrt{1+4x^6}} &= \frac{i}{2^{4/3}}\int_0^{-4^{1/3}r^2} \frac{dv}{\sqrt{v(1-v^3)}}\\ &= \frac{i}{2^{4/3}\cdot 3^{1/4}}\mathrm{sn}^{-1}\left(\sqrt{1-\frac{h(-4^{1/3}r^2)^2}{b^2}}, k\right),\end{aligned}$$

and in view of (5.91),

$$\mathrm{sn}^{-1}\left(\sqrt{1-\frac{h(-4^{1/3}r^2)^2}{b^2}}, k\right) = \mp 2^{4/3}3^{1/4} is. \tag{5.92}$$

From $\mathrm{cn}^2 x - \mathrm{sn}^2 x = 1$ (see [**42**], p. 24 (2.1.4)),

$$\mathrm{cn}^2\Big(2^{4/3}3^{1/4} is, k\Big) = \frac{h(-4^{1/3}r^2)^2}{b^2}. \tag{5.93}$$

Again, by [**42**], p. 39 (2.6.12):

$$\mathrm{cn}(iu,k) = \mathrm{nc}(u,k') = \frac{1}{\mathrm{cn}(u,k')},$$

where $k' = \sqrt{1-k^2}$, so (5.93) becomes

$$b\,\mathrm{nc}(2^{4/3}\,3^{1/4}s, k') = h(-4^{1/3}r^2), \tag{5.94}$$

with $k' = \sqrt{\bar{b}}/2$ and $\bar{b} = 2+\sqrt{3}$. Inverting h, we obtain

$$r^2(s) = 4^{-1/3}\left(\frac{1+\sqrt{3}}{2} - \frac{\sqrt{3}}{1+b\,\mathrm{nc}(2^{4/3}\,3^{1/4}s, k')}\right). \tag{5.95}$$

Thus (5.85) can be expressed as

$$
\begin{aligned}
2i\int_0^\tau r^2(s)\,ds &= \frac{1+\sqrt{3}}{4^{1/3}}(i\tau) \\
&\quad -2^{1/3}\sqrt{3}\,i\int_0^\tau \frac{ds}{1+b\,\mathrm{nc}(2^{4/3}\,3^{1/4}s,k')}.
\end{aligned}
\tag{5.96}
$$

We are left with evaluating the second term on the right side of (5.96). With $u = is$, one has

$$
\begin{aligned}
\int_0^\tau \frac{ds}{1+b\,\mathrm{nc}(2^{4/3}\,3^{1/4}s,k')} &= \frac{1}{i}\int_0^{i\tau}\frac{du}{1+b\,\mathrm{nc}(-i\,2^{4/3}\,3^{1/4}s,k')} \\
&= \frac{1}{i}\int_0^{i\tau}\frac{du}{1+b\,\mathrm{cn}(2^{4/3}\,3^{1/4}s,k)},
\end{aligned}
\tag{5.97}
$$

since nc is an even function. Consequently,

$$
2i\int_0^\tau r^2(s)\,ds = \mathbf{F}(i\tau),
\tag{5.98}
$$

where

$$
\begin{aligned}
\mathbf{F}(z) &= \frac{1+\sqrt{3}}{4^{1/3}}z - 2^{1/3}\sqrt{3}\int_0^z \frac{dv}{1+b\,\mathrm{cn}(2^{4/3}\,3^{1/4}v,k)} \\
&= \frac{1+\sqrt{3}}{4^{1/3}}z - \frac{3^{1/4}}{2}\int_0^{2^{4/3}3^{1/4}z}\frac{dv}{1+b\,\mathrm{cn}(v,k)}.
\end{aligned}
$$

Next we use [**42**], p. 93 (51):

$$
\begin{aligned}
&\int_0^u \frac{dv}{1+b\,\mathrm{cn} v} \\
&= \frac{1}{1-b^2}\,\Lambda\Big(u,\frac{ib}{\sqrt{1-b^2}},k\Big) - \frac{b}{\sqrt{(1-b^2)(k^2+k'^2b^2)}}\tan^{-1}\left(\sqrt{\frac{k^2+k'^2b^2}{1-b^2}}\,\mathrm{sd}\,u\right),
\end{aligned}
$$

with $b^2 < 1$, where

$$
\Lambda(u,i\beta,k) = \int_0^u \frac{dv}{1+\beta^2\,\mathrm{sn}^2 v}.
\tag{5.99}
$$

We recall that

$$
b = 2-\sqrt{3},\quad k = \frac{\sqrt{b}}{2} = \frac{\sqrt{6}-\sqrt{2}}{4},\quad k' = \frac{\sqrt{\bar{b}}}{2} = \frac{\sqrt{6}+\sqrt{2}}{4},\quad \beta = \frac{b}{\sqrt{1-b^2}},
$$

$$
1-b^2 = 4\sqrt{3}-6,\quad k^2+k'^2b^2 = 1-\frac{\sqrt{3}}{2},\quad \frac{k^2+k'^2b^2}{1-b^2} = \frac{1}{4}\frac{\sqrt{3}-2}{2\sqrt{3}-3},
$$

and

$$\begin{aligned}
\frac{b}{\sqrt{(1-b^2)(k^2+k'^2\alpha^2)}} &= \frac{(2-\sqrt{3})\sqrt{2}}{(\sqrt{3}-1)\sqrt{2\sqrt{3}-3}}\\
&= \frac{\sqrt{2-\sqrt{3}}\cdot\sqrt{2}}{(\sqrt{3}-1)\cdot 3^{1/4}}\\
&= \frac{\sqrt{2}\,(\sqrt{3}-1)}{2}\,\frac{\sqrt{2}}{(\sqrt{3}-1)\,3^{1/4}}\\
&= 3^{1/4},
\end{aligned}$$

$$\frac{1}{1-b^2} = \frac{1}{2}\frac{1}{2\sqrt{3}-3} = \frac{2+\sqrt{3}}{2\sqrt{3}} = \frac{1}{\sqrt{3}} + \frac{1}{2},$$

$$\sqrt{\frac{k^2+k'^2b^2}{1-b^2}} = \frac{1}{4}\frac{\sqrt{2}(\sqrt{3}-1)}{\sqrt{2\sqrt{3}-3}} = \frac{1}{4}\frac{\sqrt{2}(\sqrt{3}-1)}{3^{1/4}\sqrt{2-\sqrt{3}}} = \frac{1}{2\cdot 3^{1/4}}.$$

$$\begin{aligned}
\frac{b}{\sqrt{1-b^2}} &= \frac{2-\sqrt{3}}{\sqrt{2(2\sqrt{3}-3)}} = \frac{2-\sqrt{3}}{2^{1/2}3^{1/4}\sqrt{2-\sqrt{3}}} = \frac{1}{2^{1/2}3^{1/4}}\cdot\frac{\sqrt{2}(\sqrt{3}-1)}{2}\\
&= \frac{\sqrt{3}-1}{2\cdot 3^{1/4}}.
\end{aligned}$$

Consequently,

$$\int_0^u \frac{dv}{1+b\,\mathrm{cn}v} = \left(\frac{\sqrt{3}}{3}+\frac{1}{2}\right)\Lambda\left(u, i\frac{\sqrt{3}-1}{2\cdot 3^{1/4}}, k\right) - 3^{-1/4}\tan^{-1}\left(\frac{\mathrm{sd}\,u}{2\cdot 3^{1/4}}\right),$$

and

$$\begin{aligned}
\mathbf{F}(z) &= \frac{1+\sqrt{3}}{4^{1/3}}z - \frac{2+\sqrt{3}}{4\cdot 3^{1/4}}\Lambda\left(2^{4/3}3^{1/4}z, i\frac{\sqrt{3}-1}{2\cdot 3^{1/4}}, k\right)\\
&\quad -\frac{1}{2}\tan^{-1}\left(\frac{\mathrm{sd}(2^{4/3}3^{1/4}z)}{2\cdot 3^{1/4}}\right).
\end{aligned} \tag{5.100}$$

To evaluate the second term more explicitly, we use the following elliptic functions:

$$\begin{aligned}
\mathrm{sn}^{-1}(x,k) &= \int_0^x \frac{dt}{\sqrt{(1-t^2)(1-k^2t^2)}}, \quad 0 \le x \le 1, && \text{p.50 (3.1.2)}\\
\mathrm{cn}^{-1}(x,k) &= \int_x^1 \frac{dt}{\sqrt{(1-t^2)(k'^2+k^2t^2)}}, \quad 0 \le x \le 1, && \text{p.52 (3.2.2)}
\end{aligned}$$

$$\begin{aligned}
F(\omega,k') &= \operatorname{sn}^{-1}(\sin\omega,k') = \int_0^\omega (1-k'^2\sin^2\theta)^{-1/2}\,d\theta, && \text{p.51 (3.1.8)}\\
\mathcal{E}(u,k) &= \int_0^u \operatorname{dn}^2(v,k)\,dv, && \text{p.62 (3.4.25)}\\
D(\omega,k') &= \int_0^\omega \sqrt{1-k'^2\sin^2\theta}\,d\theta = \int_0^\omega \sqrt{\cos^2\theta - k^2\sin^2\theta}\,d\theta, && \text{p.63 (3.4.26)}\\
\mathcal{E} &= \mathcal{E}(k) = \mathcal{E}(K,k) = \int_0^{\pi/2}(1-k^2\sin^2\theta)^{1/2}\,d\theta = \int_0^K \operatorname{dn}^2 v\,dv,\\
K &= K(k) = \int_0^{\pi/2}(1-k^2\sin^2\theta)^{-1/2}\,d\theta = \int_0^1 \frac{dt}{\sqrt{(1-t^2)(1-k^2t^2)}},
\end{aligned}$$

see p.63 (3.5.4) and p.73 (3.8.1), where numbers refer to [**42**]. In particular,

$$\begin{aligned}
\Lambda(u,i\beta,k) &= \frac{k^2}{\beta^2+k^2}u + \frac{\beta}{\sqrt{(\beta^2+k^2)(\beta^2+1)}}\Big[u\big\{D(\omega,k')-F(\omega,k')\\
&\quad + \frac{\mathcal{E}}{K}F(\omega,k')\big\} - \tan^{-1}(Y/X)\Big]\\
&= \frac{k^2}{\beta^2+k^2}u + \frac{\beta}{\sqrt{(\beta^2+k^2)(\beta^2+1)}}\Big[u\big\{D(\omega,k')-F(\omega,k')\\
&\quad + \frac{\mathcal{E}}{K}F(\omega,k')\big\} + \frac{i}{2}\log\frac{\theta_4(x-iy)}{\theta_4(x+iy)}\Big],
\end{aligned}$$

where

$$\begin{aligned}
X &= 1+2\sum_{n=1}^{\infty}(-1)^n q^{n^2}\cos 2nx\,\cosh 2ny,\\
Y &= 2\sum_{n=1}^{\infty}(-1)^n q^{n^2}\sin 2nx\,\sinh 2ny,
\end{aligned}$$

with $q=\exp(-\pi K'/K)$, $x=\pi u/2K$ and $y=\pi F(\omega,k')/2K$ and $\sin\omega = \beta/\sqrt{\beta^2+k^2}$, see [**42**], p. 72 (3.7.47). θ_4 stands for Jacobi's zeta function. In our case

$$\beta = \frac{\sqrt{3}-1}{2\cdot 3^{1/4}},\quad k=\frac{\sqrt{6}-\sqrt{2}}{4},\quad k'=\frac{\sqrt{6}+\sqrt{2}}{4},\quad b=2-\sqrt{3}.$$

An algebraic computation yields

$$\frac{k^2}{\beta^2+k^2} = b\sqrt{3},\qquad \frac{\beta}{\sqrt{(\beta^2+k^2)(\beta^2+1)}} = 2\cdot 3^{1/4}\,b,\qquad \frac{\beta}{\sqrt{\beta^2+k^2}} = \sqrt{3}-1.$$

We shall show the function Λ depends only on the Jacobi's epsilon function $\mathcal{E}(u,k)$. We have

$$F(\omega,k') = \operatorname{sn}^{-1}(\sin\omega,k') = \operatorname{sn}^{-1}(\sqrt{3}-1,k').$$

From [**42**], p. 41 (2.7.14) and p. 63 (3.4.27), $\operatorname{am} v = \int_0^v \operatorname{dn}\gamma\,d\gamma$, and $E(v,k') = D(\operatorname{am} v,k')$, with $\operatorname{sn}(v)=\sin(\operatorname{am} v)$. With $\omega=\operatorname{am} v$, we have $v=\operatorname{sn}^{-1}(\sqrt{3}-1,k')$, and hence

$$D(\omega,k') = \mathcal{E}(\operatorname{sn}^{-1}(\sqrt{3}-1,k'),k').$$

If $k = \frac{\sqrt{3}-1}{2\sqrt{2}} = \frac{\sqrt{6}-\sqrt{2}}{4}$, then $K'/K = \sqrt{3}$ see [**42**], p. 86 (10). Hence

$$q = e^{-\pi\sqrt{3}}.$$

Also

$$K = K(k) = \frac{3^{1/4}}{4\sqrt{\pi}}\frac{1}{\sqrt{3}}\Gamma(1/6)\Gamma(1/3),$$

and

$$K[2\sqrt{3}E - (\sqrt{3}+1)K] = \frac{\pi}{2}.$$

Dividing by $K^2 = \frac{1}{16\pi\sqrt{3}}\Gamma^2(1/6)\Gamma^2(1/3)$, we get

$$\frac{\mathcal{E}}{K} = \frac{\pi}{4\sqrt{3}K^2} + \frac{\sqrt{3}+1}{2\sqrt{3}} = \left(\frac{2\pi}{\Gamma(1/6)\Gamma(1/3)}\right)^2 + \frac{3+\sqrt{3}}{6}.$$

Hence,

$$\begin{aligned}\Lambda(u, i\beta, k) &= b\sqrt{3}u + 2\cdot 3^{1/4}\, b\Big[u\Big\{E(am^{-1}\omega, k') - am^{-1}\omega \\ &\quad + \Big[\left(\frac{2\pi}{\Gamma(1/6)\Gamma(1/3)}\right)^2 + \frac{3+\sqrt{3}}{6}\Big] am^{-1}\omega\Big\} + \frac{i}{2}\log\frac{\theta_4(x-iy)}{\theta_4(x+iy)}\Big],\end{aligned}$$

with $am^{-1}\omega = \mathrm{sn}^{-1}(\sqrt{3}-1, k')$. Hence, the formula (5.100) becomes

$$\begin{aligned}\mathbf{F}(z) &= \frac{1+\sqrt{3}}{4^{1/3}}z - 3^{1/2}\,2^{-2/3}z \\ &\quad + \frac{1}{2}\Big[u\Big\{E(am^{-1}\omega, k') + \Big[\left(\frac{2\pi}{\Gamma(1/6)\Gamma(1/3)}\right)^2 - \frac{3-\sqrt{3}}{6}\Big] am^{-1}\omega\Big\} \\ &\quad + \frac{i}{2}\log\frac{\theta_4(x-iy)}{\theta_4(x+iy)}\Big] - \frac{1}{2}\tan^{-1}\left(\frac{\mathrm{sd}(2^{4/3}\,3^{1/4}z)}{2\cdot 3^{1/4}}\right),\end{aligned}$$

with $u = 2^{4/3}\,3^{1/4}\,z$ and $am^{-1}\omega = \mathrm{sn}^{-1}(\sqrt{3}-1, k')$. Finally, we restate the last part of Proposition 5.41 for the step 4 case ($k = 1$).

COROLLARY 5.43. *Let τ_j be the critical points of the modified complex action $f(\tau) = \tau g(\tau)$. Setting $\zeta_j = \mathbf{F}(i\tau_j)$, the lengths of the geodesics between the origin and the point (x, t), $|x| \neq 0$ are given by*

$$\ell_j^4 = \nu(\zeta_j)\big(|t| + \|x\|^4\big).$$

5.12. Modified complex action function on $\mathbf{H}_n$

In this section we use the notions and notations introduced in chapter 4. Using the classical action for the vectors (4.5) given by the Theorem 4.5,

$$S(x, t, \tau; \theta) = [t - t(0)]\theta + \sum_{j=1}^{n} a_j\theta\cot(2a_j\tau\theta)r_j^2,$$

where $r_j^2 = x_{2j-1}^2 + x_{2j}^2$. Let $g = S + \theta t(0)$ and define the modified complex action function

$$f(x, t, \tau) = -i\tau t + \sum_{j=1}^{n} a_j\tau\coth(2a_j\tau)r_j^2. \tag{5.101}$$

Recall the notation

$$x = (x', x''), \quad x'' = (x_{2p+1}, x_{2p+2}, \ldots, x_{2n}),$$

$$0 < a_1 \le a_2 \le \cdots \le a_p < a_{p+1} = \cdots = a_n.$$

The following study of the modified complex action f can be found in [**3**].

LEMMA 5.44. *For any (x,t) with $x'' \neq 0$, and $t \geq 0$, the function $f(\tau) = f(x,t,\tau)$ has finitely many critical points on the imaginary axis. There is one critical point between the origin and the first pole of f on the positive imaginary axis, and it is a local minimum for f. There are either zero or two critical points (counting multiplicity) between each pair of poles on the positive imaginary axis; of such a pair of critical points the one nearer the origin is a local minimum and the other a local maximum for f.*

PROOF. Let

$$F(\theta) = f(x,t,i\theta) = \sum_{j=1}^{n} a_j\theta\cot(2a_j\theta)r_j^2 + t\theta. \tag{5.102}$$

Then

$$F'(\theta) = t - \sum_{j=1}^{n} a_j\mu(2a_j\theta)r_j^2.$$

The lemma follows from the properties of the function μ, see chapter 1. □

LEMMA 5.45. *For any (x,t) with $x'' \neq 0$ and $t > 0$, there is exactly one branch of the set*

$$\Gamma_0 = \{\tau : Im\, f(x,t,\tau) = 0, Re\,\tau > 0, Im\,\tau > 0\} \tag{5.103}$$

that goes to ∞ in the quadrant $Re\,\tau > 0$, $Im\,\tau > 0$. On this branch $Re\, f$ increases as $\tau \to \infty$.

PROOF. Assume $\tau = s + i\theta$, with $s, \theta > 0$. Then

$$Im(\tau\coth\tau) = \frac{\theta\sinh 2s - s\sin 2\theta}{2\sinh^2 s + 2\sin^2\theta} > 0, \tag{5.104}$$

since

$$\theta\sinh 2s - s\sin 2\theta = \theta(\sinh 2s - 2s) + s(2\theta - \sin 2\theta) > 0.$$

If $\theta = Im\,\tau > 0$ fixed, then

$$\lim_{s\to+\infty} Im(\tau\coth) = \theta, \quad \text{uniformly for bounded } \theta. \tag{5.105}$$

When $\sin\theta = 0$, then

$$Im(\theta\coth\tau) = \theta\coth s \geq \frac{\theta}{s}. \tag{5.106}$$

We have the following estimation

$$\begin{aligned}\frac{\partial}{\partial s}\Big(Im\,\tau\coth\tau\Big) &= -\frac{\cos\theta\sin\theta}{\sinh^2 s+\sin^2\theta}\\ &\quad+\frac{-\theta\sinh^2 s\cos^2\theta+\theta\cosh^2 s\sin^2 s\sin^2\theta}{(\sinh^2 s+\sin^2\theta)^2}\\ &\quad+\frac{2s\cosh s\sinh s\cos\theta\sin\theta}{(\sinh^2 s+\sin^2\theta)^2}\\ &= O\Big(\frac{s+\theta}{\sinh^2 s}\Big), \qquad (5.107)\end{aligned}$$

as $s\to\infty$. We write

$$Im\, f(x,t,\tau) = -st + Im\sum_{j=1}^{n} 2a_j\tau\coth(2a_j\tau)\frac{r_j^2}{2}. \tag{5.108}$$

Choose a_j such that $r_j\neq 0$. Let θ be large enough, such that $\sin(2a_j\theta)=0$ and $\coth(2a_jr_j(\theta/2t)^{1/2})<3/2$. Using (5.105), (5.106) and (5.108), there will be at least one point $\tau=s_\theta+i\theta$, when $Im\, f(x,t,\tau)=0$. Using (5.106) and (5.108) it follows that

$$\frac{\theta}{s_\theta}\frac{r_j^2}{2}\le s_\theta t \tag{5.109}$$

and (5.105), (5.108) and (5.109) imply that

$$s_\theta t\le\frac{3}{2}\sum_{k=1}^{n}2a_k\theta\frac{r_k^2}{2}, \tag{5.110}$$

and hence s_θ belongs to the interval

$$\frac{r_j}{\sqrt{2t}}\sqrt{\theta}\le s_\theta\le\frac{3}{2}\Big(\sum_{k=1}^{n}a_k\frac{r_k^2}{t}\Big)\theta. \tag{5.111}$$

From (5.107) and (5.111) we get the estimation

$$\Big|\frac{\partial}{\partial s}Im\sum_{j=1}^{n}2a_j\tau\coth(2a_j\tau)\frac{r_j^2}{2}\Big|\le C\frac{s+\theta}{\sinh^2(2a_1s)}\le C_1\theta e^{-C_2\sqrt{\theta}},\qquad s>s_\theta, \tag{5.112}$$

where C, C_1 and C_2 are positive constants which depend only on the a_j's and on (x,t). Then for suitable large θ we have

$$\frac{\partial}{\partial s}f(x,t,s+i\theta)<0,\qquad s>s_\theta, \tag{5.113}$$

and there is only one solution s_θ of $Im\, f(x,t,s+i\theta)=0$. It follows from (5.105) that no branch of Γ_0 can escape to ∞ between two consecutive such lines $Im\,\tau=\theta$, with $\sin(2a_j\theta)=0$. Since $Im\, f=0$ on the imaginary τ-axis, a branch can escape from the quadrant through the imaginary τ-axis only at a critical point of f. By Lemma 5.44, there is no such escape for large θ. This still leaves the question of a possible branching of Γ_0 between two such consecutive $Im\,\tau=0$ lines. Two such branches must join at the two s_θ points, since they are unique. This implies the existence of a bounded region Ω on which $Im\, f$ is harmonic and non-constant and vanishes on $\partial\Omega$, which is a contradiction. Thus, for large $|\tau|$, there is exactly one branch of Γ_0 that goes to infinity within the quadrant.

In particular, f has no critical points on this single branch of Γ_0. Therefore, on this branch, $Re\, f$ must increase or decrease, it cannot have a stationary point. Now, using

$$\begin{aligned}\coth(u+iv) &= \frac{\cosh u \sinh u - i \sin v \cos v}{\sinh^2 u + \sin^2 v} \\ iv \coth(iv) &= v \cot v,\end{aligned}$$

implies that

$$\begin{aligned}&Re\{(u+iv)\coth(u+iv) - iv\coth(iv)\} \\ &= \frac{u \cosh u \sinh u + v \sin v \cos v}{\sinh^2 u + \sin^2 v} - v \cot v \\ &= \frac{\sinh^2 u}{\sinh^2 u + \sin^2 v} \cdot (u \coth u - v \cot v).\end{aligned}$$

It follows that when $\sin\theta = 0$, we have

$$Re(\tau \coth \tau) = s \coth s. \tag{5.114}$$

Relations (5.111) and (5.114) yield

$$Re\, f(x,t,s_\theta + i\theta) = \theta t + \sum_{k=1}^{n} 2a_k s_\theta \coth(2a_k s_\theta)\frac{r_k^2}{2}, \tag{5.115}$$

which increases as $\theta \to \infty$. Consequently, $Re\, f$ increases on the branch of Γ_0, which goes off to infinity, and we have derived Lemma 5.45. □

PROPOSITION 5.46. If $f'(\tau) = 0$ and $Im\, f(\tau) = 0$, then τ is imaginary.

PROOF. It follows from the next lemma. □

LEMMA 5.47. *Assume $x'' \neq 0$ and $t > 0$. Let the critical points of f on the positive imaginary axis, counted according to multiplicity, be $i\theta_1, \dots, i\theta_{2m+1}$, with*

$$\theta_1 < \theta_2 \le \theta_3 < \cdots < \theta_{2m} \le \theta_{2m+1}. \tag{5.116}$$

Let Γ be the union of Γ_0 and the closed intervals

$$[0, i\theta_1], [i\theta_2, i\theta_3], \dots, [i\theta_{2m}, i\theta_{2m+1}]. \tag{5.117}$$

Then Γ, oriented in the direction of increasing $Re\, f$, is a simple connected curve from 0 to ∞.

PROOF. The set Γ is a union of analytic arcs that meet only at critical points of f. At any critical point $\tau \neq 0$ there will be at least one branch that is oriented toward τ and one that is oriented away from τ. The real part $Re\, f$ increases along the branch of Γ that leads to ∞. It follows that any point of Γ is contained in at least one maximal oriented contour $\Gamma_1 \subset \Gamma$. Moreover, Γ_1 must begin at 0 and eventually contain the branch that leads to ∞. If Γ_2 is another such maximal oriented contour, either it coincides with Γ_1 or the union of the two contains a simple closed contour Γ_3. Suppose the latter; then Γ_3 encloses a bounded region Ω. But then $Im\, f$ is harmonic and non-constant on Ω and vanishes on the boundary, which is a contradiction. Thus there is a unique such maximal contour, and it coincides with Γ itself. □

We still assume $t > 0$ and $x'' \neq 0$. A solution of

$$|t| = \sum_{j=1}^{n} a_j \mu(2a_j\theta) r_j^2 \tag{5.118}$$

corresponds to a critical point $\tau = i\theta$ of the modified complex action function f. The square of the length of the associated geodesic is $2f(i\theta)$. Let $(i\theta_k)$ be the critical points of f on the positive imaginary axis, numbered as in (5.116). Then

$$f(i\theta_1) < f(i\theta_2) \leq f(i\theta_3) < \cdots < f(i\theta_{2m}) \leq f(i\theta_{2m+1}), \tag{5.119}$$

with strict inequality where the corresponding inequality in (5.116) is strict. This is the same as saying that the critical points occur in this order on the oriented curve Γ of Lemma 5.47. Any other ordering of the critical points along Γ would imply that Γ has self-intersections, which contradicts Lemma 5.47.

COROLLARY 5.48. The lengths of the geodesic associated with a solution of (5.118) increases with θ.

5.13. A geometric formula for the fundamental solution

In this section the complex action will be used to find the fundamental solution $K(x, t; x_0, y_0)$ for the Heisenberg operator $\Delta_H = \frac{1}{2}(X_1^2 + X_2^2)$, *i.e.*,

$$\Delta_H^{(x,t)} K(x, t, x_0, y_0) = \delta_{x-x_0} \delta_{t-t_0},$$

where δ_{a-a_0} is the Dirac distribution centered at a_0 and $X_1 = \partial_{x_1} + 2x_1\partial_t$, $X_2 = \partial_{x_2} + 2x_1\partial_t$. Due to the left invariance with respect to the Heisenberg translation it suffices to look for the fundamental solution only at the origin, because

$$K(x, t; x_0, t_0) = K\Big((x_0, t_0)^{-1} \circ_H (x, t); 0, 0\Big),$$

where $\circ_H$ is the Heisenberg group law. One shall look for a fundamental solution in the form

$$K(x, t; 0, 0) = \int_{-\infty}^{\infty} \frac{E(x,\tau) v(x,\tau)}{g(x, t, \tau)} \, d\tau,$$

see [**3**], [4]. In the above relation $g = -it + r^2 \coth(2\tau)$ is the complex action and $E = -\partial g/\partial \tau = \frac{2r^2}{\sinh^2(2\tau)}$ is the energy along the geodesics. The function $v(x, \tau)$ is called the *volume element* and satisfies the transport equation

$$\Delta_h(Ev) + \frac{\partial}{\partial \tau}(Tv + (\Delta_H g)v) = 0, \tag{5.120}$$

where

$$T = \frac{\partial}{\partial \tau} + (X_1 g)X_1 + (X_2 g)X_2 \tag{5.121}$$

is called the first order *transport operator.* The following result shows that in fact T is the differentiation along the bicharacteristics.

PROPOSITION 5.49. If f is a function and $(x(s), t(s))$ is a subRiemannian geodesic, then

$$Tf = \frac{d}{ds} f\big(x(s), t(s), s\big)_{|s=\tau}. \tag{5.122}$$

PROOF. Using $X_i g = \dot{x}_i$, $i = 1, 2$, then the chain rule yields

$$\begin{aligned}
\frac{d}{ds} f\big(x(s), t(s), s\big) &= \frac{\partial f}{\partial x_1}\dot{x}_1(s) + \frac{\partial f}{\partial x_2}\dot{x}_2(s) + \frac{\partial f}{\partial t}\dot{t}(s) + \frac{\partial f}{\partial s} \\
&= \frac{\partial f}{\partial x_1} X_1 g + \frac{\partial f}{\partial x_2} X_2 g + \frac{\partial f}{\partial t}\dot{t}(s) + \frac{\partial f}{\partial s} \\
&= \Big(\frac{\partial f}{\partial x_1} - 2x_2\frac{\partial f}{\partial t}\Big) X_1 g + \Big(\frac{\partial f}{\partial x_2} + 2x_1 \frac{\partial f}{\partial t}\Big) X_2 g \\
&\quad +2x_2 \frac{\partial f}{\partial t} X_1 g - 2x_1 \frac{\partial f}{\partial t} X_2 g + \frac{\partial f}{\partial t}\dot{t}(s) + \frac{\partial f}{\partial s} \\
&= (X_1 f)(X_1 g) + (X_2 f)(X_2 g) \\
&\quad + \frac{\partial f}{\partial t}\Big(\dot{t} + 2x_2 X_1 g - 2x_1 X_2 g\Big) + \frac{\partial f}{\partial s}.
\end{aligned}$$

Since the geodesics are horizontal curves, we obtain

$$\begin{aligned}
\frac{d}{ds} f\big(x(s), t(s), s\big) &= (X_1 f)(X_1 g) + (X_2 f)(X_2 g) \\
&\quad + \frac{\partial f}{\partial t}\big(\underbrace{\dot{t} + 2x_2\dot{x}_1 - 2x_1\dot{x}_2}_{=0}\big) + \frac{\partial f}{\partial s} \\
&= (X_1 f)(X_1 g) + (X_2 f)(X_2 g) + \frac{\partial f}{\partial s}.
\end{aligned}$$

Evaluating the previous identity at the final value $s = \tau$, we obtain (5.122). □

COROLLARY 5.50. If E is the energy along the bicharacteristics,

$$TE = 0.$$

PROOF. It comes from the fact that E is constant along the bicharacteristics. □

The transport equation (5.120) follows from a formal computation. Since $K(x, t; 0, 0)$ must satisfy

$$\Delta_H K(x, t; 0, 0) = 0, \qquad (x, t) \neq (0, 0),$$

this formally leads to

$$\begin{aligned}
0 &= \int_{-\infty}^{\infty} \Delta_H\Big(\frac{Ev}{g}\Big)\, d\tau \\
&= \int_{-\infty}^{\infty} \Big[\frac{\Delta_H(Ev)}{g} - \frac{\langle Xg, X(Ev)\rangle}{g^2} + \frac{2EvH(\nabla g)}{g^3} - \frac{(\Delta_H g)Ev}{g^2}\Big]\, d\tau, \qquad (5.123)
\end{aligned}$$

where

$$\langle Xg, X(Ev)\rangle = X_1 g\, X_1(Ev) + X_2 g\, X_2(Ev)$$

and $H(\nabla g) = \frac{1}{2}\Big[(X_1 g)^2 + (X_2 g)^2\Big]$. The Hamilton-Jacobi equation

$$\frac{\partial g}{\partial \tau} + H(\nabla g) = 0$$

yields

$$\begin{aligned}\frac{2EvH(\nabla g)}{g^3} &= -\frac{2Ev\partial g/\partial\tau}{g^3} = Ev\frac{\partial}{\partial\tau}\frac{1}{g^2}\\ &= -\frac{1}{g^2}\frac{\partial}{\partial\tau}(Ev) + \frac{\partial}{\partial\tau}\Big(\frac{Ev}{g^2}\Big).\end{aligned}$$

Assuming $\lim_{\tau\to\pm\infty} Ev/g^2 = 0$, (5.123) yields

$$0 = \int_{-\infty}^{\infty}\Big[\frac{\Delta_H(Ev)}{g} - \frac{T(Ev)+(\Delta_H g)Ev}{g^2}\Big]\,d\tau \tag{5.124}$$

with T defined by (5.121). Corollary 5.50 and an integration by parts yield

$$\begin{aligned}\frac{T(Ev)+(\Delta_H g)Ev}{g^2} &= -\Big(Tv+(\Delta_H g)v\Big)\frac{E}{g^2}\\ &= -\Big(Tv+(\Delta_H g)v\Big)\frac{\partial}{\partial\tau}\frac{1}{g}\\ &= \frac{1}{g}\frac{\partial}{\partial\tau}\Big(Tv+(\Delta_H g)v\Big) - \frac{\partial}{\partial\tau}\frac{Tv+(\Delta_H g)v}{g}.\end{aligned}$$

Under the assumption

$$\lim_{\tau\to\pm\infty}\frac{Tv+(\Delta_H g)v}{g} = 0,$$

the identity (5.124) becomes

$$0 = \int_{-\infty}^{\infty}\frac{1}{g}\Big(\Delta_H(Ev) + \frac{\partial}{\partial\tau}(Tv+(\Delta_H g)v)\Big)\,d\tau$$

and hence it suffices that v satisfies equation (5.120). In the next section we shall solve the transport equation (5.120).

5.14. The volume element on the Heisenberg group

Considering

$$X_1 = \partial_{x_1} + 2x_2\partial_t, \qquad X_2 = \partial_{x_2} - 2x_1\partial_t,$$

a computation shows

$$\begin{aligned}X_1 g &= 2x_1\coth(2\tau) - 2ix_2, \qquad X_1^2 g = 2\coth(2\tau),\\ X_2 g &= 2x_2\coth(2\tau) + 2ix_1, \qquad X_2^2 g = 2\coth(2\tau),\end{aligned}$$

where

$$g = (x_1^2+x_2^2)\coth(2\tau) - it.$$

Hence

$$\Delta_X g = 2\coth(2\tau).$$

We need to solve the transport equation

$$\Delta_X(Ev) + \frac{\partial}{\partial\tau}\Big(Tv + (\Delta_X g)v\Big) = 0.$$

We are looking for a volume element v, which satisfies simultaneously both equations

$$\begin{aligned}\Delta_X(Ev) &= 0\\ Tv + (\Delta_X g)v &= 0.\end{aligned}$$

Consider the ansatz $v = A(r^2)B(\tau)$. In the following we shall compute Tv.

$$X_1 v = \partial_{x_1} v = 2x_1 A'(r^2)B(\tau).$$
$$X_2 v = \partial_{x_2} v = 2x_2 A'(r^2)B(\tau).$$

$$\begin{aligned} Tv &= \partial_\tau v + (X_1 g)X_1 v + (X_2 g)X_2 v \\ &= A(r^2)B'(\tau) + 2(x_1\coth(2\tau) - ix_2)2x_1 A'(r^2)B(\tau) \\ &\quad +2(x_2\coth(2\tau) + ix_1)2x_2 A'(r^2)B(\tau) \\ &= A(r^2)B'(\tau) + 4r^2\coth(2\tau)A'(r^2)B(\tau). \end{aligned}$$

The equation

$$Tv + 2\coth(2\tau)v = 0$$

can be written as

$$A(r^2)B'(\tau) + 4r^2\coth(2\tau)A'(r^2)B(\tau) + 2\coth(2\tau)A(r^2)B(\tau) = 0.$$

Dividing by $A(r^2)B(\tau)\coth(2\tau)$ yields

$$\frac{B'(\tau)}{B(\tau)\coth(2\tau)} + 4r^2\frac{A'(r^2)}{A(r^2)} + 2 = 0.$$

There is a separation constant C such that

$$\begin{aligned} \frac{B'(\tau)}{B(\tau)} &= C\coth(2\tau) \\ 4r^2\frac{A'(r^2)}{A(r^2)} &= -(C+2). \end{aligned}$$

Integrating the above equations we obtain

$$\begin{aligned} B(\tau) &= K_1\sinh(2\tau)^{C/2} \\ A(r^2) &= K_2 r^{\frac{-2(C+2)}{4}}, \end{aligned}$$

where $K_1, K_2 \in \mathbb{R}$ are constants. Hence the volume element is

$$v = \mathbf{k}\frac{\sinh(2\tau)^{C/2}}{r^{\frac{2(C+2)}{4}}},$$

with $\mathbf{k}$ constant. Then we choose the constant C such that $\Delta_X(Ev) = 0$, where

$$Ev = \frac{2r^2}{\sinh^2(2\tau)}\cdot\mathbf{k}\frac{\sinh(2\tau)^{C/2}}{r^{\frac{2(C+2)}{4}}}.$$

Obviously, $\Delta_X Ev = 0$ if Ev does not depend on r, *i.e.*, when $C = 2$. In this case the volume element becomes

$$v = \mathbf{k}\frac{\sinh(2\tau)}{r^2}.$$

In the following we shall show that this is the only possible volume element, *i.e.*, we cannot choose another value of C such that $\Delta_X(Ev) = 0$.

Let

$$m = 2 - \frac{2(C+2)}{4}.$$

We choose m such that the following condition is satisfied

$$\Delta_X(r^m) = 0.$$

The following Lemma yields only one good value, which is $m = 0$, *i.e.*, $C = 2$.

LEMMA 5.51. *We have*

$$\Delta_X(r^m) = 2m^2 r^{2(m-1)}.$$

PROOF. A computation shows

$$\begin{aligned}(X_1^2+X_2^2)(r^m) &= 4r^{2(m-2)}m^2x_1^2 + 4r^{2(m-1)}m - 4r^{2(m-2)}mx_1^2 \\ &\quad +4r^{2(m-2)}m^2x_2^2 - 4r^{2(m-2)}mx_2^2 \\ &= 4mr^{2(m-1)} + 4r^{2(m-2)}\Big[m^2x_1^2 - mx_1^2 + m^2x_2^2 - mx_2^2\Big] \\ &= 4mr^{2(m-1)} + 4r^{2(m-2)}\Big[(m^2-m)\underbrace{(x_1^2+x_2^2)}_{=r^2}\Big] \\ &= 4mr^{2(m-1)} + 4r^{2(m-2)}m(m-1)r^2 \\ &= 4mr^{2(m-1)}(1+m-1) = 4m^2r^{2m-2}.\end{aligned}$$

□

5.14.0.9. *The fundamental solution.* In this section we shall compute explicitly the fundamental solution given by the integral

$$K(x,t;0,0) = \int_{-\infty}^{\infty} \frac{E(x,\tau)v(x,\tau)}{g(x,t,\tau)}\,d\tau, \tag{5.125}$$

where

$$E(x,\tau) = \frac{2|x|^2}{\sinh^2(2\tau)}, \quad g = (x_1^2+x_2^2)\coth(2\tau) - it, \quad v(x,\tau) = \mathbf{k}\frac{\sinh(2\tau)}{|x|^2}.$$

$K(x,t,0,0)$ may be thought of as the $(\text{action})^{-1}$ summed on the characteristic variety over $(0,0)$ with measure Ev.
Changing the variable of integration from τ to g yields

$$\begin{aligned}K(x,t;0,0) &= \int_{\mathbb{R}} \frac{E(x,\tau)v(x,\tau)}{g(x,\tau)}d\tau \\ &= \int_{\mathbb{R}} \frac{v}{g}\left(-\frac{\partial g}{\partial \tau}\right)d\tau = -\int_{g_-}^{g_+} v\frac{dg}{g},\end{aligned} \tag{5.126}$$

where

$$g_\pm = \lim_{\tau\to\pm\infty} g = \pm(x_1^2+x_2^2) - it.$$

In this notation we have

$$v = -\frac{e^{i\pi/2}}{4\pi^2}\frac{1}{\sqrt{(g_+-g)(g-g_-)}},$$

where we chose $k = \frac{-1}{4\pi^2}$. Then

$$K(x,t;0,0) = \int_{\mathcal{C}_-} v\frac{dg}{g},$$

where we made a branch cut

$$\mathcal{C} = (-\infty - it, g_-] \cup [g_+, -it+\infty),$$

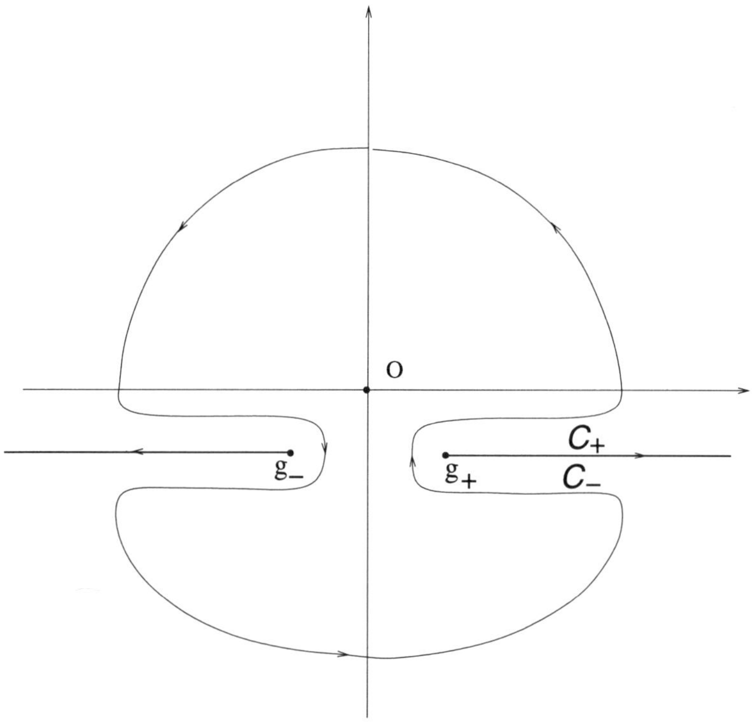

Figure 5.6: The "dumbbell" with waist at $g_\pm$.

with upper and lower directed edges denoted by $\mathcal{C}_+$ and $\mathcal{C}_-$, respectively. See Figure 5.6. Integrals of v/g vanish on upper and lower semicircles as their radii increase without bound. Therefore we can integrate on a "dumbbell" with waist at $g_\pm$. Inside this domain v/g has a simple pole at $g = 0$. Hence

$$-\frac{e^{i\pi/2}}{4\pi^2}(g_+)^{-1/2}(-g_-)^{-1/2} = \mathrm{Res}\big|_{g=0}\frac{v(g)}{g} = -\frac{1}{2\pi i}\int_{\mathcal{C}_-\cup\mathcal{C}_+} v\frac{dg}{g}.$$

From the residue calculation, we know that

$$\int_{\mathcal{C}_+} v\frac{dg}{g} = \left(e^{2\pi i}\right)^{-\frac{1}{2}}\int_{-\mathcal{C}_-} v\frac{dg}{g} = -\int_{\mathcal{C}_-} v\frac{dg}{g}.$$

Hence, one has

$$\begin{aligned} v\big|_{g=0} &= -\frac{1}{2\pi i}\int_{\mathcal{C}_-\cup\mathcal{C}_+} v\frac{dg}{g} \\ &= -\frac{1}{2\pi i}\int_{\mathcal{C}_-} v\frac{dg}{g} + \frac{1}{2\pi i}\int_{\mathcal{C}_+} v\frac{dg}{g} \\ &= -\frac{1}{2\pi i}\int_{\mathcal{C}_-} v\frac{dg}{g} - \frac{1}{2\pi i}\int_{\mathcal{C}_-} v\frac{dg}{g} \\ &= -\frac{1}{\pi i}\int_{\mathcal{C}_-} v\frac{dg}{g} = \frac{i}{\pi}K(x,t,0,0), \end{aligned}$$

and therefore we obtain

$$K(x,t;0,0) = -\frac{1}{8\pi}\frac{1}{\sqrt{|x|^4+t^2}}.$$

One may show that

$$\Delta_H^{(x,t)} K(x,t;0,0) = \delta_x \delta_t$$

in the distributions sense.

This is the Folland-Stein formula on the Heisenberg group. Unfortunately this simple formula is just a lucky coincidence of too much symmetry, and on general non-isotropic Heisenberg groups the fundamental solutions are given in the form (5.126). The reason is that the fundamental solution must include all the distances, which necessitates the use of g, and the summation over all the distances means integration in the variable τ.

REMARK 5.52. $d_C(x,t;0,0) = \sqrt{|x|^4 + t^2}$ is the Carnot-Carathéodory distance from the origin, see chapter 1.

5.15. Exercises

EXERCISE 5.1. Let $\alpha(s)$ be a differentiable function such that

$$\dot{\alpha} \sinh \alpha = 4 \sinh \alpha, \qquad \dot{\alpha} \cosh \alpha = 4 \cosh \alpha.$$

Show that $\alpha(s) = 4s + \alpha_0$, where $\alpha_0 \in \mathbb{R}$ is a constant.

EXERCISE 5.2. Show that the equation

$$\cosh(4 + x) - \cosh x = 0$$

has the unique solution $x = -2$.

EXERCISE 5.3. Consider the initial condition problem

$$\dot{r} = 2r \coth(2\tau), \quad r(0) = 0.$$

i) Show that $\int \coth x \, dx = \ln \sinh(x) + C$, for any $x > 0$.

ii) Using the method of separation of variables, show that

$$r(\tau) = C \sinh(2\tau)$$

where C is a constant.

iii) Show that for any $0 \leq s \leq \tau$ we have

$$r(s) = r(\tau) \frac{\sinh(2s)}{\sinh(2\tau)}.$$

EXERCISE 5.4. i) In polar coordinates $x_1 = r\cos\phi$, $x_2 = r\sin\phi$, the vector fields $X_1 = \partial_{x_1} + 2x_2|x|^{2k}\partial_t$, $X_2 = \partial_{x_2} - 2x_1|x|^{2k}\partial_t$, become

$$\begin{aligned} X_1 &= \cos\phi\,\partial_r - \frac{\sin\phi}{r}\partial_\phi + 2r^{2k+1}\sin\phi\,\partial_t \\ X_2 &= \sin\phi\,\partial_r + \frac{\cos\phi}{r}\partial_\phi - 2r^{2k+1}\cos\phi\,\partial_t. \end{aligned}$$

ii) Show that

$$\begin{aligned} X_1 g &= \cos\phi\,\partial_r g - 2ir^{2k+1}\sin\phi \\ X_2 g &= \sin\phi\,\partial_r g + 2ir^{2k+1}\cos\phi, \end{aligned}$$

where g is the complex action.

iii) Prove that

$$\Delta_X g = \frac{1}{2}(X_1^2 + X_2^2)g = \frac{1}{2}(\partial_r^2 g + \frac{1}{r}\partial_r g).$$

EXERCISE 5.5. Let $v = v(r, \tau)$, where $r = |x|$ and X_1, X_2 as in Exercise 5.4.

i) Then v is rotational symmetric and translation invariant along the t-axis.

ii) Show that

$$\begin{aligned} X_1^2 v &= \cos^2\phi\, \partial_r^2 v + \frac{\sin^2\phi}{r}\partial_r v, \\ X_2^2 v &= \sin^2\phi\, \partial_r^2 v + \frac{\cos^2\phi}{r}\partial_r v, \\ \Delta_X v &= \frac{1}{2}(\partial_r^2 v + \frac{1}{r}\partial_r v). \end{aligned}$$

(Note that v and g satisfy the same equation!).

EXERCISE 5.6. Find the functions A and B such that $v(r, \tau) = A(r) + B(\tau)$ satisfies the transport equation

$$E\Delta_X v + (T - \Delta_X g)\frac{\partial v}{\partial \tau} = 0,$$

following the steps:

i) Show that

$$\begin{aligned} \Delta_X v &= \frac{1}{2}(A''(r) + \frac{1}{r}A'(r)), \\ \frac{\partial v}{\partial \tau} &= B'(\tau), \\ T\Big(\frac{\partial v}{\partial \tau}\Big) &= \frac{\partial B'}{\partial \tau} + \sum_{i=1}^{2}(X_i g)(X_i B') = B''(\tau). \end{aligned}$$

ii) Choosing B such that $B''(\tau) = 0$, show there is a constant C such that

$$\begin{aligned} -\frac{1}{E}B'(\tau) &= C, \\ \frac{r^2 A''(r) + rA'(r)}{r^2\partial_r^2 g + r\partial_r g} &= C. \end{aligned}$$

iii) Using the substitution $r = e^\rho$, show that

$$\begin{aligned} r^2 A''(r) + rA'(r) &= \partial_\rho^2 A(e^\rho) \\ r^2\partial_r^2 g + r\partial_r g &= \partial_\rho^2 g(e^\rho). \end{aligned}$$

iv) Using ii) and iii) show that

$$\partial_\rho^2 A(e^\rho) = C\, \partial_\rho^2 g(e^\rho).$$

v) Show that

$$\begin{aligned} A(r) &= Cg(r) + a\ln r + b, \\ B(\tau) &= B_0 - CE\tau, \qquad B_0, C, a, b \in \mathbb{R}. \end{aligned}$$

vi) Find the function v in both cases $r(0) = 0$ and $r(0) = r_0 \neq 0$.

EXERCISE 5.7. The Heisenberg vector fields on $\mathbf{H}_1$ in polar coordinates are

$$\begin{aligned} X_1 &= \cos\phi\,\partial_r - \frac{\sin\phi}{r}\partial_\phi + 2r\sin\phi\,\partial_t, \\ X_2 &= \sin\phi\,\partial_r + \frac{\cos\phi}{r}\partial_\phi - 2r\cos\phi\,\partial_t, \end{aligned}$$

and the complex action is $g = -it + r^2\coth(2\tau)$. Find a solution of the transport equation

$$E\Delta_X v + (T - \Delta_X g)\frac{\partial v}{\partial \tau} = 0$$

under the form $v = a(r) + b(\tau)$, using the following steps:

i) Show there is a separation constant C such that

$$\begin{aligned} r^2 a''(r) + r a'(r) &= C, \\ b''(\tau) + 2\coth(2\tau)b'(\tau) + (\coth^2\tau - 1)C &= 0. \end{aligned} \tag{5.127}$$

ii) Using the substitution $r = e^\rho$ show that the equation (5.127) has the general solution

$$a(r) = \frac{1}{2}C(\ln r)^2 + C_1\ln r + C_2, \qquad C_1, C_2 \in \mathbb{R}.$$

iii) Let $v = \coth(2\tau)$. Show that

$$v' = 2(1 - v^2).$$

iv) Let $b(\tau) = \Phi(\coth(2\tau))$. Show that Φ satisfies the ODE

$$(1 - v^2)\Phi''(v) - v\Phi'(v) - \frac{C}{4} = 0. \tag{5.128}$$

v) Let $y = \Phi'$. Show that y satisfies the linear equation

$$y' - \frac{v}{1 - v^2}y = \frac{C}{4}\frac{1}{1 - v^2},$$

with the integrant factor $\mu(v) = \sqrt{v^2 - 1}$.

vi) Show that

$$y(v) = -\frac{C}{4}\frac{\ln(v + \sqrt{v^2 - 1})}{\sqrt{v^2 - 1}} + \frac{C_0}{\sqrt{v^2 - 1}}, \qquad C_0 \in \mathbb{R}.$$

vii) Show that $\cosh^{-1}(\coth 2\tau) = \ln(\coth\tau)$.

viii) Show that

$$\begin{aligned} \int \frac{\cosh^{-1}(v)}{\sqrt{v^2 - 1}}\,dv &= \frac{1}{2}\Big(\ln(v + \sqrt{v^2 - 1})\Big)^2 + C \\ &= \frac{1}{2}(\cosh^{-1}(v))^2 + C. \end{aligned}$$

ix) Show that

$$\Phi(v) = -\frac{C}{8}(\cosh^{-1}(v))^2 + C_0\cosh^{-1}(v) + C,$$

x) Show that

$$b(\tau) = -\frac{C}{8}\Big(\cosh^{-1}(\coth 2\tau)\Big)^2 + C_0\cosh^{-1}(\coth 2\tau).$$

xi) Show that the volume element is

$$v = a(r) + b(\tau) = \frac{C}{8} \ln(r^2 \tanh \tau) \ln(r^2 \coth \tau).$$

EXERCISE 5.8. Let $E(\tau)$ denote the energy along a complex geodesic $c : [0, \tau] \to \mathbb{R}^3$ of a step 2 manifold. Prove that if E satisfies
(*i*) $E(\tau)$ is real,
(*ii*) $\lim_{|\tau| \to \infty} E(\tau) = 0$,
then the momentum θ is pure imaginary.

CHAPTER 6

Quantum Mechanics on the Heisenberg group

We have seen in the first section of chapter 5 how Heisenberg operator arose from the harmonic oscillator by a certain complex quantification. In this chapter we shall emphasize more relations between the quantum harmonic oscillator and the Heisenberg operator. It turns out that the quantification of energy property is identical for both models. Before starting this discussion we shall recall in the next section a few useful things about Quantum Mechanics.

6.1. Linear Harmonic Oscillator

The potential energy of a material point of mass $m_0 = 1$ in an elastic force field is

$$U(x) = \frac{\omega^2}{2} x^2, \tag{6.1}$$

where the constant ω is the oscillator pulsation. Let E be the oscillator total energy. Classically, the particle will oscillate between $x_1 = -A$ and $x_2 = +A$, where

$$E = \frac{\omega^2}{2} A^2.$$

From the quantic point of view, the harmonic oscillator is characterized by a density probability function $\Psi(x,t)$, which satisfies the one-dimensional Schrödinger equation

$$ih\frac{\partial \Psi}{\partial t} + \frac{h^2}{2}\frac{d^2}{dx^2}\Psi = U(x)\Psi, \tag{6.2}$$

with $U(x)$ given by (6.1). The linear harmonic oscillator is one of the most important examples in elementary Quantum Mechanics. It describes vibrations in molecules and their counterparts in solids. The most eminent role of this oscillator is its linkage to the boson. The boson is one of the conceptual building blocks of microscopic physics, describing the models of the electromagnetic field, providing the basis of its quantization. It is named after the Indian physicist S. N. Bose, see [**11**].

It is a well-known fact that the energy of the linear harmonic oscillator is quantified by the relation

$$E_n = \Big(n + \frac{1}{2}\Big)h\omega, \qquad n = 0, 1, 2, \dots$$

where h is the Plank constant. Hence the harmonic oscillator, from the quantic point of view, may have only discrete values for the energy. The difference between two successive levels of energy is always the same, equal to $h\omega$. The minimum energy the linear harmonic oscillator may have is

$$E_0 = \frac{h\omega}{2}.$$

This is called the ground-state energy, see Figure 6.1.

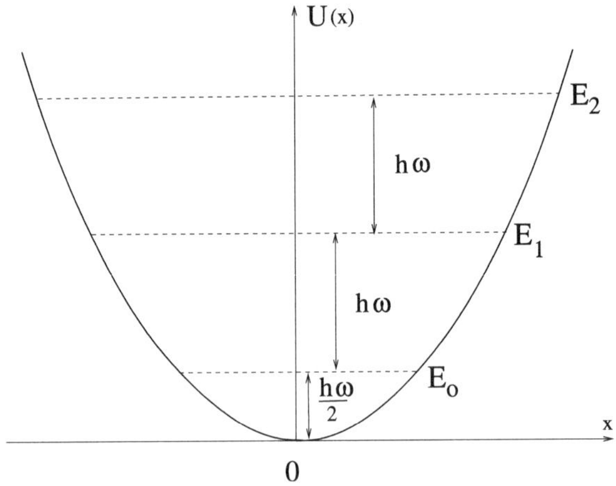

Figure 6.1: The energies of the linear harmonic oscillator are discrete.

The same quantification relation will be obtained in the Heisenberg case. The construction will be done by analogy with the elementary Quantum Mechanics construction, which uses the creation and anihilation operators.
Schrödinger's equation (6.2) can be written as

$$ih\frac{\partial \Psi}{\partial t} = \widehat{H}(\Psi),$$

with the Hamiltonian operator

$$\widehat{H} = \frac{1}{2}\widehat{p}^2 + \frac{\omega^2}{2}\widehat{x}^2, \tag{6.3}$$

where

$$\widehat{p} = \frac{h}{i}\frac{d}{dx}, \qquad \widehat{x} = x$$

are the momentum and the position operators. The uncertainty Heisenberg relation is expressed as

$$[\widehat{p}, \widehat{x}] = \frac{h}{i}\mathbb{I}, \tag{6.4}$$

where $\mathbb{I}$ is the identity operator. In the elementary Quantum Mechanics the following creation and anihilation operators are used

$$\widehat{a}^+ = \sqrt{\frac{\omega}{2h}}\widehat{x} - \frac{i}{\sqrt{2h\omega}}\widehat{p}, \tag{6.5}$$

$$\widehat{a}^- = \sqrt{\frac{\omega}{2h}}\widehat{x} + \frac{i}{\sqrt{2h\omega}}\widehat{p}. \tag{6.6}$$

They are obtained by a formal factorization of the operator $\frac{1}{h\omega}\widehat{H}$. The above operators have the following properties

$$[\widehat{a}^-,\widehat{a}^+] = \mathbb{I}, \tag{6.7}$$

$$\widehat{a}^-\widehat{a}^+ = \frac{1}{h\omega}\widehat{H}+\frac{1}{2}\mathbb{I}, \tag{6.8}$$

$$\widehat{a}^+\widehat{a}^- = \frac{1}{h\omega}\widehat{H}-\frac{1}{2}\mathbb{I}. \tag{6.9}$$

One may find more about the linear harmonic oscillator in [**45**]. In the next section we shall find similar relations on the Heisenberg group.

6.2. The operators Z and $\overline{Z}$ and the energy quantification

In this chapter we shall adopt the following notation for the Heisenberg vector fields

$$X_1 = \partial_{x_1} + 2hx_2\partial_t, \quad X_2 = \partial_{x_2} - 2hx_1\partial_t, \tag{6.10}$$

where $h > 0$ stands for the Plank's constant. The Heisenberg operator

$$\Delta_H = \frac{1}{2}X_1^2 + \frac{1}{2}X_2^2 \tag{6.11}$$

and the Hamiltonian operator (6.3) are both sum of squares of operators which do not commute. The relation

$$[X_1, X_2] = -4h\partial_t$$

is an analogue of the uncertainty relation (6.4). The vector fields X_1 and X_2 correspond to some physical fields, which cannot be observed simultaneously.
Let

$$Z = \frac{1}{2}(X_1 - iX_2), \tag{6.12}$$

$$\overline{Z} = \frac{1}{2}(X_1 + iX_2), \tag{6.13}$$

which correspond to the operators $\widehat{a}^+$ and $\widehat{a}^-$ given in (6.5)-(6.6). The following proposition shows that Z and $\overline{Z}$ have similar properties to operators $\widehat{a}^+$ and $\widehat{a}^-$ given above.

PROPOSITION 6.1. *The operators Z and $\overline{Z}$ verify the following relations*

$$1)\ [\overline{Z}, Z] = 2hi\partial_t,$$

$$2)\ \overline{Z}Z = \frac{1}{2}\Delta_H + ih\partial_t,$$

$$3)\ Z\overline{Z} = \frac{1}{2}\Delta_H - ih\partial_t.$$

PROOF. 1) It follows from 2) and 3) by subtraction.
2) We have

$$\begin{aligned}\overline{Z}Z &= \frac{1}{4}(X_1 + iX_2)(X_1 - iX_2)\\ &= \frac{1}{4}(X_1^2 + X_2^2 - i[X_1, X_2])\\ &= \frac{1}{4}(X_1^2 + X_2^2) + ih\partial_t.\end{aligned}$$

2) It is obtained from the second one by conjugation. □

COROLLARY 6.2. The Heisenberg operator can be expressed as

$$\begin{aligned} 1)\ \Delta_H &= 2Z\overline{Z} + 2ih\partial_t = 2Z\overline{Z} + [\overline{Z}, Z], \\ 2)\ \Delta_H &= 2\overline{Z}Z - 2ih\partial_t = 2\overline{Z}Z - [\overline{Z}, Z]. \end{aligned}$$

Consider the operator

$$\begin{aligned} \Delta_\lambda &= \frac{1}{2}(X_1^2 + X_2^2) - \frac{1}{2}i\lambda[X_1, X_2] \\ &= Z\overline{Z} + \overline{Z}Z + \lambda[\overline{Z}, Z]. \end{aligned} \tag{6.14}$$

Consider a quantum system described by the wave function $\psi : \mathbb{R}^3 \to \mathbb{C}$, which satisfies the equation $\Delta_\lambda\psi = 0$. The real number λ will be called the energy of ψ. Like in the case of the linear harmonic oscillator, we shall assume the existence of a *ground state* ψ_0, which is a solution of $Z\psi_0 = 0$. We shall discuss the solutions of this equation in a later section. Here we are concerned only with the values of λ. We shall find λ_0 such that $\Delta_{\lambda_0}\psi_0 = 0$. λ_0 plays the role of the *ground-state* energy. Corollary 6.2 yields

$$\begin{aligned} \Delta_H\psi_0 &= (2\overline{Z}Z - 2ih\partial_t)\psi_0 = 2\overline{Z}\overbrace{(Z\psi_0)}^{=0} - 2ih\partial_t\psi_0 \\ &= \frac{1}{2}i[X_1, X_2]\psi_0, \end{aligned}$$

and hence $\lambda_0 = +1$. $\overline{Z}$ will be considered as creation operator. We can create excited states $\psi_1, \psi_2, \psi_3, \ldots$, where

$$\psi_1 = \overline{Z}\psi_0, \quad \psi_{n+1} = \overline{Z}\psi_n = \overline{Z}^n\psi_0.$$

In this section we shall compute the energies λ_n associated with the excited states ψ_n, *i.e.*, such that $\Delta_{\lambda_n}\psi_n = 0$. Using again Corollary 6.2 and the fact that ∂_t commutes with Z and $\overline{Z}$, we have

$$\begin{aligned} \Delta_H\psi_1 &= \Delta_H(\overline{Z}\psi_0) = (2\overline{Z}Z - 2ih\partial_t)\overline{Z}\psi_0 \\ &= \overline{Z}(2Z\overline{Z})\psi_0 - 2ih\partial_t\psi_1 \\ &= \overline{Z}(2Z\overline{Z} + 2ih\partial_t - 2ih\partial_t)\psi_0 - 2ih\partial_t\psi_1 \\ &= \overline{Z}(\Delta_H\psi_0) - 2ih\partial_t\overline{Z}\psi_0 - 2ih\partial_t\psi_1 \\ &= \overline{Z}(\Delta_H\psi_0) - 4ih\partial_t\psi_1 \\ &= \overline{Z}(2\overline{Z}\underbrace{Z\psi_0}_{=0} - 2ih\partial_t\psi_0) - 4ih\partial_t\psi_1 \\ &= \frac{1}{2}i[X_1, X_2]\overline{Z}\psi_0 + i[X_1, X_2]\psi_1 \\ &= \left(1 + \frac{1}{2}\right)i[X_1, X_2]\psi_1. \end{aligned}$$

Hence

$$\Delta_H\psi_1 = \left(1 + \frac{1}{2}\right)i[X_1, X_2]\psi_1,$$

and then $\lambda_1 = 3$. We shall prove by induction that

$$\Delta_H\psi_n = \left(n + \frac{1}{2}\right)i[X_1, X_2]\psi_n. \tag{6.15}$$

Let $\psi_{n+1} = \overline{Z}\psi_n$. A similar computation yields

$$\Delta_H\psi_{n+1} = \overline{Z}(\Delta_H\psi_n) - 4ih\partial_t\psi_{n+1}.$$

Using the induction hypothesis (6.15) yields

$$\begin{aligned}\Delta_H \psi_{n+1} &= \overline{Z}\Big((n+\frac{1}{2})i[X_1,X_2]\psi_n\Big) + i[X_1,X_2]\psi_n \\ &= (n+\frac{1}{2})i[X_1,X_2]\overline{Z}\psi_n + i[X_1,X_2]\psi_n \\ &= \Big(n+1+\frac{1}{2}\Big)i[X_1,X_2]\psi_{n+1}.\end{aligned}$$

Hence we have proved by induction that $\lambda_n = 2n+1$, where $\Delta_{\lambda_n}\psi_n = 0$, for $n = 0, 1, 2, \ldots$

To conclude, we have the following comments.
1) The eigenvalue $\lambda/2$ has only discrete values, equal to $n+\frac{1}{2}$, which is the same quantification relation as in the linear harmonic oscillator case.
2) For the values $\lambda = 2n+1$ the operator Δ_λ ceases to be hypoelliptic. A fundamental solution for Δ_λ is constructed in [**4**]. The fundamental solution is extended meromorphically in λ with obstructions at the simple poles $\pm\lambda = 2n+1$.
3) The negative sign of λ is obtained if we start with a conjugate ground state ψ_0^* which satisfies

$$\overline{Z}\psi_0^* = 0,$$

and define the the conjugate excited states

$$\psi_n^* = Z\psi_{n-1}^* = Z^n\psi_0^*.$$

In a similar way, one may show that

$$\Delta_H\psi_n^* = -(n+\frac{1}{2})i[X_1,X_2]\psi_n^*, \quad n = 0,1,2,\ldots$$

i.e., $\Delta_{-(2n+1)}\psi_n^* = 0$.

6.3. The ground-state ψ_0 and the states ψ_n

The function $\psi_0 : \mathbb{R}^3 \to \mathbb{C}$ satisfies the ground-state equation

$$Z\psi_0 = 0. \tag{6.16}$$

Let $\psi_0 = u + iv$, where u and v are real-valued functions. The following result is an analog of the Cauchy-Riemann equations.

PROPOSITION 6.3. $\psi_0 = u + iv$ satisfies the ground-state equation (6.16) if and only if

$$X_1 u = -X_2 v, \quad X_1 v = X_2 u. \tag{6.17}$$

PROOF.

$$\begin{aligned}2Z\psi_0 &= (X_1 - iX_2)\psi_0 = (X_1 - iX_2)(u+iv) \\ &= X_1(u+iv) - iX_2(u+iv) \\ &= X_1 u + iX_1 v - iX_2 u + X_2 v \\ &= (X_1 u + X_2 v) + i(X_1 v - X_2 u).\end{aligned}$$

Then $2\psi_0 = 0$ if and only if both the real and the imaginary components are zero, *i.e.*, relation (6.17). □

COROLLARY 6.4. Let $\psi_0 = u + iv$ be a solution of the ground-state equation (6.16). The following relations are equivalent
i) ψ_0 is constant,
ii) v is constant,
iii) u is constant.

PROOF. We shall show that if v is constant, then ψ_0 is constant. Using relation (6.17) we have $X_1 u = X_2 u = 0$. Hence, for any horizontal vector field $W = aX_1 + bX_2$, we have $Wu = 0$.
Let $x \neq 0$. We shall show that $u(0) = u(x)$. From the connectivity theorem, there is a subRiemannian geodesic joining the origin with the point x. Let $c : [0, \tau] \to \mathbb{R}^3$ be the geodesic with $c(0) = 0$ and $c(\tau) = x$. The velocity vector $\dot{c}(s) = \dot{x}_1(s)X_1 + \dot{x}_2(s)X_2$ is horizontal and hence $\dot{c}(s)u = 0$, *i.e.*, u is constant along the geodesic $c(s)$ and hence $u(0) = u(c(s)) = u(c(\tau)) = u(x)$. □

The relationship between the states ψ_0 and ψ_1 is given by the following result.

PROPOSITION 6.5. Let $\psi_1 = \overline{Z}\psi_0$ be the first excited state, where $\psi_0 = u + iv$ is the ground-state. Then
1) $\psi_1 = X_1\psi_0 = iX_2\psi_0$,
2) $\psi_1 = 2\overline{Z}u = 2i\overline{Z}v$,
3) $Z\psi_1 = \Delta_H \psi_0$.

PROOF. *1*) Relation $2Z\psi_0 = (X_1 - iX_2)\psi_0 = 0$ yields $X_1\psi_0 = iX_2\psi_0$. Then

$$\psi_1 = \overline{Z}\psi_0 = \frac{1}{2}(X_1 + iX_2)\psi_0 = X_1\psi_0 = iX_2\psi_0.$$

2) The equations (6.17) yield

$$\begin{aligned} \psi_1 &= \overline{Z}\psi_0 = X_1\psi_0 = X_1(u+iv) = X_1 u + iX_1 v \\ &= -X_2 v + iX_1 v = i(X_1 + iX_2)v = 2i\overline{Z}v, \end{aligned}$$

and

$$\psi_1 = X_1 u + iX_1 v = X_1 u + iX_2 u = 2\overline{Z}u.$$

3) Using relation *1*) yields

$$\begin{aligned} \Delta_H \psi_0 &= \frac{1}{2}(X_1^2\psi_0 + X_2^2\psi_0) = \frac{1}{2}\Big(X_1\psi_1 + X_2(\frac{1}{i}\psi_1)\Big) \\ &= \frac{1}{2}(X_1 - iX_2)\psi_1 = Z\psi_1. \end{aligned}$$

□

Before solving the ground-state equation (6.16), we need a few results regarding the vector fields Z and $\overline{Z}$.

Let $z = x_1 + ix_2 \in \mathbb{C}$ and define the operators

$$\bar{\partial}_z = \frac{1}{2}(\partial_{x_1} - i\partial_{x_2}), \quad \partial_z = \frac{1}{2}(\partial_{x_1} + i\partial_{x_2}). \tag{6.18}$$

A straightforward computation shows

$$\bar{\partial}_z z = 1, \qquad \bar{\partial}_z \bar{z} = 0, \qquad \bar{\partial}_z |z|^2 = \bar{z}, \tag{6.19}$$

$$\partial_z z = 0, \qquad \partial_z \bar{z} = 1, \qquad \bar{\partial}_z |z|^2 = z. \tag{6.20}$$

PROPOSITION 6.6. We have

$$Z = \overline{\partial}_z + ih\overline{z}\partial_t, \quad \overline{Z} = \partial_z - ihz\partial_t. \tag{6.21}$$

PROOF.

$$\begin{aligned} Z &= \frac{1}{2}(X_1 - iX_2) = \frac{1}{2}(\partial_{x_1} + 2hx_2\partial_t) - \frac{i}{2}(\partial_{x_2} - 2hx_1\partial_t) \\ &= \frac{1}{2}(\partial_{x_1} - i\partial_{x_2}) + h(x_2 + ix_1)\partial_t = \overline{\partial}_z + ih\overline{z}\partial_t. \end{aligned}$$

The second relation is obtained by conjugation from the first, because $\overline{\partial}_z$ and ∂_z are conjugate operators.

□

Z and $\overline{Z}$ are the holomorphic and the anti-holomorphic tangent vector fields to the hypersurface

$$b\mathbb{D}_2(h) = \{\xi \in \mathbb{C}^2; Im\,\xi_2 = h|\xi_1|^2\} \subset \mathbb{C}^2,$$

where

$$\mathbb{D}_2(h) = \{\xi \in \mathbb{C}^2; Im\,\xi_2 > h|\xi_1|^2\}$$

is the Siegel upper-half space in $\mathbb{C}^2$.

$\mathbb{C}^2$ has a group of biholomorphic affine transformations, which preserves $\mathbb{D}_2(h)$ and $b\mathbb{D}_2(h)$. They are parametrized by $(z,t) \in \mathbb{C} \times \mathbb{R}$:

$$\begin{aligned} (z,t) * \xi &= \xi', \\ \xi_1' &= \xi_1 + z, \\ \xi_2' &= \xi_2 + t + ih|z|^2 + 2ih\xi_1\overline{z}. \end{aligned}$$

This group acts simply transitively on the boundary, so $b\mathbb{D}_2(h)$ can be identified with $\mathbb{H}_1(h)$ by matching a transformation (z,t) to the image of the origin

$$(z,t) \longleftrightarrow (z + t + ih|z|^2).$$

The group structure is that of the Heisenberg group $\mathbb{H}_1(h)$ with $z = x_1 + iz_2$.
In the following we shall construct a family of solutions for the equation (6.16).

PROPOSITION 6.7. 1) Let $\psi_0 = G(h|z|^2 + it)$ where $G \in \mathcal{C}^1(\mathbb{R})$ arbitrary. Then $Z\psi_0 = 0$.
2) Let $\psi_0^* = F(h|z|^2 + it)$ where $F \in \mathcal{C}^1(\mathbb{R})$ arbitrary. Then $\overline{Z}\psi_0^* = 0$.

PROOF. Using (6.19), a straightforward computation shows

$$\begin{aligned} Z\psi_0 &= G'(h|z|^2 + it)Z(h|z|^2 + it) \\ &= G'(h|z|^2 + it)(\overline{\partial}_z + ih\overline{z}\partial_t)(h|z|^2 + it) \\ &= G'(h|z|^2 + it)(h\overline{\partial}_z|z|^2 - h\overline{z}) = 0. \end{aligned}$$

$$\begin{aligned} \overline{Z}\psi_0^* &= F'(h|z|^2 - it)(\partial_z - ihz\partial_t)(h|z|^2 - it) \\ &= F'(h|z|^2 - it)(h\partial_z|z|^2 - hz) = 0. \end{aligned}$$

□

The formulas for the excited states ψ_n are given below.

PROPOSITION 6.8. For any $G \in \mathcal{C}^1(\mathbb{R})$, the function

$$\psi_n = (2hz)^n G^{(n)}(h|z|^2 + it) \tag{6.22}$$

satisfies the equation $\Delta_{\lambda_n}\psi_n = 0$ with $\lambda_n = 2n+1$.

PROOF. We shall prove the relation by induction over n.

$$\begin{aligned}
\psi_1 &= \overline{Z}\psi_0 = \overline{Z}G(h|z|^2 + it) \\
&= G'(h|z|^2 + it)(\partial_z - ihz\partial_t)(h|z|^2 + it) \\
&= G'(h|z|^2 + it)(h\partial_z|z|^2 + hz) \\
&= 2hzG'(h|z|^2 + it),
\end{aligned}$$

and hence (6.22) is verified for $n = 1$. Assume (6.22) verified for some $n \geq 1$. Then

$$\begin{aligned}
\psi_{n+1} &= \overline{Z}\psi_n = 2^n h^n \Big(\overline{Z}(z^n)G^{(n)}(h|z|^2 + it) + z^n\overline{Z}G^{(n)}(h|z|^2 + it)\Big) \\
&= 2^n h^n \Big(nz^{n-1}\underbrace{\overline{Z}(z)}_{=0}G^{(n)}(h|z|^2 + it) + z^n G^{(n+1)}(h|z|^2 + it)\overline{Z}(h|z|^2 + it)\Big) \\
&= 2^n h^n z^n G^{(n+1)}(h|z|^2 + it)(h\partial_z|z|^2 + hz) \\
&= 2^{n+1}h^{n+1}z^{n+1}G^{(n+1)}(h|z|^2 + it),
\end{aligned}$$

which is formula (6.22) for $n+1$.

□

An important space in Quantum Mechanics is the space of functions vanishing at infinity

$$\mathcal{N}_0 = \{f : \mathbb{R}^2_x \times \mathbb{R}_t \to \mathbb{R};\ \lim_{|(x,t)|\to\infty} f(x,t) = 0\}.$$

The function G should be chosen such that $\psi_n \in \mathcal{N}_0$, for any $n \geq 0$.

For instance, if we choose $G(u) = \frac{1}{u}$, then

$$\phi_0 = \frac{1}{h|z|^2 + it} \in \mathcal{N}_0,$$

and Proposition 6.8 yields

$$\psi_n = \frac{(-2hz)^n n!}{(h|z|^2 + it)^{n+1}} \in \mathcal{N}_0.$$

6.4. The operator $\partial_\tau - Z$

In this section we shall investigate the non-commutativity relations aroused from the operators Z and $\overline{Z}$. The calculus developed is a consequence of the Heisenberg non-commutativity relation

$$[Z, \overline{Z}] = \frac{i}{2}[X_1, X_2] = -2ih\partial_t. \tag{6.23}$$

The 1-parameter group of diffeomorphisms associated to Z and the exponential function will be studied.

LEMMA 6.9. *For any integer $n \geq 1$, we have*

$$[Z, \overline{Z}^n] = n\overline{Z}^{n-1}[Z, \overline{Z}] = n[Z, \overline{Z}]\overline{Z}^{n-1}.$$

PROOF. Relation (6.23) implies that $[Z, \overline{Z}]$ commutes with Z and $\overline{Z}$. This solves the second identity. For the first identity we shall use induction. The case $n = 1$ is obvious. Assuming $[Z, \overline{Z}^n] = n\overline{Z}^{n-1}[Z, \overline{Z}]$, we have

$$\begin{aligned}
[Z, \overline{Z}^{n+1}] &= Z\overline{Z}^n\overline{Z} - \overline{Z}^{n+1}Z \\
&= ([Z, \overline{Z}^n] + \overline{Z}^n Z)\overline{Z} - \overline{Z}^{n+1}Z \\
&= [Z, \overline{Z}^n]\overline{Z} + \overline{Z}^n Z\overline{Z} - \overline{Z}^{n+1}Z \\
&= [Z, \overline{Z}^n]\overline{Z} + \overline{Z}^n[Z, \overline{Z}] \\
&= n\overline{Z}^{n-1}[Z, \overline{Z}]\overline{Z} + \overline{Z}^n[Z, \overline{Z}] \\
&= n\overline{Z}^n[Z, \overline{Z}] + \overline{Z}^n[Z, \overline{Z}] \\
&= (n+1)\overline{Z}^n[Z, \overline{Z}],
\end{aligned}$$

which proves the induction step. □

THEOREM 6.10. *Let f be an analytic function on $\mathbb{R}$. Then*

$$[Z, f(\overline{Z})] = f'(\overline{Z})[Z, \overline{Z}] = [Z, \overline{Z}]f'(\overline{Z}).$$

PROOF. The function f can be written as a convergent power series

$$f(u) = \sum_{n \geq 0} \frac{f^{(n)}(0)}{n!} u^n.$$

Lemma 6.9 yields

$$\begin{aligned}
[Z, f(\overline{Z})] &= [Z, \sum \frac{f^{(n)}(0)}{n!}\overline{Z}^n] = \sum \frac{f^{(n)}(0)}{n!}[Z, \overline{Z}^n] \\
&= \sum \frac{f^{(n)}(0)}{n!} n\overline{Z}^{n-1}[Z, \overline{Z}] = f'(\overline{Z})[Z, \overline{Z}].
\end{aligned}$$

The second identity comes from the fact that $[Z, \overline{Z}]$ commutes with the powers of Z and $\overline{Z}$. □

Starting with a ground-state ψ_0, we have constructed the excited states $\psi_n = \overline{Z}^n\psi_0$. We can construct more general states of the form $f(\overline{Z})\psi_0$, with f analytic function.

PROPOSITION 6.11. Let ψ_0 be a solution of the ground-state equation $Z\psi_0 = 0$ and let f be an analytic function.
1) We have

$$Zf(\overline{Z})\psi_0 = f'(\overline{Z})[Z, \overline{Z}]\psi_0 = [Z, \overline{Z}]f'(\overline{Z})\psi_0.$$

2) In general, for any $n \in \mathbb{N}$, we have

$$Z^n f(\overline{Z})\psi_0 = f^{(n)}(\overline{Z})[Z, \overline{Z}]^n\psi_0 = [Z, \overline{Z}]^n f^{(n)}(\overline{Z}).$$

PROOF. 1) We have

$$Zf(\overline{Z}) = [Z, f(\overline{Z})] + f(\overline{Z})Z.$$

Using $Z\psi_0 = 0$, Theorem 6.10 yields

$$Zf(\overline{Z})\psi_0 = [Z, f(\overline{Z})]\psi_0 = f'(\overline{Z})[Z, \overline{Z}]\psi_0.$$

2) As $[Z, \overline{Z}]$ commutes with Z, using relation 1) and Theorem 6.10 yields

$$\begin{aligned} Z^2 f(\overline{Z})\psi_0 &= Zf'(\overline{Z})[Z,\overline{Z}]\psi_0 = Z[Z,\overline{Z}]f'(\overline{Z})\psi_0 \\ &= [Z,\overline{Z}]Zf'(\overline{Z})\psi_0 = [Z,\overline{Z}]^2 f''(\overline{Z})\psi_0 \\ &= f''(\overline{Z})[Z,\overline{Z}]^2\psi_0. \end{aligned}$$

By iteration we get formula 2). □

COROLLARY 6.12. We have

$$Z^n f(\overline{Z})\psi_0 = f^{(n)}(\overline{Z})(-2ih\partial_t)^n\psi_0 = (-2ih\partial_t)^n f^{(n)}(\overline{Z})\psi_0.$$

COROLLARY 6.13. Let $\psi_m = \overline{Z}^m\psi_m$ be the m-th excited state. Then

$$Z^n\psi_m = \frac{m!}{(m-n)!}(-2ih\partial_t)^n\psi_{m-n}, \quad 0 \le n \le m.$$

In particular,

$$Z^n\psi_n = n!(-2ih\partial_t)^n\psi_0.$$

PROOF. Choose $f(\overline{Z}) = \overline{Z}^m$ in Proposition 6.11 and get

$$\begin{aligned} Z^n\psi_m &= Z^n\overline{Z}^m\psi_0 = m(m-1)\dots(m-n+1)\overline{Z}^{m-n}(-2ih\partial_t)^n\psi_0 \\ &= \frac{m!}{(m-n)!}(-2ih\partial_t)^n\underbrace{\overline{Z}^{m-n}\psi_0}_{=\psi_{m-n}}. \end{aligned}$$

□

Consider the operator $e^{\tau Z}$ with $\tau \in \mathbb{R}$, defined by

$$e^{\tau Z} = \sum_{n\ge 0}\frac{\tau^n}{n!}Z^n. \tag{6.24}$$

THEOREM 6.14. *Let f be an analytic function and let $v = f(\overline{Z})\psi_0$. Then the solution of the Cauchy problem*

$$\begin{aligned} \partial_\tau u &= Zu \quad \text{on } [0,\infty)\times\mathbb{C} \\ u_{|\tau=0} &= v, \end{aligned}$$

is given by

$$u = f(\overline{Z} + [Z,\overline{Z}]\tau)\psi_0.$$

PROOF. The solution can be written formally as $u = e^{\tau Z}v$. We shall compute it using Proposition 6.11

$$\begin{aligned} u &= e^{\tau Z}v = e^{\tau Z}f(\overline{Z})\psi_0 = \sum_{n\ge 0}\frac{\tau^n}{n!}Z^n f(\overline{Z})\psi_0 \\ &= \sum_{n\ge 0}\frac{\tau^n}{n!}f^{(n)}(\overline{Z})[Z,\overline{Z}]^n\psi_0 \\ &= \sum_{n\ge 0}\frac{(\tau[Z,\overline{Z}])^n}{n!}f^{(n)}(\overline{Z})\psi_0 \\ &= f(\overline{Z} + \tau[Z,\overline{Z}])\psi_0. \end{aligned}$$

□

In the theory of the linear harmonic oscillator the excited states $(\psi_n)_{n\geq 0}$ form a complete orthonormal system and the function v can be expanded always as

$$v = \sum_{n\geq 0} v_n\psi_n = \sum_{n\geq 0} v_n\overline{Z}^n\psi_0 = f(\overline{Z})\psi_0.$$

DEFINITION 6.15. $\rho_Z(\tau) = e^{\tau Z}v$ is called the 1-parameter group of diffeomorphisms associated to the vector field Z.

Z belongs to the complexified Lie algebra of the Lie group $\mathbf{H}_1$. The above definition is motivated by the following properties.

PROPOSITION 6.16. The following relations hold
1) $\rho_Z(0) = v$,
2) $\rho_Z(\tau_1 + \tau_2) = \rho_Z(\tau_1)\,\rho_Z(\tau_2), \quad \forall\tau_1, \tau_2 \in \mathbb{R}$,
3) $\frac{d}{d\tau}\rho_Z(\tau) = Z$.

PROOF. Relations 1) and 3) follow from the Definition 6.15. We still need to show 2). Using Theorem 6.14

$$\begin{aligned}
\rho_Z(\tau_1)\,\rho_Z(\tau_2) &= \rho_Z(\tau_1)e^{\tau_2 Z}v = \rho_Z(\tau_1)e^{\tau_2 Z}f(\overline{Z})\psi_0\\
&= \rho_Z(\tau_1)f(\overline{Z} + \tau_2[Z,\overline{Z}])\psi_0\\
&= e^{\tau_1 Z}f(\overline{Z} + \tau_2[Z,\overline{Z}])\psi_0\\
&= f(\overline{Z} + \tau_2[Z,\overline{Z}] + \tau_1[Z,\overline{Z}])\psi_0\\
&= f(\overline{Z} + (\tau_1+\tau_2)[Z,\overline{Z}])\psi_0\\
&= e^{(\tau_1+\tau_2)Z}f(\overline{Z})\psi_0 = e^{(\tau_1+\tau_2)Z}v\\
&= \rho_Z(\tau_1+\tau_2).
\end{aligned}$$

□

The missing direction $[Z, \overline{Z}]$ plays an important role in the non-commutativity relations. The usual multiplication rule for exponentials $e^{a+b} = e^a e^b$ does not work if a and b are replaced by non-commutative operators. This rules are present in Quantum Mechanics in many places and there are consequences of the uncertainty principle. For instance, we have

$$e^Z e^{\overline{Z}}\psi_0 = e^{\overline{Z}+[Z,\overline{Z}]}\psi_0.$$

This is obtained from the following result making $s = \tau = 1$.

PROPOSITION 6.17. For any $\tau, s \in \mathbb{R}$ we have

$$e^{\tau Z}e^{s\overline{Z}}\psi_0 = e^{s\overline{Z}+s\tau[Z,\overline{Z}]}\psi_0.$$

PROOF. Let $f(\overline{Z}) = e^{s\overline{Z}}$ and $v = f(\overline{Z})\psi_0$. Then

$$e^{\tau Z}e^{s\overline{Z}}\psi_0 = e^{\tau Z}v = f(\overline{Z} + \tau[Z,\overline{Z}])\psi_0 = e^{s\overline{Z}+s\tau[Z,\overline{Z}]}\psi_0.$$

□

Because the operators e^Z and $e^{\overline{Z}}$ do not commute, we have another formula when their position is switched.

THEOREM 6.18. *For any $\alpha, \beta \in \mathbb{C}$ we have*

$$e^{\alpha\overline{Z}}e^{\beta Z} = e^{\alpha\overline{Z}+\beta Z-\frac{1}{2}\alpha\beta[Z,\overline{Z}]}. \tag{6.25}$$

PROOF. For any $\tau \in \mathbb{R}$ let

$$T(\tau) = e^{\tau\alpha\overline{Z}} e^{\tau\beta Z}. \tag{6.26}$$

We have

$$\frac{d}{d\tau} e^{\tau\alpha Z} = \alpha Z e^{\tau\alpha Z} = e^{\tau\alpha Z}\alpha Z, \tag{6.27}$$

$$[Z, e^{\tau\alpha\overline{Z}}] = \tau\alpha e^{\tau\alpha\overline{Z}}[Z, \overline{Z}] = [Z, \overline{Z}]\tau\alpha e^{\tau\alpha\overline{Z}}, \tag{6.28}$$

because of Theorem 6.10 with $f(\overline{Z}) = e^{\tau\alpha\overline{Z}}$. Differentiating in relation (6.26) and using relations (6.27) and (6.28) yields

$$\begin{aligned}
\frac{d}{d\tau}T(\tau) &= \alpha\overline{Z}e^{\tau\alpha\overline{Z}}e^{\tau\beta Z} + e^{\tau\alpha\overline{Z}}\beta Z e^{\tau\beta Z} \\
&= (\alpha\overline{Z} + \beta Z - \beta Z)e^{\tau\alpha\overline{Z}}e^{\tau\beta Z} + e^{\tau\alpha\overline{Z}}\beta Z e^{\tau\beta Z} \\
&= (\alpha\overline{Z} + \beta Z)T(\tau) - \beta Z e^{\tau\alpha\overline{Z}}e^{\tau\beta Z} + e^{\tau\alpha\overline{Z}}\beta Z e^{\tau\beta Z} \\
&= (\alpha\overline{Z} + \beta Z)T(\tau) + \Big(- \beta Z e^{\tau\alpha\overline{Z}} + e^{\tau\alpha\overline{Z}}\beta Z\Big) e^{\tau\beta Z} \\
&= (\alpha\overline{Z} + \beta Z)T(\tau) + [e^{\tau\alpha\overline{Z}}, \beta Z]e^{\tau\beta Z} \\
&= (\alpha\overline{Z} + \beta Z)T(\tau) - \beta[Z, e^{\tau\alpha\overline{Z}}]e^{\tau\beta Z} \\
&= (\alpha\overline{Z} + \beta Z)T(\tau) - \tau\alpha\beta[Z, \overline{Z}]e^{\tau\alpha\overline{Z}}e^{\tau\beta Z} \\
&= (\alpha\overline{Z} + \beta Z)T(\tau) - \tau\alpha\beta[Z, \overline{Z}]T(\tau).
\end{aligned}$$

Hence $T(\tau)$ satisfies the following linear ODE

$$\frac{d}{d\tau}T(\tau) + \Big(\tau\alpha\beta[Z, \overline{Z}] - \alpha\overline{Z} - \beta Z\Big)T(\tau) = 0. \tag{6.29}$$

An integrant factor is

$$\mu(\tau) = e^{\frac{1}{2}\tau^2\alpha\beta[Z,\overline{Z}] - \tau(\alpha\overline{Z} + \beta Z)}.$$

Multiplying by μ, yields the exact equation

$$\frac{d}{d\tau}(T(\tau)\mu(\tau)) = 0.$$

Hence

$$T(\tau) = C\mu^{-1}(\tau) = Ce^{\tau(\alpha\overline{Z}+\beta Z) - \frac{1}{2}\tau^2\alpha\beta[Z,\overline{Z}]},$$

where C is a constant. Making $\tau = 0$ yields

$$C = T(0) = e^{0\cdot\alpha\overline{Z}}e^{0\cdot\beta Z} = 1,$$

and hence

$$T(\tau) = e^{\tau(\alpha\overline{Z}+\beta Z) - \frac{1}{2}\tau^2\alpha\beta[Z,\overline{Z}]}.$$

Making $\tau = 1$ yields relation (6.25).

□

Relation (6.25) written as

$$e^{\alpha\overline{Z}}e^{\beta Z} = e^{\alpha\overline{Z} + \beta Z + \frac{1}{2}\alpha\beta[\overline{Z}, Z]}$$

is an analog of the Campbell-Hausdorff formula from the Lie groups theory. This says that if X, Y are left invariant vector fields and $\alpha, \beta \in \mathbb{R}$, then

$$e^{\alpha X}e^{\beta Y} = e^{\alpha X + \beta Y + \frac{1}{2}\alpha\beta[X,Y] + \textit{higher level brackets}}.$$

For the step 2 Lie groups there are no higher level brackets. We expect formula (6.25) to look more complicated in the case when there are many missing directions.

6.5. Heisenberg derivative

In the elementary Quantum Mechanics the following Heisenberg differentiation rule holds

$$\frac{d\widehat{A}}{d\mathbf{t}} = \frac{i}{h}[\widehat{H}, \widehat{A}],$$

where $\widehat{H} = -\frac{h^2}{2}\Delta + U(x)$ is the Hamiltonian operator and $\widehat{A}$ is an operator which characterizes a real physical system. The operator is called conservative if $\frac{d\widehat{A}}{d\mathbf{t}} = 0$, *i.e.*,

$$[\widehat{H}, \widehat{A}] = \widehat{H}\widehat{A} - \widehat{A}\widehat{H} = 0.$$

If the operator $\widehat{A}$ depends explicitly on the time $\mathbf{t}$, the differentiation rule becomes

$$\frac{d\widehat{A}}{d\mathbf{t}} = \frac{\partial \widehat{A}}{\partial \mathbf{t}} + \frac{i}{h}[\widehat{H}, \widehat{A}].$$

This is a similar formula of the following relation from symplectic mechanics

$$\frac{df}{d\mathbf{t}} = \frac{\partial f}{\partial \mathbf{t}} + \sum_i \Big(\frac{\partial f}{\partial q_i}\frac{\partial H}{\partial p_i} - \frac{\partial f}{\partial p_i}\frac{\partial H}{\partial q_i}\Big) = \frac{\partial f}{\partial \mathbf{t}} + \{H, f\},$$

where H is the Hamiltonian function and $\{\ ,\ \}$ is the Poisson bracket. The operator

$$\{\widehat{H}, \widehat{A}\} = \frac{i}{h}[\widehat{H}, \widehat{A}]$$

is called the quantum Poisson bracket.

In the following we shall compute the quantum Poisson bracket and find the conservative operators of the type $\widehat{A} = f(\overline{Z})$, where f is an analytic function. The Hamiltonian operator in our case is

$$\widehat{H} = -\frac{h^2}{2}(X_1^2 + X_2^2) = -h^2\Delta_H.$$

PROPOSITION 6.19. *If f is an analytic function, we have*

$$\begin{aligned}[\Delta_H, f(\overline{Z})] &= 2\overline{Z}f'(\overline{Z})[Z,\overline{Z}]\\ &= 2f'(\overline{Z})\overline{Z}\,[Z,\overline{Z}]\\ &= 2[Z,\overline{Z}]f'(\overline{Z})\overline{Z}.\end{aligned}$$

PROOF. Using $\Delta_H = 2Z\overline{Z} + [\overline{Z}, Z]$, yields

$$\begin{aligned}[\Delta_H, f(\overline{Z})] &= \Delta_H f(\overline{Z}) - f(\overline{Z})\Delta_H\\ &= \Big(2Z\overline{Z} + [\overline{Z}, Z]\Big)f(\overline{Z}) - f(\overline{Z})\Big(2Z\overline{Z} + [\overline{Z}, Z]\Big)\\ &= 2Z\overline{Z}f(\overline{Z}) + [\overline{Z}, Z]f(\overline{Z}) - 2f(\overline{Z})Z\overline{Z} - f(\overline{Z})[\overline{Z}, Z]\\ &= 2\Big(Z\overline{Z}f(\overline{Z}) - f(\overline{Z})Z\overline{Z}\Big) = 2\Big(Zf(\overline{Z})\overline{Z} - f(\overline{Z})Z\overline{Z}\Big)\\ &= 2\Big(Zf(\overline{Z}) - f(\overline{Z})Z\Big)\overline{Z} = 2[Z, f(\overline{Z})]\overline{Z}\\ &= 2f'(\overline{Z})[Z,\overline{Z}]\overline{Z} = 2\overline{Z}f'(\overline{Z})[Z,\overline{Z}],\end{aligned}$$

where we have used that $[Z,\overline{Z}]$ commutes with $\overline{Z}$ and $f(\overline{Z})$. □

In our case

$$\partial_t = \frac{i}{2h}[Z, \overline{Z}].$$

This means that the bracket $[Z, \overline{Z}]$ corresponds to the time operator. An operator which does not depend explicitly on time in our case will not depend on $[Z, \overline{Z}]$. The operator $f(\overline{Z})$ does not depend explicitly on $[Z, \overline{Z}]$. Hence $f(\overline{Z})$ is conservative if and only if it commutes with Δ_H, *i.e.*

$$2\overline{Z}f'(\overline{Z})[Z, \overline{Z}] = 0,$$

or $f'(\overline{Z}) = 0$, *i.e.* f is a constant function.

6.6. The fundamental solution for Δ_H

The infinite matrix formalism in Quantum Mechanics was introduced in 1925 by W. K. Heisenberg. Using this idea we shall look for a fundamental solution in the form

$$K = \langle M\psi, \psi^* \rangle = \sum_{n,k\geq 0} M_{nk}\psi_n\psi_k^*,$$

where M is an infinite transition probability matrix, *i.e.*, $M_{kn} = M_{nk} \in [0, 1]$. In this way, we can take into account all possible transitions between the excited states and the conjugated states.

Let $A = h|z|^2 + it$ and $\overline{A} = h|z|^2 - it$. We shall consider the following ground-states functions

$$\psi_0 = \frac{1}{A^\alpha}, \qquad \psi_0^* = \frac{1}{\overline{A}^\alpha}.$$

Proposition 6.8 yields

$$\psi_n = (-2hz)^n \alpha(\alpha + 1) \dots (\alpha + n - 1)\frac{1}{A^{\alpha+n}},$$

$$\psi_n^* = (-2h\overline{z})^n \alpha(\alpha + 1) \dots (\alpha + n - 1)\frac{1}{\overline{A}^{\alpha+n}}.$$

Assuming M diagonal matrix,

$$\begin{aligned} K &= K(z, t; 0, 0) = \sum \widetilde{M}_{nk}\frac{z^n}{A^{\alpha+n}}\frac{\overline{z}^n}{\overline{A}^{\alpha+k}} = \frac{1}{|A|^{2\alpha}}\sum \widetilde{M}_{nk}\frac{z^n\overline{z}^k}{A^n\overline{A}^k} \\ &= \frac{1}{|A|^{2\alpha}}\sum_{n\geq 0} \widetilde{M}_{nn}\frac{|z|^{2n}}{|A|^{2n}}. \end{aligned}$$

The operator Δ_H is homogeneous with respect to the dilations

$$(z, t) \to (\lambda z, \lambda^2 t).$$

The term

$$\frac{|z|^{2n}}{|A|^{2n}} = \Big(\frac{|z|^2}{h^2|z|^4 + t^2}\Big)^n$$

is not homogeneous with respect to the above dilations, unless $n = 0$. Hence

$$K = \frac{\widetilde{M}_{00}}{|A|^{2\alpha}} = \widetilde{M}_{00}\psi_0\psi_0^*.$$

We shall determine α such that

$$\Delta_H K = 0, \quad \text{for } (z,t) \neq (0,0).$$

In the computation we shall use the following relations

$$\begin{aligned} \overline{Z}A &= 2hz, \quad \overline{Z}\,\overline{A} = 0, \\ ZA &= 0, \qquad Z\overline{A} = 2h\overline{z}. \end{aligned}$$

Then

$$\begin{aligned} \overline{Z}\psi_0^* &= 0, \qquad\qquad Z\psi_0 = 0, \\ \overline{Z}\psi_0 &= \frac{-2\alpha hz}{A^{\alpha+1}}, \quad Z\psi_0^* = \frac{-2\alpha h\overline{z}}{\overline{A}^{\alpha+1}}, \end{aligned}$$

and then

$$Z(z\psi_0^*) = \psi_0^* - \frac{2\alpha h|z|^2}{\overline{A}^{\alpha+1}}, \quad \overline{Z}(\overline{z}\psi_0) = \psi_0 - \frac{2\alpha h|z|^2}{A^{\alpha+1}}.$$

Using $\Delta_H = Z\overline{Z} + \overline{Z}Z$, yields

$$\begin{aligned} \Delta_H(\psi_0\psi_0^*) &= Z\overline{Z}(\psi_0\psi_0^*) + \overline{Z}Z(\psi_0\psi_0^*) = Z((\overline{Z}\psi_0)\psi_0^*) + \overline{Z}((Z\psi_0)\psi_0^*) \\ &= Z\Big(\frac{-2\alpha hz}{A^{\alpha+1}}\frac{1}{\overline{A}^\alpha}\Big) + \overline{Z}\Big(\frac{1}{A^\alpha}\frac{-2\alpha h\overline{z}}{\overline{A}^{\alpha+1}}\Big) \\ &= \frac{-2\alpha h}{A^{\alpha+1}}Z\Big(\frac{z}{\overline{A}^\alpha}\Big) + \frac{-2\alpha h}{\overline{A}^{\alpha+1}}\overline{Z}\Big(\frac{\overline{z}}{A^\alpha}\Big) \\ &= \frac{-2\alpha h}{A^{\alpha+1}}Z\Big(z\psi_0^*\Big) + \frac{-2\alpha h}{\overline{A}^{\alpha+1}}\overline{Z}\Big(\overline{z}\psi_0\Big) \\ &= \frac{-2\alpha h}{A^{\alpha+1}}\Big(\psi_0^* - \frac{2\alpha h|z|^2}{\overline{A}^{\alpha+1}}\Big) + \frac{-2\alpha h}{\overline{A}^{\alpha+1}}\Big(\psi_0 - \frac{2\alpha h|z|^2}{A^{\alpha+1}}\Big) \\ &= -2\alpha h\left(\frac{1}{A^{\alpha+1}\overline{A}^\alpha} - \frac{2\alpha h|z|^2}{|A|^{2(\alpha+1)}} + \frac{1}{\overline{A}^{\alpha+1}A^\alpha} - \frac{2\alpha h|z|^2}{|A|^{2(\alpha+1)}}\right) \\ &= \frac{-2\alpha h}{|A|^{2\alpha}}\left(\frac{1}{A} - \frac{2\alpha h|z|^2}{|A|^2} + \frac{1}{\overline{A}} - \frac{2\alpha h|z|^2}{|A|^2}\right) \\ &= \frac{-2\alpha h}{|A|^{2\alpha}}\left(\frac{\overline{A}+A}{A\overline{A}} - \frac{4\alpha h|z|^2}{A\overline{A}}\right) \\ &= \frac{-2\alpha h}{|A|^{2\alpha}} \cdot \frac{2h|z|^2 - 4\alpha h|z|^2}{A\overline{A}} = 0 \end{aligned}$$

for $\alpha = \frac{1}{2}$. Hence the fundamental solution has the form

$$K = \frac{C}{(A\overline{A})^{1/2}} = \frac{C}{(h^2|z|^4 + t^2)^{1/2}}.$$

6.7. The fundamental solution for $\Delta_\lambda = \Delta_H - \frac{i}{2}\lambda[X_1, X_2]$

In this case we shall look for a fundamental solution at the origin of the form

$$K(z,t;0,0) = C\psi_0\psi_0^*,$$

with

$$\psi_0 = \frac{1}{A^\alpha}, \quad \psi_0^* = \frac{1}{\overline{A}^\beta}.$$

Δ_λ is a linear combination of $Z\overline{Z}$ and $\overline{Z}Z$

$$\begin{aligned}\Delta_\lambda &= Z\overline{Z} + \overline{Z}Z + \lambda[\overline{Z}, Z] \\ &= Z\overline{Z} + \overline{Z}Z + \lambda\overline{Z}Z - \lambda Z\overline{Z} \\ &= (1+\lambda)Z\bar{Z} + (1-\lambda)\overline{Z}Z.\end{aligned}$$

A straightforward computation shows

$$\begin{aligned}\Delta_\lambda(\psi_0\psi_0^*) &= \frac{-2(1+\lambda)\alpha h}{A^{\alpha+1}\overline{A}^\beta} + \frac{4\alpha\beta(1+\lambda)|z|^2h^2}{A^{\alpha+1}\overline{A}^{\beta+1}} \\ &\quad - \frac{2(1-\lambda)\beta h}{\overline{A}^{\beta+1}A^\alpha} + \frac{4\alpha\beta(1-\lambda)|z|^2h^2}{\overline{A}^{\beta+1}A^{\alpha+1}} \\ &= -\frac{2h}{A^\alpha\overline{A}^\beta}\left(\frac{(1+\lambda)\alpha}{A} + \frac{(1-\lambda)\beta}{\overline{A}}\right) + \frac{8\alpha\beta|z|^2h^2}{A^{\alpha+1}\overline{A}^{\beta+1}} \\ &= \frac{h}{A^{\alpha+1}\overline{A}^{\beta+1}}\Big(8\alpha\beta h|z|^2 - 2[\alpha+\beta+\lambda(\alpha-\beta)]h|z|^2 \\ &\quad -2it[\beta-\alpha-\lambda(\beta+\alpha)]\Big).\end{aligned}$$

$$\begin{aligned}\mathit{Im}\,\Delta_\lambda(\psi_0\psi_0^*) = 0 &\implies \beta - \alpha = \lambda(\beta+\alpha), \\ \mathit{Re}\,\Delta_\lambda(\psi_0\psi_0^*) = 0 &\implies 4\alpha\beta = \alpha+\beta+\lambda(\alpha-\beta) = (\alpha+\beta)(1-\lambda^2).\end{aligned}$$

The first relation of the above equations yields

$$\beta = \alpha\frac{1+\lambda}{1-\lambda}, \quad \lambda \neq 1,$$

and substituting in the second equation, yields

$$4\alpha^2\frac{1+\lambda}{1-\lambda} = \alpha + \alpha\frac{1+\lambda}{1-\lambda} + \lambda\Big(\alpha - \alpha\frac{1+\lambda}{1-\lambda}\Big).$$

Multiplying by $\frac{1-\lambda}{\alpha(1+\lambda)}$, $\lambda \neq -1$, yields

$$\begin{aligned}4\alpha &= \frac{1-\lambda}{1+\lambda} + 1 + \frac{\lambda(1-\lambda)}{1+\lambda} - \lambda \\ &= \frac{2(1-\lambda^2)}{1+\lambda} = 2(1-\lambda).\end{aligned}$$

Thus $\alpha = \frac{1-\lambda}{2}$ and $\beta = \frac{1-\lambda}{2}\frac{1+\lambda}{1-\lambda} = \frac{1+\lambda}{2}$. Hence

$$\begin{aligned}K(z,t;0,0) &= CA^{\frac{1-\lambda}{2}}\overline{A}^{\frac{1+\lambda}{2}} \\ &= \frac{C}{(h|z|^2+it)^{\frac{1-\lambda}{2}}(h|z|^2-it)^{\frac{1+\lambda}{2}}}.\end{aligned}$$

6.8. An application to the eigenfunctions of Δ_H

In this section se shall study the relation between the eigenfunctions of Δ_H and the associated sub-elliptic Schrödinger equation.

PROPOSITION 6.20. Let ψ be an eigenfunction for Δ_H, *i.e.*, there is a constant $E \in \mathbb{R}$ such that

$$\Delta_H \psi = E\psi.$$

1) Then $\Psi = \overline{Z}\psi$ satisfies the following Schrödinger equation

$$4ih\partial_t \Psi + \Delta_H \Psi = E\Psi.$$

2) Then $\Psi^* = Z\psi$ satisfies the equation

$$-4ih\partial_t \Psi^* + \Delta_H \Psi^* = E\Psi^*.$$

PROOF. 1) Using the commutativity between $\overline{Z}$ and ∂_t yields

$$\begin{aligned}
\Delta_H &= \Delta_H \overline{Z}\psi = (2\overline{Z}Z - 2ih\partial_t)\overline{Z}\psi \\
&= (2\overline{Z}Z\overline{Z} - 2ih\partial_t \overline{Z})\psi \\
&= \overline{Z}(2Z\overline{Z})\psi - 2ih\partial_t \overline{Z}\psi \\
&= \overline{Z}(2Z\overline{Z} + 2ih\partial_t - 2ih\partial_t)\psi - 2ih\partial_t \overline{Z}\psi \\
&= \overline{Z}(\Delta_H - 2ih\partial_t)\psi - 2ih\partial_t \overline{Z}\psi \\
&= \overline{Z}(E\psi - 2ih\partial_t \psi) - 2ih\partial_t \overline{Z}\psi \\
&= E\overline{Z}\psi - 2ih\partial_t \overline{Z}\psi - 2ih\partial_t \overline{Z}\psi \\
&= E\Psi - 4ih\partial_t \Psi.
\end{aligned}$$

2) Comes from 1) by conjugation. □

Choosing $E = 0$, we have

COROLLARY 6.21. If $\Delta_H \varphi = 0$ on U, then $\Phi = \overline{Z}\varphi$ satisfies

$$4ih\partial_t \Phi + \Delta_H \Phi = 0 \tag{6.30}$$

on U.

EXAMPLE 6.22. Choosing

$$\varphi = \frac{1}{\sqrt{h^2|z|^4 + t^2}},$$

we have $\Delta_H \varphi = 0$ on $U = \mathbb{R}^3 \backslash \{(0,0)\}$. Then

$$\Phi = \overline{Z}\varphi = \frac{-h\overline{z}}{(h|z|^2 + it)\sqrt{h^2|z|^4 + t^2}}$$

satisfies equation (6.30) on U.

6.9. The Schrödinger kernel

Propagators. We shall find a Green function for the operator

$$L = ih\partial_u + \frac{h^2}{2}(X_1^2 + X_2^2), \tag{6.31}$$

where h is the Plank's constant. The Green function $P(x, t, u; x_0, t_0, u_0)$ is also referred to in Quantum Mechanics as a *propagator.* If Ψ is a wave function, which satisfies the equation

$$ih\partial_u\Psi + h^2\Delta_H\Psi = 0, \tag{6.32}$$

with the initial condition $\Psi_0(x, t) = \Psi(x, t, u_0)$, then the solution at any instant of time $u > u_0$ is given by

$$\Psi(x, t, u) = \int P(x, t, u; x_0, t_0, u_0)\Psi_0(x_0, t_0)\, dx_0\, dt_0.$$

In Quantum Mechanics the propagator is denoted by $\Psi(x, t, u|x_0, t_0, u_0)$. Equation (6.32) is a subelliptic Schrödinger equation with zero potential energy. It describes a free particle in the Heisenberg quantum framework. One may look for solutions of (6.32) of the type

$$\Psi = \sum_j \Phi_j(x, t)\varphi_j(u).$$

Substituting in (6.32) we arrive at

$$ih\Phi_j(x, t)\varphi_j'(u) + h^2\Delta_H\Phi_j(x, t)\,\varphi_j(u) = 0.$$

Dividing by $\Phi_j\varphi_j$ yields

$$ih\frac{\varphi_j'(u)}{\varphi(u)} + h^2\frac{\Delta_H\Phi_j(x, t)}{\Phi_j(x, t)} = 0.$$

There is a separation constant E_j, which has the significance of energy, such that

$$\begin{aligned} ih\varphi_j'(u) &= E_j\,\varphi(u), \\ h^2\Delta_H\Phi_j &= -E_j\Phi_j. \end{aligned}$$

Then

$$\varphi_j(u) = e^{-\frac{i}{hu}E_j\,u^2}.$$

From the Classical Mechanics, $Eu^2 = gu$, where g is the action. Using this idea, we let $f_j = E_j u^2$. Then

$$\Psi = \sum_j \Phi_j(x, t)e^{-\frac{if_j}{hu}}.$$

In the next section se shall show that the modified complex action $f = \tau g$ is replacing all f_j's, while the sum is replaced by an integral. This is an important application of the fact that the modified complex action f can produce all sub-Riemannian distances $d_j^2 = 2\tau E_j$. This shows how useful complex Hamiltonian mechanics techniques are for describing the Schrödinger equation.

We shall show that the formula for the Schrödinger's kernel at $x_0 = 0, t_0 = 0, u_0 = 0$ is

$$P(x, t, u) = \Big(\frac{1}{2\pi ihu}\Big)^2 \int_{-\infty}^{+\infty} e^{-\frac{if(x,t,\tau)}{hu}}V(\tau)\,d\tau,$$

where $V(\tau)$ satisfies a certain transport equation.

This shows that the propagator P depends on each subRiemannain distance d_j. This is a well known fact of Quantum Mechanics stated by R. Feynman. He explained the double slit experiment assuming the fact that the electron takes all possible paths between any two points. Hence each path has a contribution to the propagator, which is defined as a path integral. For macro objects, like airplanes, planets or bullets, only the classical path is important and their motion is explained in the context of Riemannian geometry. The picture for the sub-atomic particles is completely different and the correct framework to study their motion is the subRiemannian geometry.

The Green function at the origin . A Green function P for the operator L is a distribution on $\mathbb{R}^2_x \times \mathbb{R}_t$ with the following properties

$$1)\ \partial_u P - \Delta_H P = 0, \text{ for } u > 0, \tag{6.33}$$

$$2)\ \lim_{u \searrow 0} P(x,t,u) = \delta(x)\delta(t), \tag{6.34}$$

where δ stands for the Dirac distribution.
The following result will be useful.

LEMMA 6.23. *For any smooth function φ we have*

$$\Delta_H e^{\varphi} = e^{\varphi}\Big(\Delta_H \varphi + \frac{1}{2}|\nabla_X \varphi|^2\Big),$$

where $|\nabla_X \varphi|^2 = (X_1\varphi)^2 + (X_1\varphi)^2$ and $\Delta_H = \frac{1}{2}(X_1^2 + X_2^2)$.

PROOF. We have $X_j e^{\varphi} = e^{\varphi} X_j \varphi$ and

$$X_j^2 e^{\varphi} = X_j(e^{\varphi} X_j \varphi) = e^{\varphi}(X_j\varphi)^2 + e^{\varphi} X_j^2 \varphi.$$

Hence

$$\Delta_H e^{\varphi} = \frac{1}{2}\Big(X_1^2\varphi + X_2^2\varphi\Big) = e^{\varphi}\Big(\Delta_H\varphi + \frac{1}{2}|\nabla_X\varphi|^2\Big)$$

□

The above lemma works for any vector fields $X_1, X_2, \dots, X_{2m}$.

Let $g = g(x,t,\tau) = -it + h\coth(2h\tau)|x|^2$ be the complex action from the origin. g satisfies the Hamilton-Jacobi equation

$$\frac{\partial g}{\partial \tau} + \frac{1}{2}|\nabla_X g|^2 = 0. \tag{6.35}$$

The modified complex action $f = \tau g = -i\tau t + \tau h\coth(2h\tau)|x|^2$ satisfies

$$\frac{\partial f}{\partial \tau} = g + \tau\frac{\partial g}{\partial \tau} \Longrightarrow \tau\frac{\partial f}{\partial \tau} = f + \tau^2\frac{\partial g}{\partial \tau}. \tag{6.36}$$

Equations (6.35) and (6.36) yield

$$|\nabla_X f|^2 = \tau^2|\nabla_X g|^2 = -2\tau^2\frac{\partial g}{\partial \tau} = 2\Big(f - \tau\frac{\partial f}{\partial \tau}\Big). \tag{6.37}$$

Let $\varphi = \frac{-if}{hu}$. Then using (6.37) we have

$$\frac{1}{2}|\nabla_X\varphi|^2 = -\frac{1}{2h^2u^2}|\nabla_X f|^2 = -\frac{1}{h^2u^2}\Big(f - \tau\frac{\partial f}{\partial \tau}\Big). \tag{6.38}$$

As $\Delta_H\varphi = \frac{-i}{hu}\Delta_H f$, Lemma 6.23 yields

$$\Delta_H e^{\varphi} = e^{\varphi}\Big(-\frac{i}{hu}\Delta_H f - \frac{1}{h^2u^2}\Big(f - \tau\frac{\partial f}{\partial \tau}\Big)\Big). \tag{6.39}$$

Hence

$$h^2\Delta_H\Big(\frac{1}{u^2}e^{\varphi}V(\tau)\Big) = \frac{1}{u^3}e^{\varphi}V(\tau)\bigg(-ih\Delta_H f - \frac{1}{u}\Big(f - \tau\frac{\partial f}{\partial \tau}\Big)\bigg), \tag{6.40}$$

and

$$ih\frac{\partial}{\partial u}\Big(\frac{1}{u^2}e^{\varphi}V(\tau)\Big) = -\frac{e^{\varphi}}{u^3}V(\tau)\Big(\frac{f}{u} + 2ih\Big). \tag{6.41}$$

Relations (6.40) and (6.41) yield

$$\begin{aligned}
(h^2\Delta_H - ih\partial_u)\Big(\frac{1}{u^2}e^{\varphi}V(\tau)\Big) &= \frac{1}{u^3}e^{\varphi}V(\tau)\Big(ih(2-\Delta_H f) + \frac{\tau}{u}\frac{\partial f}{\partial \tau}\Big)\\
&= \frac{ih}{u^3}e^{\varphi}V(\tau) + \frac{ih}{u^3}e^{\varphi}V(\tau)(1-\Delta_H f)\\
&\quad + \frac{ih}{u^3}e^{\varphi}V(\tau)\frac{\tau}{u}\Big(\frac{-i}{h}\Big)\frac{\partial f}{\partial \tau}\\
&= \frac{ih}{u^3}e^{\varphi}V(\tau) + \frac{ih}{u^3}e^{\varphi}V(\tau)(1-\Delta_H f)\\
&\quad + \frac{ih}{u^3}\frac{\partial}{\partial \tau}\Big(e^{\varphi}V(\tau)\tau\Big)\\
&\quad - \frac{ih}{u^3}e^{\varphi}\frac{dV}{d\tau}\tau - \frac{ih}{u^3}e^{\varphi}V(\tau)\\
&= -\frac{ih}{u^3}e^{\varphi}\Big(\tau\frac{dV}{d\tau} + (\Delta_H f - 1)V(\tau)\Big)\\
&\quad + \frac{\partial}{\partial \tau}\Big(\frac{ih}{u^3}e^{\varphi}V(\tau)\tau\Big).
\end{aligned}$$

Let $V(\tau)$ satisfy the equation

$$\tau\frac{dV}{d\tau} + (\Delta_H f - 1)V(\tau) = 0.$$

If we set $V(\tau) = \tau v(\tau)$ then v satisfies

$$\tau\frac{d(\tau v)}{d\tau} + \Big(\Delta_H(\tau g) - 1\Big)\tau v = 0,$$

which can be written as

$$\frac{dv}{d\tau} + (\Delta_H g)v = 0, \tag{6.42}$$

and it is called the *transport equation.* A straightforward computation provides $\Delta_H g = 2h\coth(2h\tau)$. Hence the transport equation (6.42) becomes

$$\frac{dv}{d\tau} + 2h\coth(2h\tau)v = 0,$$

and it can be solved by the method of separation. The solution is

$$v(\tau) = \frac{2hC}{\sinh(2h\tau)},$$

where C is a constant. If we employ $V(0) = 1$, then

$$V(\tau) = \tau v(\tau) = \frac{2h\tau}{\sinh(2h\tau)}.$$

Therefore

$$(h^2\Delta_H - ih\partial_u)\Big(\frac{1}{u^2}e^{\varphi}V(\tau)\Big) = \frac{\partial}{\partial\tau}\Big(\frac{ih}{u^3}e^{\varphi}\frac{2h\tau^2}{\sinh(2h\tau)}\Big).$$

As the expression under ∂/∂_τ vanishes at $\tau = \pm\infty$, integrating yields

$$(h^2\Delta_H - ih\partial_u)\int_{-\infty}^{\infty}\frac{1}{u^2}e^{\varphi}V(\tau)\,d\tau = 0, \quad \forall u > 0.$$

Let

$$P(x,t,u) = \frac{C}{u^2}\int_{-\infty}^{\infty}\frac{1}{u^2}\,e^{\varphi}V(\tau)\,d\tau,$$

where

$$\varphi = \frac{if(x,t,\tau)}{hu}.$$

Hence the first relation of (6.33) is satisfied. To prove the second relation of (6.33) we note first that if $0 < |2h\gamma| < \pi$, then

$$Re\ f(x,t,\tau + i\gamma\,\text{sgn}t) \geq \epsilon(\gamma)(|x|^4 + t^2)^{1/2}$$

for some $\epsilon(\gamma) > 0$, see (2.7.7) of [**5**]. Consequently P can be put in the form

$$P(x,t,u) = \frac{C}{u^2}\int_{-\infty+i\gamma\text{sgn}t}^{+\infty+i\gamma\text{sgn}t} e^{-\frac{if(x,t,\tau)}{hu}}\,V(\tau)\,d\tau.$$

This yields

$$\lim_{u\to 0}\Big(\frac{d}{du}\Big)^m P(x,t,u) = 0,$$

uniformly in (x,t) on compact subsets of $\mathbb{R}^3\backslash\{0\}$ for each fixed n, $n = 0,1,2,\ldots$ To derive 2) we need to compute

$$\lim_{R\to\infty}\int_{\mathbb{R}} P(x,t,\frac{1}{R})\,dx\,dt.$$

This was done in [**5**], see (2.8.13)-(2.8.15). In particular,

$$\lim_{R\to\infty}\int_{\mathbb{R}} P(x,t,\frac{1}{R})\,dx\,dt = (2\pi ih)^2C = 1,$$

which yields the following result.

THEOREM 6.24. *The Schrödinger kernel, i.e., the kernel of* $e^{-ihu\Delta_H}$ *at the origin is given by*

$$P(x,t,u) = \frac{1}{(2\pi hiu)^2}\int_{-\infty}^{+\infty} e^{\frac{-if(x,t,\tau)}{hu}}\,\frac{2h\tau}{\sinh(2h\tau)}\,d\tau,$$

where $f = \tau g = -i\tau t + h\tau\coth(2\tau h)(x_1^2 + x_2^2)$ *is the complex modified action function.*

We have proved that $P(x,t,u)$ is the kernel relative to the origin. It follows from the left invariance of Δ_H that the full Schrödinger kernel is obtained by left translations. The Heisenberg convolution with P gives the solutions of the usual initial-value problem for the Schrödinger operator on the Heisenberg group $\mathbf{H}_1$.

The Green function at any point. In order to get a closed form expression for the Schrödinger kernel, instead of using the Heisenberg convolution, we shall use direct computation. The computation and the propagator formula will be very similar to the one in the case when $x_0 = 0, t_0 = 0$.

In the general case the complex action between the points (x_0, y_0) and (x, y) within time τ is

$$\begin{aligned} g = g(x,t,x_0,t_0,\tau) &= -it + h\coth(2h\tau)(|x|^2 + |x_0|^2) \\ &\quad -2h|x|\,|x_0|\frac{1}{\sinh(2h\tau)}. \end{aligned} \tag{6.43}$$

THEOREM 6.25. *The complex action g given by* (6.43) *satisfies the Hamilton-Jacobi equation*

$$\frac{\partial g}{\partial \tau} + \frac{1}{2}(X_1 g)^2 + \frac{1}{2}(X_2 g)^2 = 0.$$

PROOF. By direct computation. Let $r = |x|$ and $r_0 = |x_0|$. We have

$$\begin{aligned} \partial_t g &= -i, \\ \partial_{x_1} g &= 2hx_1\Big(\coth(2h\tau) - \frac{r_0}{r}\frac{1}{\sinh(2h\tau}\Big), \\ \partial_{x_2} g &= 2hx_2\Big(\coth(2h\tau) - \frac{r_0}{r}\frac{1}{\sinh(2h\tau}\Big). \end{aligned}$$

Therefore

$$\begin{aligned} &(X_1 g)^2 + (X_2 g)^2 \\ =\ & \big(\partial_{x_1} g + 2hx_2\partial_t g\big)^2 + \big(\partial_{x_2} g - 2hx_1\partial_t g\big)^2 \\ =\ & 4h^2x_1^2\Big(\coth(2h\tau) - \frac{r_0}{r}\frac{1}{\sinh(2h\tau)}\Big)^2 - 4h^2x_2^2 \\ & -8h^2x_1x_2 i\Big(\coth(2h\tau) - \frac{r_0}{r}\frac{1}{\sinh(2h\tau)}\Big) \\ & +4h^2x_2^2\Big(\coth(2h\tau) - \frac{r_0}{r}\frac{1}{\sinh(2h\tau)}\Big)^2 - 4h^2x_1^2 \\ & +8h^2x_1x_2 i\Big(\coth(2h\tau) - \frac{r_0}{r}\frac{1}{\sinh(2h\tau)}\Big) \\ =\ & 4h^2r^2\Big(\coth(2h\tau) - \frac{r_0}{r}\frac{1}{\sinh(2h\tau)}\Big)^2 - 4h^2r^2 \\ =\ & 4h^2r^2\left(\coth^2(2h\tau) + \frac{r_0^2}{r^2}\frac{1}{\sinh^2(2h\tau)} - 2\frac{r_0}{r}\frac{\coth(2h\tau)}{\sinh(2h\tau)} - 1\right) \\ =\ & 4h^2r^2\left(1 + \frac{1}{\sinh^2(2h\tau)} + \frac{r_0^2}{r^2}\frac{1}{\sinh^2(2h\tau)} - 2\frac{r_0}{r}\frac{\cosh(2h\tau)}{\sinh^2(2h\tau)} - 1\right) \\ =\ & \frac{4h^2r^2}{\sinh^2(2h\tau)}\left(1 + \frac{r_0^2}{r^2} - 2\frac{r_0}{r}\cosh(2h\tau)\right) \\ =\ & \frac{4h^2}{\sinh^2(2h\tau)}\Big(r^2 + r_0^2 - 2rr_0\cosh(2h\tau)\Big). \end{aligned}$$

Hence

$$\begin{aligned} H(\nabla g) &= \frac{1}{2}(X_1 g)^2 + \frac{1}{2}(X_2 g)^2 \\ &= \frac{2h^2}{\sinh^2(2h\tau)}\Big(r^2 + r_0^2 - 2rr_0\cosh(2h\tau)\Big). \end{aligned}$$

On the other hand

$$\begin{aligned} \frac{\partial g}{\partial \tau} &= h(r^2 + r_0^2)\frac{\partial}{\partial \tau}\coth(2h\tau) - 2rr_0 h\frac{\partial}{\partial \tau}\frac{1}{\sinh(2h\tau)} \\ &= -2h^2(r^2 + r_0^2)\frac{1}{\sinh^2(2h\tau)} + 4h^2 rr_0\frac{\cosh(2h\tau)}{\sinh^2(2h\tau)} \\ &= \frac{-2h^2}{\sinh^2(2h\tau)}\Big(r^2 + r_0^2 - 2rr_0\cosh(2h\tau)\Big) \end{aligned}$$

and hence

$$\frac{\partial g}{\partial \tau} = -H(\nabla g).$$

□

Following the same computation as in the proof of Theorem 6.24, the volume element $v(\tau)$ satisfies

$$\frac{dv}{d\tau} + (\Delta_H g)v = 0.$$

Using

$$X_1^2 g = X_2^2 g = 2h\Big(\coth(2h\tau) - \frac{r_0}{r}\frac{1}{\sinh(2h\tau)}\Big),$$

yields

$$\Delta_H g = 2h\coth(2h\tau) - 2h\frac{r_0}{r}\frac{1}{\sinh(2h\tau)}.$$

The transport equation becomes

$$\begin{aligned} \frac{dv}{v} &= \Big(-2h\coth(2h\tau) + 2h\frac{r_0}{r}\frac{1}{\sinh(2h\tau)}\Big)\,d\tau \Longleftrightarrow \\ \ln|v| &= -2h\frac{1}{2h}\ln|\sinh(2h\tau)| + 2h\frac{r_0}{r}\frac{1}{2h}\ln|\tanh(h\tau)| + C \Longleftrightarrow \\ v(\tau) &= \frac{C\tanh(h\tau)^{r_0/r}}{\sinh(2h\tau)} \\ &= C\frac{\sinh(h\tau)^{r_0/r}}{\cosh(h\tau)^{r_0/r}}\frac{1}{2\sinh(h\tau)\cosh(h\tau)} \\ &= \frac{C}{2}\frac{\sinh(h\tau)^{\frac{r_0}{r}-1}}{\cosh(h\tau)^{\frac{r_0}{r}+1}}. \end{aligned}$$

We shall choose the constant C such that $V(0) \to 1$ for $r_0 \to 0$. Hence $C = 2h$ and

$$V(\tau) = \tau v(\tau) = h\tau\frac{\sinh(h\tau)^{\frac{r_0}{r}-1}}{\cosh(h\tau)^{\frac{r_0}{r}+1}}.$$

The propagator's formula is given by the following theorem.

THEOREM 6.26. *The Schrödinger kernel is given by*

$$P(x,t,x_0,t_0,u) = \frac{h}{(2\pi hiu)^2}\int_{-\infty}^{+\infty} e^{\frac{-if(x,t,x_0,t_0,\tau)}{hu}} \tau \frac{\sinh(h\tau)^{\frac{r_0}{r}-1}}{\cosh(h\tau)^{\frac{r_0}{r}+1}}\,d\tau,$$

where

$$f = \tau g = -i\tau t + \tau h\coth(2h\tau)(|x|^2+|x_0|^2) - 2\tau h|x|\,|x_0|\frac{1}{\sinh(2h\tau)}.$$

is the complex modified action function.

6.10. The heat kernel on $\mathbf{H}_1$

In this section v denotes the time variable and we shall consider the heat operator

$$\partial_v - \Delta_H = \partial_v - \frac{1}{2}(X_1^2 + X_2^2). \tag{6.44}$$

The fundamental solution at the origin is the function $P(x,t;v)$ defined on $\mathbf{H}_1 \times \mathbb{R}_+$ such that the following relations are satisfied

$$\partial_v P - \Delta_H P = 0, \quad v > 0,$$

$$\lim_{v\searrow 0} P(y,t;v) = \delta(y)\delta(t).$$

The heat kernel has been obtained in [**6**], [**26**], [**34**]. For a variational approach using Hamiltonian formalism, see [**3**]. In the following we shall use the results proved in the previous section.

With the transformation

$$v = u/(ih), \qquad y_j = x_j/(ih)$$

the heat operator (6.44) becomes the Schrödinger's operator

$$ih\partial_u + \frac{h^2}{2}(X_1^2 + X_2^2).$$

Then the Theorem 6.26 becomes

THEOREM 6.27. *The heat kernel at the origin is given by*

$$P(y,t,v) = \frac{1}{(2\pi v)^2}\int_{-\infty}^{\infty} e^{-f(y,t,\tau)/v} V(\tau)\,d\tau, \tag{6.45}$$

where

$$f(y,t,\tau) = -i\tau t + h\tau\coth(2\tau h)(y_1^2 + y_2^2),$$

is the modified complex action and

$$V(\tau) = \frac{2h\tau}{\sinh(2h\tau)}$$

is the volume element.

The small time behavior when $x \neq 0$. We shall use the steepest descends method in order to compute the small time asymptotics for the heat kernel.

Consider the Laplace integral in the complex plane

$$F(\lambda) = \int_C V(z)e^{\lambda h(z)}\, dz,$$

where the $V(z), h(z)$ are holomorphic functions on a domain D, which contains the smooth piece-wise curve C. The following theorem will provide asymptotic expansions for $F(\lambda)$ when the parameter $\lambda \to \infty$. If $z = z(\tau)$, $t_1 \leq \tau \leq t_2$ is a parametrization of the curve C, we write

$$F(\lambda) = \int_{t_1}^{t_2} V(z(\tau))z'(\tau)e^{\lambda[\phi(\tau)+i\psi(\tau)]}\, d\tau,$$

where $h = \phi + i\psi$, and $\phi(\tau) = \phi(z(\tau)), \psi(\tau) = \phi(z(\tau))$. The following result can be found for instance in Iacob [**35**], vol. III, p. 83:

THEOREM 6.28. *Let C be a piece-wise smooth curve and the functions $V(z)$ and $h(z)$ satisfying*

1) $V(z)$, $h(z)$ are holomorphic at each point of C,

2) the maximum of the function $\phi(z) = Re\, h(z)$ on the curve C is reached at a point z_0 interior to C,

3) $V(z_0) \neq 0$, $h'(z_0) = 0$, $h''(z_0) \neq 0$,

4) $\int_C |V(z)|e^{\lambda\phi(z)}\, dz < \infty$, for $\lambda > 0$.

Then for $\lambda \to \infty$ we have the following asymptotic expansion

$$F(\lambda) = e^{\lambda h(z_0)} \sum_{n=0}^{\infty} a_n \lambda^{-n-\frac{1}{2}},$$

with the coefficients

$$a_n = \frac{\sqrt{2\pi}(2n)!}{n!2^{2n}} b_{2n}, \qquad V(z(\tau))z'(\tau) = \sum_{n=0}^{\infty} b_n \tau^n.$$

The dominant term is

$$F(\lambda) = \sqrt{\frac{-2\pi}{\lambda h''(z_0)}} e^{\lambda h(z_0)} \Big[V(z_0) + \mathcal{O}\Big(\frac{1}{\lambda}\Big)\Big].$$

6.10.0.10. *The small time asymptotics for the heat kernel.* We shall assume $x \neq 0$. The contour for the integral (6.45) can be moved to the line $Im\, \tau = \theta_c$:

$$P(y, t; u) = \frac{1}{(2\pi u)^2} \int_{Im\, \tau = \theta_c} e^{-f(y,t,\tau)/u} V(\tau)\, d\tau.$$

In order to apply Theorem 6.28, we shall let $h = -f$, $\lambda = 1/u$, and $C = \{Im\, \tau = \theta_c\}$. We know that the unique critical point of f with respect to τ in the strip $\{|Im \tau| < \pi/2h\}$ is the point $\tau_c = i\theta_c$, where θ_c satisfies

$$t = h\mu(2h\theta)\|y\|^2. \tag{6.46}$$

Then on the contour $C = \{Im\, \tau = \theta_c\}$, the function $Re\, f$ has a strict minimum at $\tau_c = i\theta_c$, and its value is $d^2(y,t)/2$. We have that $\phi = Re\, h$ has a strict maximum at $z_0 = \tau_c$, which is in the interior of C.

Since z_0 is a strict maximum, $h'(z_0) = 0$ and $h''(z_0) < 0$. It is obvious that $V(z_0) \neq 0$. We have

$$V(z_0) = \frac{2hz_0}{\sinh(2hz_0)} = \frac{2hi\theta_c}{\sinh(2hi\theta_c)} = \frac{2h\theta_c}{\sin(2h\theta_c)} \neq 0.$$

The condition 4) of the Theorem 6.28 is satisfied, because

$$\begin{aligned}\int_C |V(\tau)|e^{\lambda\phi(\tau)}\,d\tau &= \int_{Im\,\tau=\theta_c} |V(\tau)e^{-f/u}|\,d\tau \\ &\leq e^{-d^2(x,t)/(2u)} \int_R |V(s+i\theta_c)|\,ds < \infty.\end{aligned}$$

We also note that while the first derivative of f vanishes at $\tau = i\theta_c$, the second derivative is

$$f''(i\tau_c) = \frac{\partial^2 f}{\partial \tau^2}\Big|_{\tau=i\theta_c} = \frac{4h^2}{\sin^2(2h\theta_c)}[1 - 2h\theta_c\cot(2h\theta_c)]\|y\|^2.$$

Applying Theorem 6.28, we have

$$\begin{aligned}P(y,t;u) &= \frac{1}{(2\pi u)^2}F(\lambda) = \frac{1}{(2\pi u)^2}\sqrt{\frac{-2\pi}{\lambda h''(z_0)}}e^{\lambda h(z_0)}[V(z_0) + \mathcal{O}(1/\lambda)] \\ (6.47) \qquad &= \frac{1}{(2\pi u)^2}\sqrt{\frac{2\pi u}{f''(i\theta_c)}}e^{-d^2(x,t)/(2u)}[V(i\theta_c) + \mathcal{O}(u)].\end{aligned}$$

Using

$$\sqrt{\frac{2\pi u}{f''(i\theta_c)}}V(i\theta_c) = \sqrt{2\pi u}\cdot\frac{\sin 2h\theta_c}{2h(1 - ah\theta_c\cot(2a\theta_c))^{1/2}\|y\|}\cdot\frac{2h\theta_c}{\sin 2h\theta_c},$$

then (6.47) yields:

THEOREM 6.29. *Given a fixed point (x,t), $x \neq 0$, let θ_c denote the solution of (6.46) in the interval $[0, \pi/2h)$. Then the heat kernel has the following asymptotic expansion*

$$P(y,t;u) = \frac{1}{(2\pi u)^2}e^{-d(x,t)^2/(2u)}\Big[\frac{\theta_c\sqrt{2\pi u}}{(1 - 2h\theta_c\cot(2h\theta_c))^{1/2}\|y\|} + \mathcal{O}(u)\Big], \quad \textit{as } u \searrow 0.$$

The small time behavior when $x = 0$. In this section we shall assume $h = 1$.

THEOREM 6.30. *At points $(0,t)$ with $t > 0$, we have the following expansion*

$$P(0,t;u) = \frac{1}{4u^2}\sum_{k=1}^{\infty}(-1)^{k+1}ke^{-\ell_k^2(0,t)/2u}$$

as $u \to 0^+$, where $\ell_k(0,t)$ is the length of the k-th geodesic connecting the origin and the point $(0,t)$.

PROOF. As we know, the heat kernel for the operator Δ_H has the form

$$P(x,t,u) = \frac{1}{4\pi^2u^2}\int_{-\infty}^{+\infty} e^{-f(x,t,\tau)/u}V(\tau)d\tau$$

where

$$f(x,t,\tau) = -i\tau t + \tau\coth(2\tau)(x_1^2 + x_2^2)$$

is the modified complex action and

$$V(\tau) = \frac{2\tau}{\sinh(2\tau)}$$

is the volume element. Since $x = 0$, we have to look at the integral

$$\frac{1}{4\pi^2 u^2} \int_{-\infty}^{+\infty} e^{\frac{it\tau}{u}} \frac{2\tau}{\sinh(2\tau)} d\tau = \frac{1}{8\pi^2 u^2} \int_{-\infty}^{+\infty} e^{\frac{it\tau}{2u}} \frac{\tau}{\sinh(\tau)} d\tau.$$

This integral can be evaluated explicitly by residue calculus. Let us consider the following contour integral first. See Figure 6.2.

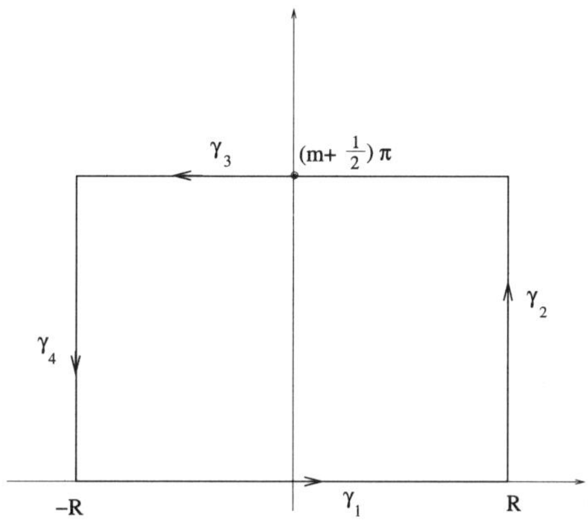

Figure 6.2: *The paths* γ_i, $i = \overline{1,4}$.

Let $\gamma_2 : \big[0, \big(m+\frac{1}{2}\big)\pi\big] \to \mathbb{C}$ and $\gamma_2(y) = R + iy$.

$$\int_{\gamma_2} e^{i\frac{tz}{2u}} \frac{z}{\sinh(z)} dz = i \int_0^{(m+\frac{1}{2})\pi} e^{i\frac{t}{2u}[R+iy]} \frac{R+iy}{\sinh(R+iy)} dy.$$

But

$$\begin{aligned} \sinh(R+iy) &= \frac{e^{R+iy} - e^{-R-iy}}{2} \\ &= \frac{e^R\big[\cos(y) + i\sin(y)\big] - e^{-R}\big[\cos(y) - i\sin(y)\big]}{2} \\ &= \cos(y)\frac{e^R - e^{-R}}{2} + i\sin(y)\frac{e^R + e^{-R}}{2}. \end{aligned}$$

Hence,

$$\sinh(R+iy) = \cos(y)\sinh(R) + i\sin(y)\cosh(R).$$

It follows that

$$\big|\sinh(R+iy)\big| = \sqrt{\cos^2(y)\sinh^2(R) + \sin^2(y)\cosh^2(R)} = \sqrt{\sinh^2(R) + \sin^2(y)}.$$

Since $R > 0$, we have

$$\big|\sinh(R+iy)\big| \geq \big|\sinh(R)\big| = \sinh(R)$$

and hence

$$\big|\sinh(R+iy)\big| \geq \big|\sin(y)\big|.$$

Therefore, we get the following bound for the modulus of $\int_{\gamma_2}$.

$$\Big|\int_{\gamma_2}\Big| \leq \int_0^{(m+\frac{1}{2})\pi} e^{-\frac{ty}{2u}} \frac{\sqrt{R^2+y^2}}{|\sinh(R+iy)|}dy.$$

Since u, t, $y > 0$, this implies that $e^{-\frac{ty}{2u}} \leq 1$. Besides, $0 \leq y \leq (m+\frac{1}{2})\pi$, and we have

$$\sqrt{R^2+y^2} \leq \sqrt{R^2+\big(m+\frac{1}{2}\big)^2\pi^2}.$$

Hence,

$$\Big|\int_{\gamma_2}\Big| \leq \int_0^{(m+\frac{1}{2})\pi} \frac{\sqrt{R^2+\big(m+\frac{1}{2}\big)^2\pi^2}}{|\sinh(R)|}dy \leq \big(m+\frac{1}{2}\big)\pi\frac{\sqrt{R^2+\big(m+\frac{1}{2}\big)^2\pi^2}}{|\sinh(R)|}.$$

But

$$\lim_{R\to\infty}\frac{\sqrt{R^2+\big(m+\frac{1}{2}\big)^2\pi^2}}{|\sinh(R)|} = 0,$$

we have

$$\int_{\gamma_2} e^{i\frac{ty}{2u}}\frac{z}{\sinh(z)}dz \to 0 \quad \text{as } R\to+\infty.$$

Similarly, one can show that

$$\int_{\gamma_4} e^{i\frac{ty}{2u}}\frac{z}{\sinh(z)}dz \to 0 \quad \text{as } R\to+\infty.$$

Now we need to evaluate $\int_{\gamma_3}$. Here $\gamma_3 : \big[R,-R\big] \to \mathbb{C}$, $\gamma_3(x) = x+i\big(m+\frac{1}{2}\big)\pi$.

$$\begin{aligned}\int_{\gamma_3} e^{i\frac{tz}{2u}}\frac{z}{\sinh(z)}dz &= -\int_{-\gamma_3} e^{i\frac{tz}{2u}}\frac{z}{\sinh(z)}dz\\ &= -\int_{-R}^{R} e^{i\frac{t}{2u}\big[x+i(m+\frac{1}{2})\pi\big]}\frac{x+i\big(m+\frac{1}{2}\big)\pi}{\sinh[x+i(m+\frac{1}{2})\pi]}dx.\end{aligned}$$

We shall split

$$\int_{-R}^{R} = \int_{-R}^{-1}+\int_{-1}^{1}+\int_{1}^{R}.$$

When $|x|\geq 1$, we shall bound

$$\Big|\sinh\big[x+i(m+\frac{1}{2})\pi\big]\Big| \geq |\sinh(x)|.$$

When $|x|\leq 1$, we shall bound

$$\Big|\sinh\big[x+i(m+\frac{1}{2})\pi\big]\Big| \geq \big|\sin\big(m+\frac{1}{2}\big)\pi\big| = |\cos(m\pi)| = 1.$$

Also,

$$\sinh(x) = \frac{e^x-e^{-x}}{2} \geq \frac{e^x}{4} \Leftrightarrow 2e^x-2e^{-x}\geq e^x \Leftrightarrow x\geq\frac{\log 2}{2},$$

so, for $x \geq 1$, we have $\sinh(x) \geq \frac{e^x}{4}$. Therefore,

$$\begin{aligned}
\Big|\int_{\gamma_3}\Big| &\leq e^{-\frac{t}{2u}\left(m+\frac{1}{2}\right)\pi}\Big[\int_{-1}^{1}\frac{\sqrt{x^2+\left(m+\frac{1}{2}\right)^2\pi^2}}{\left|\sinh\left[x+i(m+\frac{1}{2})\pi\right]\right|}dx+\int_{-R}^{-1}+\int_{1}^{R}\Big] \\
&\leq e^{-\frac{t}{2u}\left(m+\frac{1}{2}\right)\pi}\Big[\int_{-1}^{1}\sqrt{1+(m+\frac{1}{2})^2\pi^2}dx \\
&\quad+\int_{-R}^{-1}\frac{\sqrt{x^2+\left(m+\frac{1}{2}\right)^2\pi^2}}{|\sinh(x)\pi|}dx+\int_{1}^{R}\frac{\sqrt{x^2+\left(m+\frac{1}{2}\right)^2\pi^2}}{|\sinh(x)\pi|}dx\Big] \\
&= e^{-\frac{t}{2u}\left(m+\frac{1}{2}\right)\pi}\Big[2\sqrt{1+(m+\frac{1}{2})^2\pi^2}+2\int_{1}^{R}\frac{\sqrt{x^2+\left(m+\frac{1}{2}\right)^2\pi^2}}{|\sinh(x)\pi|}dx\Big] \\
&\leq e^{-\frac{t}{2u}\left(m+\frac{1}{2}\right)\pi}\Big[2\sqrt{1+(m+\frac{1}{2})^2\pi^2}+8\int_{1}^{R}\frac{\sqrt{x^2+\left(m+\frac{1}{2}\right)^2\pi^2}}{e^x}dx\Big].
\end{aligned}$$

So by Cauchy's theorem, we have

$$\int_{-R}^{R}e^{i\frac{t}{2u}\tau}\frac{\tau}{\sinh(\tau)}d\tau+\sum_{j=1}^{3}\int_{\gamma_j}=2\pi i\sum_{k=1}^{m}\operatorname{Res}\Big(e^{i\frac{t}{2u}z}\frac{z}{\sinh(z)},\ k\pi i\Big)$$

As $R\to+\infty$, we have

$$\begin{aligned}
&\int_{-\infty}^{+\infty}e^{i\frac{t}{2u}\tau}\frac{\tau}{\sinh(\tau)}d\tau \\
=\;& 2\pi i\sum_{k=1}^{m}\operatorname{Res}\Big(e^{i\frac{t}{2u}z}\frac{z}{\sinh(z)},\ k\pi i\Big) \\
&+\mathcal{O}\Big(e^{-\frac{t}{2u}\left(m+\frac{1}{2}\right)\pi}\Big[2\sqrt{1+(m+\frac{1}{2})^2\pi^2}+8\int_{1}^{R}\frac{\sqrt{x^2+\left(m+\frac{1}{2}\right)^2\pi^2}}{e^x}dx\Big]\Big) \\
=\;& 2\pi i\sum_{k=1}^{m}e^{-\frac{t}{2u}k\pi}\frac{2}{2}(-1)^k\cdot k\pi i \\
&+\mathcal{O}\Big(e^{-\frac{t}{2u}\left(m+\frac{1}{2}\right)\pi}\Big[2\sqrt{1+(m+\frac{1}{2})^2\pi^2}+8\int_{1}^{R}\frac{\sqrt{x^2+\left(m+\frac{1}{2}\right)^2\pi^2}}{e^x}dx\Big]\Big) \\
=\;& -2\pi^2\sum_{k=1}^{m}(-1)^k k\big(e^{-\frac{t}{2u}\pi}\big)^k \\
&+\mathcal{O}\Big(e^{-\frac{t}{2u}\left(m+\frac{1}{2}\right)\pi}\Big[2\sqrt{1+(m+\frac{1}{2})^2\pi^2}+8\int_{1}^{R}\frac{\sqrt{x^2+\left(m+\frac{1}{2}\right)^2\pi^2}}{e^x}dx\Big]\Big).
\end{aligned}$$

Now let $m \to \infty$ and by Theorem 1.55, one has

$$\begin{aligned}
\frac{1}{8\pi^2u^2}\int_{-\infty}^{+\infty} e^{i\frac{t\tau}{2u}}\frac{\tau}{\sinh(\tau)}d\tau &= \frac{2\pi}{8\pi^2u^2}\sum_{k=1}^{m}\mathrm{Res}\Big(e^{i\frac{tz}{2u}}\frac{z}{\sinh(z)}, k\pi i\Big) \\
&= \frac{2\pi^2}{8\pi^2u^2}\sum_{k=1}^{\infty}(-1)^{k+1}ke^{-\frac{k\pi t}{2u}} \\
&= \frac{1}{4u^2}\sum_{k=1}^{\infty}(-1)^{k+1}ke^{-\frac{s_k^2}{2u}}.
\end{aligned}$$

where $s_k = \ell(0,t)$ is the length of the k-th geodesic connecting the origin and the point $(0,t)$.

In order to show that the series is convergent, we shall use the formula

$$-\frac{1}{(1+z)^2} = \sum_{k=1}^{\infty}(-1)^k kz^k$$

with $z = e^{-\frac{t}{2u}\pi}$, and $0 < z < 1$. Then

$$\begin{aligned}
\sum_{k=1}^{\infty}(-1)^k k\big(e^{-\frac{t}{2u}\pi}\big)^k &= -\frac{e^{-\frac{t}{2u}\pi}}{\big(1+e^{-\frac{t}{2u}\pi}\big)^2} = -\frac{1}{e^{\frac{t}{2u}\pi}\big(1+2e^{-\frac{t}{2u}\pi}+e^{-2\frac{t}{2u}\pi}\big)} \\
&= -\frac{1}{e^{\frac{t}{2u}\pi}+2+e^{-\frac{t}{2u}\pi}} = -\frac{1}{\big(e^{\frac{t}{4u}\pi}+e^{-\frac{t}{4u}\pi}\big)^2} \\
&= -\frac{1}{4\cosh^2\big(\frac{\pi t}{2u}\big)}.
\end{aligned}$$

This completes the proof of the theorem. □

6.11. Maxwell's equations

The first group of Maxwell's equations is

$$\text{(6.48)} \qquad \mathrm{curl}(E) = -\frac{1}{c}\frac{\partial B}{\partial t}, \qquad \mathrm{div}(B) = 0,$$

where $E : \mathcal{D}\times\mathbb{R}_t \to \mathbb{R}^3$ is the electric field, and $B : \mathcal{D}\times\mathbb{R}_t \to \mathbb{R}^3$ is the magnetic field on a domain $\mathcal{D}\subset\mathbb{R}^3$.

If the electromagnetic field is static, then E and B are time independent. In this case

$$\text{(6.49)} \qquad \mathrm{curl}(E) = 0, \qquad \mathrm{div}(B) = 0.$$

The equations (6.48) and (6.49) are written in vectorial form. In the following we shall write them in differential form. For this we need to define the Hoge operator. Let $\Lambda^k(\mathbb{R}^n)$ be the space of k-forms on $\mathbb{R}^n$.

DEFINITION 6.31. The operator

$$* : \Lambda^k(\mathbb{R}^n) \to \Lambda^{n-k}(\mathbb{R}^n),$$

defined by

$$*\eta^k = \sum_I \mathrm{sgn}\begin{bmatrix}1\ldots n\\ I\ \ J\end{bmatrix}\eta_I\, dx^J,$$

where $J = (j_1 < \cdots < j_{n-k})$ is the complementary multi-index to $I = (i_1 < \cdots < i_k)$, is called the Hoge operator.

One may verify the following property:

$$**\eta^k = (-1)^{k(n-k)}\eta^k, \qquad \eta^k \in \Lambda^k(\mathbb{R}^n).$$

Let $dv = dx^1 \wedge \cdots \wedge dx^n$ be the volume form on $\mathbb{R}^n$.

DEFINITION 6.32. If V is a vector field, define $i_V : \Lambda^k(\mathbb{R}^n) \to \Lambda^{k-1}(\mathbb{R}^n)$ by

$$(i_V\eta^k)(W_1, \dots, W_{k-1}) = \eta^k(V, W_1, \dots, W_{k-1}).$$

One may associate with each vector field V a 1-form η^1_V defined by the equation

$$*\eta^1_V = i_V(dv).$$

A computation in local coordinates is provided below. If $V = \sum_i V^i \partial_{x_i}$ is the vector field, using the relation $**\eta^1 = (-1)^{n-1}\eta^1$, yields

$$\begin{aligned}\eta^1_V &= (-1)^{n-1} * \Big(V^1\, dx^2 \wedge \cdots \wedge dx^n - V^2\, dx^1 \wedge dx^3 \wedge \cdots \wedge dx^n \pm \dots\Big)\\ &= V^1\, dx^1 + V^2\, dx^2 + \cdots + V^n\, dx^n.\end{aligned}$$

DEFINITION 6.33. The divergence of a vector field V is defined by the equation

$$d(*\eta^1_V) = d(i_V(dv)) := \operatorname{div}(V)\, dv. \tag{6.50}$$

$d(*\eta^1_V)$ and $dv = dx^1 \wedge dx^2 \wedge dx^3$ are 3-forms on $\mathbb{R}^3$ and hence, they are proportional. The proportionality function is $\operatorname{div}(V)$. One may work out the divergence in local coordinates and get

$$\operatorname{div} V = \sum_{i=1}^{3} \frac{\partial V^i}{\partial x^i}.$$

The following relations will be useful

$$(*\eta^1_V)(u, w) = (i_V dv)(u, w) = dv(v, u, w) = \det(v, u, w).$$

As $*\eta^1_V \in \Lambda^2(\mathbb{R}^3)$, let

$$*\eta^1_V = A_1\, dx^2 \wedge dx^3 + A_2\, dx^1 \wedge dx^3 + A_3\, dx^1 \wedge dx^2. \tag{6.51}$$

A computation yields

$$\begin{aligned}A_1 &= (*\eta^1_V)(\partial_{x_2}, \partial_{x_3}) = \det(V, \partial_{x_2}, \partial_{x_3}) = V^1,\\ A_2 &= (*\eta^1_V)(\partial_{x_1}, \partial_{x_3}) = \det(V, \partial_{x_1}, \partial_{x_3}) = -V^2,\\ A_3 &= (*\eta^1_V)(\partial_{x_1}, \partial_{x_2}) = \det(V, \partial_{x_1}, \partial_{x_2}) = V^3,\\ *\eta^1_V &= V^1\, dx^2 \wedge dx^3 + V^2\, dx^3 \wedge dx^1 + V^3\, dx^1 \wedge dx^2.\end{aligned}$$

DEFINITION 6.34. The 2-form associated with the magnetic vector field $B = (B^1, B^2, B^3)$ is defined by

$$\Omega = *\eta^1_B.$$

Then

$$\Omega = B^1\, dx^2 \wedge dx^3 + B^2\, dx^3 \wedge dx^1 + B^3\, dx^1 \wedge dx^2.$$

Formula (6.50) becomes

$$d\Omega = \operatorname{div}(B)\, dv, \tag{6.52}$$

and hence, Maxwell's equation $\operatorname{div}(B) = 0$ can be written in differential form as

$$d\Omega = 0.$$

The case of concern is when the magnetic field comes from a potential

$$\Omega = d\omega,$$

where $\omega = \sum_{j=1}^{3} \omega_j\, dx_j$, and ker $\omega = \mathcal{H}$ is the horizontal distribution. The equation of motion for a particle with charge e, speed c and mass m is given by the Lorentz equation

$$m\frac{d\vec{v}}{dt} = \frac{e}{c}\vec{v} \times \vec{B}, \tag{6.53}$$

where $\vec{B} = (\Omega_1, \Omega_2, \Omega_3)$ is the magnetic vector field. The equation (6.53) is the Euler-Lagrange equation for the following Lagrangian

$$L(x, \dot{x}) = \frac{m}{2}\dot{x}^\alpha \dot{x}^\beta \delta_{\alpha\beta} - e\dot{x}^i \omega_i(x) = \frac{1}{2}m|\dot{x}|^2 - e\,\omega(\dot{x}).$$

In the case of a plane curve with velocity $\overrightarrow{v} = (\dot{x}_1, \dot{x}_2, 0)$ and magnetic field $\overrightarrow{B} = (0, 0, \Phi(x))$, the right side of (6.53) becomes

$$\begin{aligned} (\dot{x}_1, \dot{x}_2, 0) \times (0, 0, \Phi(x)) &= \Phi(x)(\dot{x}_2 \partial_{x_1} - \dot{x}_1 \partial_{x_2}) \\ &= \Phi(x)\mathcal{J}(\dot{x}_1 \partial_{x_1} + \dot{x}_2 \partial_{x_2}) \\ &= \Phi(x)\mathcal{J}(\overrightarrow{v}), \end{aligned}$$

while the left side of (6.53) is $m(\ddot{x}_1, \ddot{x}_2, 0)$. For different values of Φ we recover the models discussed in the previous sections.

• *Step 2 case: Heisenberg group*

Consider the vector fields

$$X_1 = \frac{\partial}{\partial x_1} + 2x_2\frac{\partial}{\partial t}, \quad X_2 = \frac{\partial}{\partial x_2} - 2x_1\frac{\partial}{\partial t},$$

which belong to *ker* ω, where $\omega = dt + 2(x_1\, dx_2 - x_2\, dx_1)$. The magnetic field is given by the 2-form $\Omega = d\omega = 4dx_1 \wedge dx_2$. The corresponding Lagrangian is

$$L = \frac{1}{2}(\dot{x}_1^2 + \dot{x}_2^2) + \theta(\dot{t} - 2(\dot{x}_1 x_2 + x_1 \dot{x}_2)). \tag{6.54}$$

The Euler-Lagrange equations for the Lagrangian (6.54) are

$$\ddot{x}_1 = 4\theta\dot{x}_2, \quad \ddot{x}_2 = -4\theta\dot{x}_1, \quad \theta = \text{constant}. \tag{6.55}$$

Let $\phi(s) = (x_1(s), x_2(s))$ be the projection of a solution on the x-plane. Then (6.55) becomes

$$\frac{d\dot{\phi}(s)}{ds} = 4\theta\mathcal{J}\dot{\phi}(s), \tag{6.56}$$

with $\mathcal{J}^2 = -I$, which is the Lorentz equation (6.53) with mass $m = 1$, charge $e = \theta$ and constant magnetic vector field $\vec{B} = (0, 0, 4)$. The projections on the (x_1, x_2)-space are circles, called Larmour orbits (see [**53**]). These are the trajectories of charged particles in constant magnetic fields. The trajectory is a circle perpendicular on the magnetic field (see Figure 3.1). The velocity $\vec{v} = \dot{\phi}(s)$, the Lorentz force $\ddot{\phi}(s)$ and the magnetic field $\vec{\Omega} = (0, 0, 4)$ form a right oriented orthonormal frame. In the case of Heisenberg group (step 2) the magnetic field is constant and hence does not vanish. As a consequence, the trajectory of the particle will always be a circle, even if it starts from outside of the origin (see Figure 6.3). This is true only for step 2 and it is not working for superior steps.

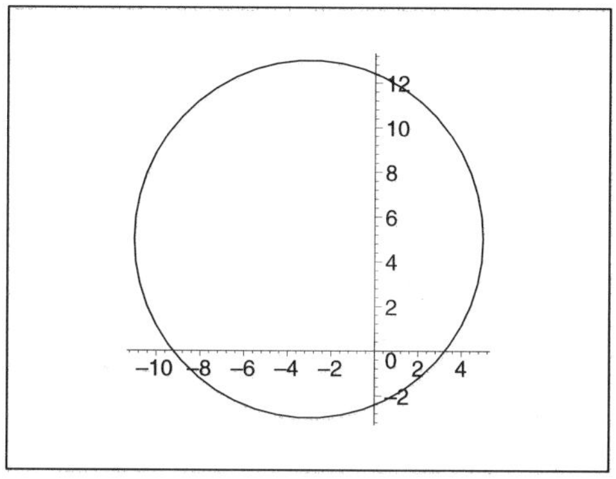

Figure 6.3: Larmour orbits outside the origin.

- *Step* 4 *case*
  Consider the vector fields

$$X_1 = \frac{\partial}{\partial x_1} + 4x_2|x|^2\frac{\partial}{\partial t}, \quad X_2 = \frac{\partial}{\partial x_2} - 4x_1|x|^2\frac{\partial}{\partial t}.$$

We have $span\{X_1, X_2\} = ker\,\omega$, where $\omega = dt + 4|x|^2(x_1\,dx_2 - x_2\,dx_1)$. The magnetic field is given by the 2-form $\Omega = d\omega = 16|x|^2 dx_1 \wedge dx_2$. The corresponding Lagrangian is

$$L = \frac{1}{2}(\dot{x}_1^2 + \dot{x}_2^2) + \theta\big(\dot{t} + 4|x|^2(x_1\dot{x}_2 - x_2\dot{x}_1)\big). \tag{6.57}$$

The Euler-Lagrange equations for the Lagrangian (6.57) are

$$\ddot{x}_1 = 16\,\theta|x|^2\dot{x}_2, \quad \ddot{x}_2 = -16\,\theta|x|^2\dot{x}_1, \quad \theta = \text{constant}. \tag{6.58}$$

Let $\phi(s) = (x_1(s), x_2(s))$. Then (6.58) becomes

$$\frac{d\dot{\phi}(s)}{ds} = \theta\dot{\phi}(s) \times \vec{\Omega},$$

which is the Lorentz equation (6.53) with mass $m = 1$, charge $e = \theta$ and magnetic field oriented along the t-axis $\vec{\Omega} = \big(0, 0, 16|x|^2\big)$. This magnetic vector field vanishes at the origin and it has spherical symmetry. The trajectories for the charged particles depend whether the particle passes through the origin. Figure 3.2 represents the trajectory of a charged particle which passes through the origin. The trajectory is a union of loops of angle $\pi/3$ at the origin.

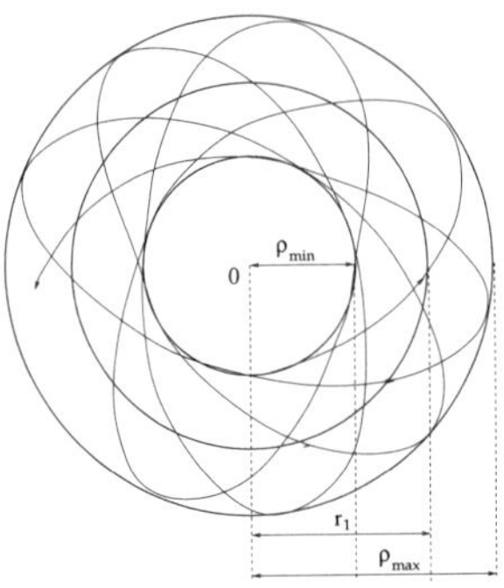

Figure 6.4: Trajectories outside of the origin.

If the trajectory is not passing through the origin, it looks like in Figure 6.4. The particle remains all the time between two circles. Using the results of section §3.10, the particle will never leave the circular crown.

- *Step 2k case*

Consider the vector fields

$$X_1 = \frac{\partial}{\partial x_1} + 2kx_2|x|^{2(k-1)}\frac{\partial}{\partial t}, \qquad X_2 = \frac{\partial}{\partial x_2} - 2kx_1|x|^{2(k-1)}\frac{\partial}{\partial t}.$$

We have $span\{X_1, X_2\} = ker\,\omega$, where $\omega = dt + 2k|x|^{2(k-1)}(x_1\,dx_2 - x_2\,dx_1)$. The magnetic field is given by the 2-form $\Omega = d\omega = 4k^2|x|^{2(k-1)}dx_1 \wedge dx_2$. The corresponding Lagrangian is

$$L = \frac{1}{2}(\dot{x}_1^2 + \dot{x}_2^2) + \theta\big(\dot{t} + 2k|x|^{2(k-1)}(x_1\dot{x}_2 - x_2\dot{x}_1)\big). \tag{6.59}$$

The Euler-Lagrange equations for the Lagrangian (6.59) are

$$\ddot{x}_1 = 4k^2\,\theta|x|^{2k}\dot{x}_2, \quad \ddot{x}_2 = -4k^2\,\theta|x|^{2k}\dot{x}_1, \quad \theta = \text{constant}. \tag{6.60}$$

Let $\phi(s) = (x_1(s), x_2(s))$. Then (6.60) becomes

$$\frac{d\dot{\phi}(s)}{ds} = \theta\dot{\phi}(s) \times \vec{\Omega},$$

which is the Lorentz equation (6.53) with mass $m = 1$, charge $e = \theta$ and magnetic field oriented along the t-axis $\vec{\Omega} = \big(0, 0, 4k^2|x|^{2(k-1)}\big)$. The magnetic field vanishes at the origin. Figures 3.2 and 6.3 show the trajectories through the origin in the cases $k = 2$ and $k = 3$. The trajectory is periodic and consists of loops with an angle of $\frac{\pi}{2k-1}$ at the origin. The number of loops is $N = 2k - 1$. For $k = 1$ (step 2 case) there is only one loop, which is a circle (Larmour orbit), see Figure 6.2.

For $k = 2$ (step 4 case) there are 3 loops, see Figure 3.2. In the cases $k = 3$ and $k = 4$ above, there are 5 and 7 loops, respectively (see Figure 3.3 and Figure 3.4). If the particle is not passing through the origin, the trajectory belongs to a circular crown, like in Figure 6.4 or Figure 6.5.

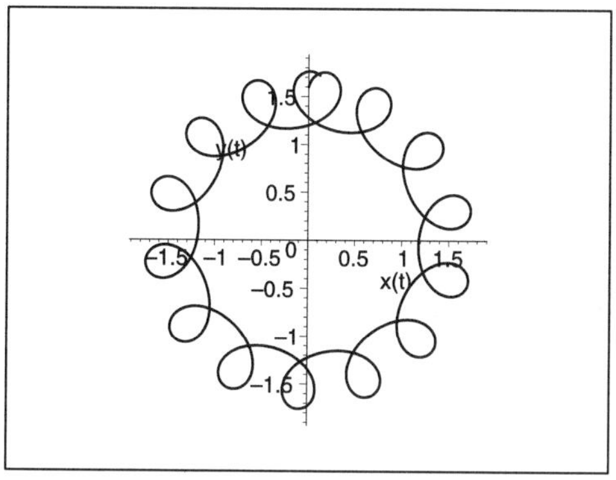

Figure 6.5: Trajectories outside of the origin.

6.12. Quantization of energy

The lengths are the same for all parametrizations. In this section we shall consider geodesics $\gamma : [0,1] \to \mathbb{R}^3$ joining the origin and the point $(0,0,t)$. As the Hamiltonian is preserved along the solutions, the velocity has constant length in the subRiemannian metric (in which X_1 and X_2 are orthonormal), and hence we have identity in Cauchy's inequality

$$\ell(\gamma) = \int_0^1 |\dot\gamma(s)|\,ds = \Big(\int_0^1 ds\Big)^{1/2}\Big(\int_0^1 |\dot\gamma(s)|^2\,ds\Big)^{1/2} = \sqrt{2E}.$$

Hence the energies of a unit mass particle moving along the geodesic are given by

$$E_m = \frac{1}{2}\ell_m^2, \quad m = 1,2,3,\dots$$

Theorem 3.27 yields (with $k+1$ changed into k):

THEOREM 6.35. *The particle has discrete energies $0 < E_1 < E_2 < \dots$, given by*

$$(E_m)^k = \frac{(2k-1)m^{2k-1}(2F^2)^k|t|}{Q}, \quad m = 1,2,3\dots$$

For $k = 1$,

$$E_m = \frac{m\pi}{2}|t|,$$

and for $k = 2$,

$$E_m^2 = \frac{|t|}{2}\left(\frac{m\sqrt{\pi}\,\Gamma\big(\frac{1}{6}\big)}{3\Gamma\big(\frac{2}{3}\big)}\right)^3, \quad m = 1,2,3,\dots$$

E_1 is the ground-state energy, *i.e.*, the lowest energy of the particle. All the other energy levels correspond to excited states. If the electron is in ground-state, then it can be excited up to the level of energy E_m if it receives energy from a photon with energy $h\nu_m = E_m - E_1$, where h is Plank's constant and ν_m is the frequency of the photon. The difference $E_{m+1} - E_m$ gives the distance between the energy levels. In the step 2 case, the energy levels are equidistant (see Figure 6.6)

$$E_2 - E_1 = E_3 - E_2 = \dots = E_{m+1} - E_m = \frac{\pi}{2}|t|.$$

In the step 2 case ($k = 1$) the smallest energy is $\pi|t|/2$. This corresponds to the ground-state energy $h\omega/2$ for a linear harmonic oscillator with pulsation $\omega = \pi|t|/h$, studied in this chapter, section §6.1

For step $k > 2$, the energy levels become more and more distant as m increases (see Figure 6.7)

$$0 < E_2 - E_1 < E_3 - E_2 < \cdots < E_{m+1} - E_m < \ldots$$

This is characterizing the energy spectrum for a charged particle in a static, non-constant, rotational invariant magnetic field.

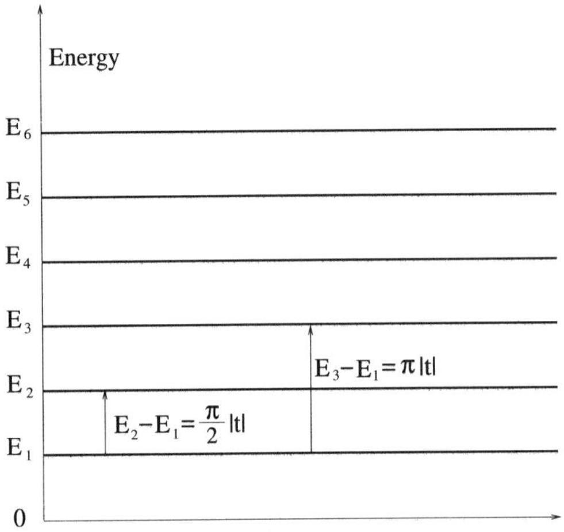

Figure 6.6. Energy levels for $k = 1$.

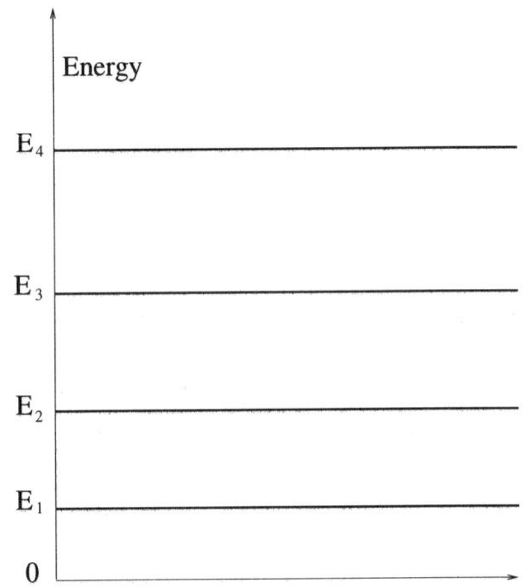

Figure 6.7. Energy levels for $k = 2$.

Let $f_k(x) = C_k x^{2-\frac{1}{k}}$, where the constant

$$C_k = 2M\left(\frac{(2k-1)|t|}{Q}\right)^{1/k}.$$

Then

$$E_{m+1} - E_m = f_k(m+1) - f_k(m) = f_k'(\xi_m),$$

with $m < \xi_m < m+1$. Hence, the difference between two consecutive energy levels is estimated as

$$\left(2 - \frac{1}{k}\right) C_k m^{1-\frac{1}{k}} < E_{m+1} - E_m < \left(2 - \frac{1}{k}\right) C_k (m+1)^{1-\frac{1}{k}}.$$

- *Case $k = 2$:* We have

$$\frac{3}{2} C_2 \sqrt{m} < E_{m+1} - E_m < \frac{3}{2} C_2 \sqrt{m+1},$$

with

$$C_2 = \sqrt{\frac{|t|}{2}} \left(\frac{\sqrt{\pi}\,\Gamma\left(\frac{1}{6}\right)}{3\Gamma\left(\frac{2}{3}\right)}\right)^{3/2} \approx 2.676\sqrt{|t|}.$$

Using $3C_2/2 \approx 4.014\sqrt{|t|}$, yields

$$4\sqrt{m|t|} < E_{m+1} - E_m, \quad m = 1, 2, 3, \ldots$$

6.13. Kepler-type laws

From Kepler's first law, all the planets have plane trajectory, which are ellipses with the sun in one of the focuses. The second Kepler's law says that the areal velocity α is constant along the motion. If T is the orbital period and r is the semi-major axis of the orbit, then the third Kepler's law says that $T^{2/3}$ and r are proportional.

We shall work out an analogy between Kepler's laws and our models.

• *The third law*
Integrating in $\dot{t} = \partial H/\partial\theta$, yields

$$t(\phi) - t(\phi_0) = -\frac{2k}{2k-1} r_{max}^{2k} \Psi(\phi - \phi_0),$$

where

$$\Psi(u) = \int_0^{(2k-1)u} \sin^{\frac{2k}{2k-1}} v \, dv.$$

Let

$$T = \left| t(\phi_0 + \frac{\pi}{2k-1}) - t(\phi_0) \right| \tag{6.61}$$

and

$$Q_k = \frac{2k}{2k-1} \Psi\Big(\frac{\pi}{2k-1}\Big) > 0.$$

If $\theta > 0$, ϕ is decreasing and t is increasing. Then (6.61) becomes

$$T = Q_k r_{max}^{2k}. \tag{6.62}$$

T is the period, *i.e.*, the time needed for the particle to come back at the origin. Hence

$$T^{1/2k} = c \cdot r_{max}$$

is an analog of the third law of Kepler, with $c = Q_k^{1/2k}$. For $k = 1$, the step 2 case, one has

$$T^{1/2} = \sqrt{\pi} \cdot r_{max},$$

where r_{max} is the diameter of the circle in the x-plane. For $k = 2$, the step 4 case, one has

$$T^{1/4} = \left(\frac{\sqrt{\pi}}{3} \frac{\Gamma(1/6)}{\Gamma(2/3)} \right)^{\frac{1}{4}} \cdot r_{max}.$$

• *The second law*
Let $x(s) = (x_1(s), x_2(s))$ be a curve in the plane. Consider polar coordinates $x_1 = r\cos\phi$, $x_2 = r\sin\phi$, with $r = |x|$. Let $\phi_0 = \phi(0)$ and $\phi_1 = \phi(1)$. The vectorial radius $\overrightarrow{OP}$ will sweep an area equal to

$$\mathcal{A} = \int_{\phi_0}^{\phi_1} \frac{1}{2} r^2 \, d\phi.$$

The above formula can be also written in differential form

$$d\mathcal{A} = \frac{1}{2} r^2 \, d\phi.$$

See Figure 1.1.

The areal velocity is defined as

$$\alpha = \frac{d\mathcal{A}}{ds}.$$

Let $c = (x_1, x_2, t)$ be a subRiemannian geodesic in step 2 case. Then the rate of change of the t-component is equal to 4α, *i.e.*, $\dot{t} = 4\alpha$. For the proof, see Theorem 1.7.

6.14. Exercises

EXERCISE 6.1. Let $P(x,t;u) = \frac{1}{(2\pi u)^2}\int_{-\infty}^{\infty} e^{-f(x,t,\tau)/u}V(\tau)$ be the heat kernel for Δ_H.

i) Show that

$$\int_{\mathbb{R}^3} P(x,t;u)\,dx\,dt = 1.$$

ii) Prove that the function P is smooth on $\mathbf{H}_1 \times \mathbb{R}$ and satisfies

$$\lim_{u\to 0}\left(\frac{d}{du}\right)^m P(x,t;u) = 0,$$

uniformly in (x,t) in compact subsets on $\mathbf{H}_1 \searrow 0$, for each $m = 1,2,3,\ldots$.

EXERCISE 6.2. Find the full heat kernel $P(x,t,x_0,t_0,u)$ for Δ_H using

i) left translations;

ii) direct computation starting from the modified complex action function.

EXERCISE 6.3. Show that the heat kernel $P(x,t,u)$ satisfies the estimate

$$P(x,t,u) \le C\,\frac{e^{-d(x,t)^2}/2u}{u^2}\min\left\{1, \sqrt{\frac{u}{\|x\|d(x,t)}}\right\}, \qquad (x,t,u)\in \mathbf{H}_1\times\mathbb{R}_+.$$

EXERCISE 6.4. Let $f(x,t,\tau) = -i\tau t + a\tau\coth(2a\tau)\|x\|^2$ be the modified complex action function. Show that

$$\frac{\partial^2 f}{\partial\tau^2}\bigg|_{\tau=i\theta_c} \ge \frac{2\|ax\|^2}{3},$$

where θ_c is the solution of

$$t = a\mu(2a\theta)\|x\|^2$$

in the strip $|Re\,\theta| < \pi/2a$, and $\mu(\phi) = \frac{\phi}{\sin^2\phi} - \cot\phi$.

EXERCISE 6.5. Prove that at points $(0,t)$, $t\neq 0$, the heat kernel

$$P(x,t,u) = \frac{1}{4au^2}e^{-d(0,t)^2/2u}\Big\{1 + O\Big(e^{-d^2(0,t)/2u}\Big)\Big\},$$

as $u\to 0^+$.

EXERCISE 6.6. Find the small time behavior for the kernel of the Schrödinger operator

a) in the case $x = 0$;

b) in the case $x \neq 0$.

EXERCISE 6.7. Consider Hankel's function of first type

$$H_n^{(1)}(\lambda) = \frac{1}{i\pi}\int_{-\infty}^{\infty+i\pi} e^{\lambda\sinh z - nz}\,dz.$$

Using Theorem 6.28 show that

$$H_n^{(1)}(\lambda) = \sqrt{\frac{2}{\pi\lambda}}e^{i(\lambda-\frac{n\pi}{2}-\frac{\pi}{4})} + O(\lambda^{-3/2})$$

as $\lambda \to \infty$.

EXERCISE 6.8. Consider the function $F(\lambda) = \int_a^b f(t)e^{\lambda h(t)}\,dt$, $\lambda \in \mathbb{R}$. Let $M = \max_{t\in[a,b]}|h(t)| < \infty$ and assume there is $\lambda_0 > 0$ such that

$$\int_a^b |f(t)|e^{\lambda_0 h(t)}\,dt < \infty.$$

Show that

$$|F(\lambda)| \le Ce^{\lambda M}, \qquad \forall\lambda > \lambda_0,\ C = \text{constant}.$$

EXERCISE 6.9 (Watson). Let $\alpha > 0, \beta > -1$ and $g(t)$ be a infinitely differentiable function on the interval $[0, T]$. The asymptotic expansion of the integral

$$G(\lambda) = \int_0^T e^{-\lambda t^\alpha} t^\beta g(t)\,dt$$

is

$$G(\lambda) = \frac{1}{\alpha}\sum_{n=0}^{\infty} g^{(n)}(0)\frac{\Gamma\big((\beta+n+1)/\alpha\big)}{n!\lambda^{(\beta+n+1)/\alpha}},$$

as $\lambda \to \infty$.

EXERCISE 6.10. *i*) Show that $V(z) \sim 1 - \frac{2}{3}h^2z^2 + \frac{14}{45}h^2 + \mathcal{O}(z^6)$, where

$$V(z) = \frac{2hz}{\sinh(2hz)}.$$

ii) Find the coefficient a_1 of the asymptotic expansion of the heat kernel.

Bibliography

[1] R. Abraham and J. E. Marsden. *Foundations of Mechanics.* Addison Wesley, 1994.

[2] V. I. Arnold. *Mathematical Models of Clasical Mechanics.* Springer Verlag, 1989.

[3] R. Beals, B. Gaveau, and P. C. Greiner. *Hamilton-Jacobi theory and the heat kernel on Heisenberg groups.* Number 79. J. Math. Pure Appl., 2000, pp. 633-689.

[4] R. Beals, B. Gaveau, and P.C. Greiner. *On a Geometric Formula for the Fundamental Solution of Subelliptic Laplacians.* Number 181. Math. Nachr., 1996, pp. 81-163.

[5] R. Beals, B. Gaveau, and P.C. Greiner. *Complex Hamiltonian mechanics and parametrices for subelliptic Laplacians, I, II, III.* Number 21. Bull. Sci. Math., 1997, pp. 1-36, 97-149, 195-259.

[6] R. Beals and P.C. Greiner. *Calculus on Heisenberg Manifolds.* Ann. Math. Studies, Vol. 119, Princetown University Press, 1988.

[7] A. Bellaïche. *Sub-Riemannian Geometry.* Progr. Math. 144, Birkhäuser, Basel, 1996.

[8] C.M. Bender, S. Boettcher, and P. N. Meisinger. J. Math. Phys., 1999.

[9] C. Berenstein, D.C. Chang, and J. Tie. *Laguerre Calculus and Its Applications on the Heisenberg Group*, volume 22. AMS/IP, Studies in Advanced Math., 2001.

[10] J. M. Bismut. *Large Deviations and the Malliavin Calculus.* Progr. in Math. 144, Birkhäuser, Boston, 1984.

[11] S. N. Bose. *Plancks gesetz und lichquantentenhypothese*, volume 26. Zeitschrift fur Physik, 1924.

[12] O. Calin. *Ph.D Thesis, University of Toronto.* 2000.

[13] O. Calin. *Geodesics on a certain step 2 sub-Riemannian manifold.* Number 22. Annals of Global Analysis and Geometry, 2002, pp. 317-339.

[14] O. Calin and D. C. Chang. *Geometric Mechanics on Riemannian Manifolds, Applications to Partial Differential Equations.* Applied and Numerical Analysis, Birkhäuser, 2004.

[15] O. Calin and D. C. Chang. *Geometric Mechanics on a step 4 subRiemannian manifold.* Taiwanese Journal of Mathematics, June, 2005.

[16] O. Calin, D. C. Chang, and P. C. Greiner. *On a Step* $2(k+1)$ *sub-Riemannian manifold*, volume 14. Journal of Geometric Analysis, 2004, pp. 1-18.

[17] O. Calin, D. C. Chang, and P. C. Greiner. *Real and complex Hamiltonian mechanics on some SubRiemannian manifolds*, volume 18. Asian Journal of Mathematics, 2004, pp. 137-160.

[18] O. Calin, D. C. Chang, and P. C. Greiner. *Geometric mechanics on the Heisenberg group.* Bulletin of the Institute of Mathematics, Academia Sinica, 2005.

[19] O. Calin and V. Mangione. *Variational calculus on sub-Riemannian manifolds*, volume 8. Balcan Journal of Geometry and Applications, 2003.

[20] A. Calogelo, G. Carcano, and R. Pini. *Optimization in Heisenberg group.* to appear in Optimization, A journal of mathematical programming and operations research.

[21] E. Cartan. *Les groupes de transformations continus, infinis, simples.* Ann. École Norm. Sup.

[22] W.L. Chow. *Uber Systeme von linearen partiellen Differentialgleichungen erster Ordnung.* Number 117. Math. Ann., 1939, pp. 98-105.

[23] A. Erdélyi et al. *Higher Transcendental Functions*, volume I,IV. Bateman Manuscript project, McGraw-Hill, New York, 1955.

[24] G. Folland and E.M. Stein. *Estimates for the* $\overline{\partial}_b$*, Complex and Analysis on the Heisenberg Group.* Number 27. Comm. Pure Appl. Math., 1974, pp. 429-522.

[25] G.B. Folland. *Introduction to PDEs, second edition.* Princeton University Press, New Jersey, 1995.

[26] B. Gaveau. *Principe de moindre action, propagation de la chaleur et estimées sous-elliptiques sur certains groupes nilpotents*, volume 139. Acta Math, 1977, pp. 95-153.

[27] M. Giaquinta and S. Hildebrand. *Calculus of Variations, I, II*, volume 310. Springer, 1977.

[28] P.C. Greiner and E.M. Stein. *Estimates for the $\bar{\partial}$-Neumann Problem*, volume 19. Math. Notes Series, Princeton University Press, Princeton, New Jersey, 1977.

[29] M. Gromov. *Metric Structures for Riemannian and Non-Riemannian Spaces, with appendices by M. Katz, P. Pansu and S. Semmes*, volume 152. Progr. Math., Birkkäuser, Boston, MA, 1999.

[30] U. Hamenstädt. *Some regularity theorems for Carnot-Carathéodory metrics*, volume 32. J. Differential Geom., 1990, pp. 819-850.

[31] P. Hartman. *Ordinary Differential equations.* Wiley, 1984.

[32] S. Hawking. *A Brief History of Time.* Bantam, 1998.

[33] L. Hörmander. *Hypoelliptic second order differential equations*, volume 119. Acta Math., 1967.

[34] A. Hulanicki. *The distribution of energy in the Brownian motion in the Gaussian field and aanalytic-hypoellipticity of certain subelliptic operators on the Heisenberg group.* Number 56. Studia Math., 1976, pp. 165-173.

[35] C. Iacob, D. Homentcovschi, N. Marcov, and A. Nicolau. *Matematici clasice si moderne.* Ed. Tehnica, Bucuresti., 1983.

[36] A. Kanpp. *Lie groups, Lie algebras and cohomology.* Math. Notes, Princetown University Press, New Jersey, 1988.

[37] S. Kobayashi and K. Nomizu. *Foundations of Differential Geometry*, volume I-II. Wiley Interscience, 1996.

[38] A. Koranyi. *Geometric properties of Heisenberg groups.* Advances in Math., 1985, pp. 28-38.

[39] A. Koranyi. *Geometric aspects of analysis on the Heisenberg group.* Topics in Modern Harmonic Analysis, May-June 1982, pp. 209-258.

[40] A. Koranyi and H. M. Riemann. *Quasiconformal mappings in the Heisenberg group.* Invent. Math., 1985, pp. 309-338.

[41] S. Krantz. *Function of several complex variables.* Wadsworth & Brooks/Cole, Pacific Grove, CA, 1992.

[42] D. F. Lawden. *Elliptic Functions and Applications*, volume 80. Applied Mathematical Sciences, Springer-Verlag New York Inc., 1989.

[43] E. B. Lee and L. Markus. *Foundations of Optimal Control Theory.* Wiley, New York, 1968.

[44] W. Liu and H. J. Sussmann. *Shortest paths for sub-Riemannian metrics on rank two distributions*, volume 118. Mem. Amer. Math. Soc., 1995.

[45] A. Messiah. *Quantum Mechanics.* Dover Publications, 2000.

[46] R. S. Millman and G. D. Parker. *Elements of Differential Geometry.* Prentice Hall, Inc., 1977.

[47] A. Nagel, F. Ricci, and E.M. Stein. *Singular Integrals with flag kernel and analysis on quadratic manifolds.* Number 181. Journal of Functional Analysis, 2001, pp. 29-118.

[48] D.H. Phong and E.M. Stein. *Hilbert integrals, singular integrals, and Radon transforms II.* Number 86. Inventiones Math., 1986, pp. 75-113.

[49] D.H. Phong and E.M. Stein. *Hilbert integrals, singular integrals, and Radon transforms I.* Number 157. Acta Math., 1986, pp. 99-157.

[50] R. Strichartz. *Subriemannian geometry.* Number 24. J. Differential Geom., 1986, pp. 221-263.

[51] R. Strichartz. *Corrections to "Subriemannian geometry".* Number 38. J. Differential Geom., 1989, pp. 595-596.

[52] H. J. Sussmann. *A cornucopia of four-dimensional abnormal sub-Riemannian minimizers in A. Bellaïche (ed.), Sub-Riemannian Geometry.* Progr. Math. Birkkäuser, Basel, 1996, pp. 341-364.

[53] W. Thirring. *A Course in Mathematical Physics, vol 1: Classical Dynamical Systems.* Springer-Verlag, Berlin · Heidelberg · New York., 1978.

[54] A. L. Jr. Xavier and M. A. M. de Aguiar. *Semiclassical approximations to the coherent-state propagator for a particle in a box*, volume 54. Physical Review A, 1996, pp. 1808-1819.

Index